Complexity and Ecosystem Management

Complexity and Ecosystem Management

The Theory and Practice of Multi-Agent Systems

Edited by

Marco A. Janssen

Associate Research Scientist, Center for the Study of Institutions, Population, and Environmental Change (CIPEC), Indiana University, USA

IN ASSOCIATION WITH THE INTERNATIONAL SOCIETY FOR ECOLOGICAL ECONOMICS

Edward Elgar

Cheltenham, UK • Northampton, MA, USA

Published by
Edward Elgar Publishing Limited
Glensanda House
Montpellier Parade
Cheltenham
Glos GL50 1UA
UK

Edward Elgar Publishing, Inc.
136 West Street
Suite 202
Northampton
Massachusetts 01060
USA

A catalogue record for this book
is available from the British Library

Library of Congress Cataloguing in Publication Data
International Society for Ecological Economics. Meeting (2000: Canberra, A.C.T.)
Complexity and ecosystem management: the theory and practice of multi-agent systems/ edited by Marco A. Janssen; in association with the International Society for Ecological Economics.
p. cm.
Edited papers from the Biannual Meeting of the Society for Ecological Economists held in Canberra, Australia, 2000.
Includes bibliographical references (p.).
1. Ecosystem management—Congresses. I. Janssen, Marco, 1969– II. Title.

QH75.A1 I58 2000
333.95—dc21

2002034676

ISBN 1 84376 061 4

Printed and bound in Great Britain by MPG Books Ltd, Bodmin, Cornwall

Contents

List of figures

List of tables

List of contributors

Nick Abel, CSIRO Sustainable Ecosystems, Canberra, Australia
Georges Amon, Unit : Dynamiques et Fonctions des Territoires Ruraux, Cemagref, France
J. Marty Anderies, CSIRO Sustainable Ecosystems, Canberra, Australia
Patrick d'Aquino, Cirad, St Louis, Sénégal
Sigried Aubert, Cirad, Saint-Denis, La Reunion (France)
Didier Babin, Cirad, Montpellier, France
Alfons Balmann, Department of Agricultural Economics and Social Sciences, Humboldt University of Berlin, Germany
Olivier Barreteau, Cemagref, Montpellier, France
Stanislas Boissau, Communication & Innovation studies, Wageningen University, Wageningen, The Netherlands
François Bousquet, Cirad-IRRI, IRRI-Thailand Office, Bangkok, Thailand
Jean-Paul Bousset, Unit: Dynamiques et Fonctions des Territoires Ruraux, Cemagref, France
Roger Bradbury Centre for Resource & Environmental Studies, Australian National University, Canberra, Australia
Jean-Christophe Castella, IRD-IRRI, SAM-Regional Program, Vietnam Agricultural Science Institute, Ha Noi, Vietnam.
Guillaume Deffuant, Laboratoire d'Ingénierie des Systèmes Complexes, Cemagref, France
Michel Etienne, INRA, Unité d'Ecodéveloppement, Avignon, France
Tom Evans, Department of Geography, Indiana University, USA
Kathrin Happe, Department of Farm Economics, University of Hohenheim, Stuttgart, Germany
Jérôme Henriot, Unit: Dynamiques et Fonctions des Territoires Ruraux, Cemagref, France
Matthew Hoffmann, Department of Political Science and International Relations, University of Delaware, USA
Mark Howden, CSIRO Sustainable Ecosystems, Canberra, Australia

Sylvie Huet, Laboratoire d'Ingénierie des Systèmes Complexes, Cemagref, France
John Ive, CSIRO Sustainable Ecosystems, Canberra, Australia
Wander Jager, Department of Marketing, University of Groningen, the Netherlands
Marco A. Janssen, CIPEC, Indiana University, USA
Konrad Kellermann, Department of Agricultural Economics and Social Sciences, Humboldt University of Berlin, Germany
Hugh Kelley, Department of Economics, Department of Cognitive Science Indiana University, USA
Anne Kleingarn, Department of Agricultural Economics and Social Sciences, Humboldt University of Berlin, Germany
Art Langston, CSIRO Sustainable Ecosystems, Canberra, Australia
Christophe Le Page, Cirad,, Montpellier, France
Tim Lynam, Institute of Environmental Studies University of Zimbabwe, Zimbabwe
Steven M. Manson, Department of Geography, University of Minnesota, USA
Mark Stafford Smith, CSIRO Sustainable Ecosystems, Alice Springs, Australia
Jacqui Stol, CSIRO Sustainable Ecosystems, Canberra, Australia
Bill Tatnell, Department of Land and Water Conservation, Buronga, Australia
Brian H. Walker, CSIRO Sustainable Ecosystems, Canberra, Australia
Gérard Weisbuch, Laboratoire de Physique Statistique, Ecole Normale Supérieure, France

Preface

During the biannual meeting of the International Society for Ecological Economics 2000 (ISEE 2000) held in Canberra, Australia, a number of papers dealt with complexity and ecosystem management. ISEE kindly invited me to edit a volume on this topic. This volume is compiled of papers which were presented in their initial form at ISEE 2000, and other complimentary papers. The resulting volume consists of a comprehensive overview of the rapidly developing area of the use of multi-agent systems to study the interactions of people and the environment.

I want to thank all of those who have contributed to the realization of the volume. First, I would like to thank all the authors for providing interesting and high quality contributions. I would also like to thank those who have helped to review the chapters. These include some of the authors who have cross-reviewed the chapters, and other external reviewers such as Thomas Berger, Olivier Chikumbo, Mike Common, Graeme Cumming, Peter Deadman, Govert Geldof, Hasse Goosen, Jeljer Hoekstra, Martijn van der Heide, Marco Huigen, Robert Imeson, Louis Lebel, Arend Ligtenberg, Kevin Lim, Jon Norberg, Dawn Parker, Rusty Pritchard, Jan Sendzimir, Daniel Stow, and John Wolfenden. I am also grateful to Jeroen van den Bergh who stimulated me to compile this volume, support from the Center for the Study of Institutions, Population, and Environmental Change at Indiana University through National Science Foundation grants SBR9521918 and SES0083511 and Edward Elgar Publishing for the editorial assistance.

Marco A. Janssen, CIPEC, Indiana University, USA

1. Introduction

Marco A. Janssen

1.1 THE ISSUE

Our world view on the structure of systems is changing rapidly. According to the traditional Newtonian paradigm, we can assume a system is in equilibrium, and study it in isolated parts. Scholars in the field of complexity adopt a paradigm of complex adaptive systems, which assumes that systems are emergent structures on a macro scale due to interactions between micro-level agents who adapt themselves to their environment. Unlike the Newtonian paradigm, which assumes that systems have a predictable behavior, the behavior of complex adaptive systems cannot be predicted accurately. However, complex adaptive systems can provide us with insights under which circumstances simple local rules can lead to emergent macro-level structures.

Examples of complex adaptive systems are immune systems, nervous systems, economies and ecologies. These systems can be studied by a number of new computation-based modeling tools, including genetic algorithms, cellular automata, neural networks, multi-agent systems, and artificial life forms. In this volume the interactions between people and their environment is addressed. Both theoretical and practical examples are presented of developing and applying methodology of studying the management of complex adaptive social-ecological systems.

Economists study the management of ecosystems in terms of harvesting ecosystem services from renewable resources. Substantial progress has been made during the last 30 years. Prior to 1970, model analysis was mainly static, such as the seminal work on renewable resource harvesting by Gordon (1954). After 1970 the trend shifted towards dynamic systems for the economics of renewable resources. The resulting optimization problem was addressed by dynamic programming, game theory and equilibrium analysis (Clark, 1990; Dasgupta and Heal, 1979; Mäler, 1974). Irreversibility and uncertainty have been addressed since the early 1970s (Arrow and Fisher, 1974; Henry, 1974) and remain among the main focuses of environmental economics (e.g. Chichilnisky, 2000).

In mainstream environmental economics it is usual to analyse a representative agent who has perfect knowledge and maximize its utility of consumption for an infinitive time horizon. Such an approach resulted in interesting insights but is of limited use if systems are characterized by non-convex dynamics, structural uncertainty, heterogeneity among agents and spatial heterogeneity. The question is how to analyse ecosystem management problems with spatial explicit non-convex dynamics influenced by multiple stakeholders who consume different types of ecosystem services. We need new tools, and multi-agent systems are promising new tools in the toolkit for the social scientist.

Multi-agent systems, also referred to as agent-based (computational) modeling, consist of a number of interacting autonomous agents (Conte *et al.*, 1997; Epstein and Axtell, 1996; Gilbert and Troitzsch, 1999; Weiss, 1999). Agents can represent animals, people or organizations; can be reactive or proactive; may sensor the environment; communicate with other agents; learn, remember, move and have emotions. The main components of multi-agent systems are cellular automata, and models of the agents. Each agent is represented as a computerized independent entity capable of acting locally in response to stimuli or to communication from other agents. Therefore, the first task is to build architectures for intelligent agents and secondly to design an organization of interacting agents to accomplish a task.

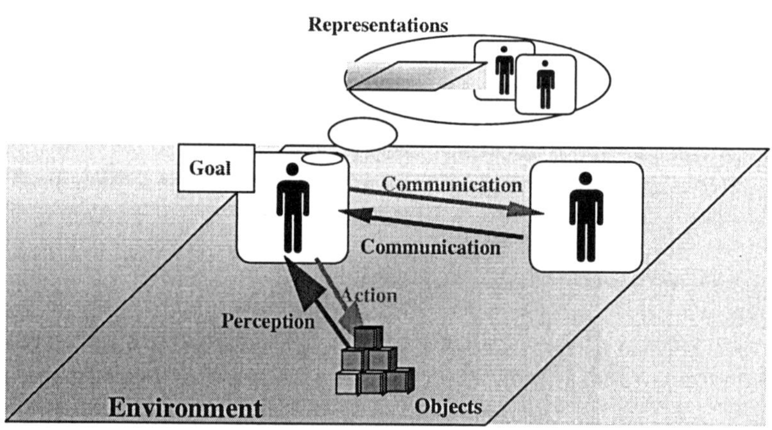

Figure 1.1: Multi-agent system general organization and principles (from Ferber, 1999)

A multi-agent system perspective of ecosystem management is essentially different from traditional environmental economic approaches. Three different types of agents can be distinguished: *humans* who differ in mental maps, goals, locations, and abilities; *non-humans* such as animals and plants; and *passive* agents such as non-living entities. We mainly focus on the human agents who are more complicated than other types of agents.

Wooldridge (1999) argues that intelligent agents are able to act flexibly and autonomously. By flexibility we mean that agents are goal-directed (satisfying or maximizing their utility), reactive (responding to changes in the environment) and capable of interacting with other agents. One of the difficulties is balancing reactive and goal-directed behavior. Developing models with agents who have only reactive behavior is relatively simple. Individual-based ecological modeling addresses problems by simulating non-human agents as reactive objects (e.g. DeAngelis and Gross, 1992). Econophysics (www.econophysics.org), an emerging field that applies methods from statistical physics and non-linear dynamics to macroeconomic modeling and financial market analysis, approaches human behavior as behavior of particles.

However, humans combine reactive and goal-directed behavior. Conventional economics makes use of fully rational actors to study human behavior. The rational actors are self-regarding individuals maximizing their own wellbeing. However, the powerful concept of the rational actor seems to be invalid according to experimental research in economics and psychology (e.g. Thaler, 1992; Gintis, 2000b).

Deliberation about an economic decision is a costly activity in terms of time and cognitive effort, and many social scientists argue that people often employ simpler decision rules, aimed at satisfying rather than optimizing (see Chapter 6). Models of bounded rationality have been used as an alternative in economics (Simon, 1955; Sargent, 1993). Still other important dimensions of the economic agents have been excluded, such as emotions, motivations, and perceptions. In order to include this dimension of behavior we have to enter the domain of psychology.

Psychology and many other social sciences were originally focused on experimental research of individual and group behavior. Since the early 50s social scientists have used computers to simulate behavioral and social processes, although the real breakthrough became in the late 80s due to the development of new simulation techniques like cellular automata, genetic algorithms and neural networks, and the widespread availability of personal computers. Computers became laboratories, allowing simulating behavior theories in virtual environments. Those multi-agent systems are used to study the stock-market dynamics (Palmer *et al.*, 1994), evolution of cooperation (e.g., Axelrod, 1984), evolution of language (Cangelosi and Parisi, 2001), rise and fall of (ancient) societies (Kohler and Gumerman, 2000), pedestrian behavior (Jiang, 1999), land use and land cover change (Parker *et al.*, 2001), etc. Simulation of multi-agent systems is now recognized as a promising methodology for social science as shown by the colloquium on this topic organized in October 2001 by the National Academy of Sciences of the USA.

In this book we apply multi-agent systems to study the interactions between humans and ecosystems. Until recently nature was being seen by most individuals as an infinite resource of materials and energy to satisfy human needs, as well as an infinite sink for human wastes and pollutants. In the last century humankind became aware of its large domination of the Earth by transforming land surface, by altering biogeochemical cycle, and by adding or removing species and genetically distinct populations in most of the earth's ecosystems (Vitousek *et al.*, 1997). The expected growth of the human population and its associated economic activities are likely to accelerate the scale and intensity of human-induced changes.

The concept of ecosystem management was introduced during the last ten years as an approach to focus on long-term sustainability of ecosystems. Various interpretations on ecosystem management are being used and are discussed in, amongst others, Grumbine (1994, 1997) and Christensen *et al.* (1996). Key elements of ecosystem management are the focus on long-term sustainability, inclusion of humans as ecosystem components, the inherent change and evolution of ecosystems, the importance of structural complexity and connectedness to maintain resilience of systems, and the acknowledgement of our incomplete knowledge of ecosystems which require adaptive management (Christensen *et al.*, 1996).

We are living in a world with fast-changing ecological and social systems. Our current institutional arrangements are based on past experience. Change is too fast to depend on learning by doing alone. Computational laboratories can explore possible futures. The resulting better understanding of the functioning of institutions can contribute to a more sustainable co-evolution of people and nature.

1.2 MULTI-AGENT MODELING

Most multi-agent models applied to ecosystem management consist of two elements: agents and cellular automata. We discuss briefly both elements. For a more detailed discussion on the methodology of multi-agent modeling we refer to Ferber (1999), Weiss (1999) and Parker *et al.* (2001).

1.2.1 Cellular Automata

Originally, the cellular automata (CA) approach was introduced by John von Neumann and Stanislaw Ulam at the end of the 1940s, mainly to give a reductionist model of life and self-reproduction. The *Game of Life*, invented by John Conway in 1970, popularized the CA approach (Gardner, 1970). This game consists of cells on a checkerboard which can have two states, 'alive' and 'dead'. Time goes by in discrete steps. According to some deterministic rules, which are the same for each cell, the state of a cell in the next time step depends on its own present state and the states of all its surrounding cells in the present period. The resulting surprising complex dynamics which evolved from this simple game attracted the attention of many people. Since the early 1970s CA have been used by many disciplines to study complex dynamic behavior of systems.

The basic features of a CA are (Hegselmann, 1998):

- There is a D-dimensional lattice.
- Time is advancing in *discrete* steps.
- There are a *finite* number of states. At each site of the lattice we have a cell, which is in one of the possible states.
- The cells change their states according to *local* rules, both in space and in time.
- The transition rules usually used are deterministic, but non-deterministic rules are allowed too.
- The system is *homogeneous* in the sense that the set of possible states is the same for each cell and the same transition rule applies to each cell.
- The updating procedure can consist of applying the transition rule to all cells *synchron* or *asynchron*.

With regard to our interest for ecosystem management, the application of CA can be rather straightforward. In fact, CA can be used to produce a dynamic GIS. The lattice represents a map of a certain area. Each possible state of a cell represents a possible land use. Due to physical restrictions, cells on some locations may be restricted to a limited number of states. For example, a secondary forest cannot turn back to a primary forest. Transition rules determine when a certain land use of a cell changes into another land use. Cell changes can be influenced by local rules; for example, if the cell is a forest-cell and if one of the neighbor cells is on fire then the cell turns to fire. However, global rules are also possible since land use changes can be influenced by demand for certain land on a higher level of scale. For example, demand for extra agricultural land can be translated as changing those cells to agriculture which are the most suitable. Note that we do not have to use CA for representing the environment, since not all problems are spatially explicit, or different spatial explicit models can be used.

One of the main drawbacks of the attention to CA only is the limited inclusion of social processes. It must be noted that social agents can also be represented as CA. Within the field of social simulation CA are used to simulate social phenomena. One of the first papers in the field is the analysis of segregation processes by Schelling (1971). In the CA approach for social processes each cell represents an agent, which interacts with its neighbors. The state of the cells relates to different characteristics of the agents such as social class, attitude, social orientation, etc. An overview of this field is given in Hegselmann (1998).

A drawback of using CA for representing social agents is its simplicity. For example, social networks are probably more complex than the local neighbors on a lattice. The possible states in which a social agent can be might be too large to be efficiently represented as a CA. The study of agents has been a topic of research for a long time in computer science which has developed its own tools and frameworks.

1.2.2 Agents

The architecture of agents in multi-agent systems has been much influenced by work in Artificial Intelligence (AI). In this field a popular wave is the autonomous agents research or behavior-based AI, which studies the behavior of adaptive autonomous agents in the physical world (robots) or in the cyberspace (software agents). This field in AI is highly inspired by biology. The phenomena of interest are those traditionally covered by ethnology and ecology (in the case of animals) or psychology and sociology (in the case of humans). The agents often consist of sensors to derive information from the environment and intelligent functions such as perception, planning, learning, etc. Behavior is defined as regularity observed in the interaction dynamics between characteristics and processes of a system and the characteristics and processes of an environment. Examples include: a theory at the behavior level that explains the formation of paths in an ant society in terms of a set of behavioral rules without reference to how they are neurophysiologically implemented. Another example is the study of behavioral rules implemented in robots who have to survive (they need to reload energy every now and then) in a physical environment with other robots as a way to explore emergent behavior in such a group. An overview of this field can be found in, for example, Steels (1995) and Maes (1995).

Distributed artificial intelligence is a relatively recent development of artificial intelligence studies (Bond and Gasser, 1988). It concerns the properties of sets of intercommunicating agents coexisting in a common environment. The aim may be to study the properties of such systems in an abstract way, or to design systems of immediate practical use, or to use such a programmed multi-agents system as a model of a human or other real-world system.

The main dilemma concerning the architecture of agents with regard to the study of ecosystem management is the degree of complexity embodied in the agent. Since the roots of agent research lie in computer science, the agents are designed for certain tasks but do not necessary represent theoretical insights from behavioral science. The fields of social simulation (e.g. Conte *et al.*, 1997) and agent-based computational economics (e.g. Tesfatsion, 2001) use the techniques of multi-agent systems together with concepts from sociology, psychology and economics to design more comprehensive agents from a social science point of view. Nevertheless, empirical validation of the agent architecture remains an activity which receives too little attention. More cooperation with experimental research and computational experimentalists is necessary to improve the construction of the agents.

1.3 APPLYING MULTI-AGENT SYSTEMS FOR ECOSYSTEM MANAGEMENT

In this section I briefly discuss what type of questions can be addressed by multi-agent systems, and what the possible problems are. As Roger Bradbury clearly discusses in Chapter 4, accepting the paradigm of complexity also means the recognition that the future cannot be predicted. Using multi-agent systems as a tool for the study of complex social ecological systems implies that we are not interested in predicting the state of the system in a certain moment in the future, like Newtonian science can for the orbits of the planets and stars in our universe. The focus of multi-agent systems is to understand persistent emergent phenomena. Emergence arises from the context-dependent, non-linear interaction of the objects and the rules (Holland, 1998). Non-linear interaction means that the behavior of the system cannot be obtained by summing or averaging the behavior of the constituents.

An example is an ant colony where ants with limited individual capabilities are able to construct comprehensive societies. A colony of a million army ants is a sophisticated 'super-organism' (Gordon, 1999). The colony carries out its raids and can even keep nest temperatures constant to within a degree. An ant colony seems endowed with intelligence far beyond that of any individual ant: It seems that intelligence, natural or artificial, is an emergent property of collective communication. Camazine *et al.* (2001) provide a state of the art volume on self-organization in biological systems.

They show that the combination of laboratory research, field experiments and modeling provide a fruitful combination to understand which micro-level behavior leads to emerging patterns at the macro-level.

An example of an emergent property in economics is the equilibrium of demand and supply. In conventional economics it is assumed that the economic system is in equilibrium, but within agent-based computational economics, the economic agents are simulated from the bottom up (Tesfatsion, 2001).

A prominent example of using multi-agent systems for ecosystem management is the Bali irrigation system studied by Lansing (2000). A large number of subaks, and each subak consists of about 50 farmers, have to coordinating planting and irrigation. Synchronous planting has the benefit of effectively controlling rice pests, but leads to water bottleneck, since they all use the water from the same river. Lansing equipped the agents with simple heuristics. Each year a subak compares its yields with the yield of its closest neighbors, and copies the cropping pattern of its (best) neighbor. The model, using real data of a watershed in southern Bali, produces similar spatial cropping patterns as observed in the field.

The Bali example shows the fruitful use of a multi-agent system perspective for the study of ecosystem management. Lansing (2002) discusses the role of 'artificial societies' in a historical debate in social science on methodology. It is clear that the mainstream statistical approach is confronted with an alternative, and this alternative is criticized. It is therefore necessary that models of multi-agent systems are applied to real case studies. This volume shows that there is progress in that respect by presenting a number of real-world case studies, and a number of methodological challenges such as validation, experimental research and the use of models in a participatory process.

1.4 A GUIDE TO THIS BOOK

This section provides a brief outline about the different contributions, and how they are linked together. Chapters 2 and 3 discuss multi-agent systems as rule-based systems. Marty Anderies proposes in Chapter 2 a general framework to describe each system in terms of rules and agents. The proposed framework can provide a common language for multi-agent systems, and Anderies applies his framework to the several contributions of this volume. In Chapter 3 the evolution of institutional rules is discussed. How are they coded, created, selected and remembered? Marco Janssen argues that linguistics and immunology provide interesting metaphors and computational tools to improve the methodology of simulating institutional rules.

The difference between Newtonian and complexity paradigms as mentioned at the start of this chapter are also the starting points for Chapters 4 and 5. Roger Bradbury describes clearly in Chapter 4 why we cannot predict the future of social-ecological systems, and argues that we can only study under which conditions emergent phenomena hold. In Chapter 5, Steve Manson discusses the problem of validating multi-agent systems. He presents a number of practical guidelines and tools, and discusses how differently applied studies in this volume address validation.

One of the crucial problems in multi-agent modeling is the validity of the design of the agents. Chapter 6 by Wander Jager and Marco Janssen is an attempt to test their artificial agents to experimental data. Their artificial agents provide new explanations of observed behavior of subjects in laboratory experiments. A next step would be to test these new hypotheses in the laboratory. This chapter illustrates that more interaction between experimental and modeling research is needed.

A traditional differential equation model of an Australian rangeland system is the starting point in Chapter 7 by Janssen *et al.* In the original model, space explicit dynamics were ignored, but by using a multi-agent system approach, it is shown that spatial heterogeneity of grazing pressure can lead to quite different conclusions compared to the original model.

The rest of this volume presents applications of multi-agent models. Applications vary from agricultural systems to forests and rangelands. Chapters 8 and 9 analyse the effects of agricultural policies in Europe. Balmann *et al.* developed an agent-based model that includes spatial interaction of some 2600 individually behaving heterogeneous farms. They apply this model to analyse the effects of a limitation of livestock density per hectare on a selected intensive production region in the southwest of Germany.

In Chapter 9, Deffuant *et al.* study the dynamics of organic farming practices in the Allier départment in France. The agent-based model is the result of an iterative process between modeling researchers and specialists of agri-environment. It represents explicitly the propagation of information and the dynamics of discussions among farmers, as well as its impact on the farmers' decision process. Deffuant *et al.* found that the initial heterogeneity in perceptions of farmers towards organic farming was a crucial factor to explain observed adoption rates.

Tim Lynam, Chapter 10, uses spatially explicit, multi-agent simulation to study the relation of household characteristics and long-term production of the Masoka agro-ecosystem in Zimbabwe. Field research provided statistics of the households which were used to generate characteristics of the population in the model. Assuming different types of decision-making strategies, such as income maximizers and need satisfaction variance minimizers, and rainfall variability, a detailed sensitivity analysis was performed. Lynam concludes that although the model is incomplete, better understanding of the relation of natural resources and the wellbeing of households was derived.

Matthew Hoffmann *et al.* present in Chapter 11 a first version of a multi-agent model to study deforestation and reforestation patterns in Indiana (USA). Prior to European colonization, Indiana was nearly entirely forested. By 1900, only 5–10% of the state was forested, but since then forestland has rebounded to approximately 20% of the state's land cover. The multi-agent model of land cover change simulates the observed process of deforestation and afforestation in Indiana over time.

The last two chapters deal with interactions with stakeholders in order to solve collective-action problems. Bousquet *et al.* provide an overview of the work of CIRAD on their use of multi-agent systems for ecosystem co-management by performing role games. By using role games the required knowledge can be obtained, the model can be validated and used in decision-making processes. The discussion of the different applications shows how models can be used in support of collective decision-making processes in natural resource management.

In the last chapter, Nick Abel and colleagues discuss a large project in the rangelands of New South Wales, Australia. Participatory analysis of the land-use values of five stakeholder groups showed a great potential for multiple land-use as an efficient way of reducing land-use conflicts and satisfying stakeholders' needs, but legal and administration hurdles prevented the system change. Participatory policy making generated 160 proposals for legal, policy and administrative changes designed to simplify the complexity, realize win–win opportunities, resolve conflicts and promote regional resilience. The project has been influential within the region, but at State level there has been little change so far.

In sum, the contributions in this volume show that models of multi-agent systems can be applied for the study of ecosystem management. Applications focused on diffusion processes, explaining puzzling historical observations, and using models together with stakeholders in an attempt to solve contemporary management problems. The continuation of this endeavor can be improved when a number of methodological issues are addressed, notably validation, and the link between the behavioral rules of agents and experimental data.

PART I

Methods and Concepts

2. The transition from local to global dynamics: a proposed framework for agent-based thinking in social-ecological systems

J. Marty Anderies

2.1 INTRODUCTION

Over the years as resource management science has evolved, ever more sophisticated quantitative techniques have been developed. This is no different than in other areas of science, and it is especially evident in those where large numbers of individual objects (members of a population, spins on a lattice, etc.) interact locally to generate global dynamics. For example, individual people (economic agents) take actions concerning production, consumption, and the gathering and transfer of information that produce economic phenomena at a larger (e.g. national) scale. That is, these local individual actions generate emergent global dynamics.

Usually it is the global dynamics of a particular system that are of interest. Examples of global phenomena include spatial distributions of competing plant populations, phase transitions, population dynamics of predator prey systems (or more general 'enemy–victim' systems (Keeling *et al.*, 2000)), or macro-economic indicators such as GDP, unemployment, or income distributions. The difficulty is that most often only local dynamics, e.g. the behavioral attributes of individuals, or the local interactions between agents, are well (or better) understood. The challenge is to move from local understanding to global understanding. This transition is an important step in improving our understanding of the way social ecological systems behave at scales relevant for the development of policy.

The use of modern computing equipment has allowed researchers to explore the behavior of systems comprised of large numbers of interacting individual units. The construction of such models on the computer is, in principle, extremely easy; simply define the individuals and the rules by which they interact and allow the model to evolve. Repeated simulation of such models generates a sample space of trajectories which can help characterize the system's qualitative behavior. Having many sample trajectories (or more interestingly, long-run 'attractors' that define the long-run system configuration for a given set of assumptions) is a considerable improvement over having only one observation, namely the current state of the system. The potential for increased understanding of how changes in individual behavior affect the behavior of social-ecological systems at larger scales is part of the motivation of what has come to be called multi-agent modeling in the resource management/ecological economics literature.

Although the term multi-agent (or agent-based, or individual-based) is quite recent, 'multi-agent' models, or at least 'multi-agent' thinking is not. An obvious example is Boltzmann's statistical mechanics of the end of the 19th century which offers an 'individual-based' alternative to the very aggregated models of Boyle and Charles of the behavior of gases. The importance of microscopic (individual) dynamics in determining macroscopic (aggregate) dynamics has long been recognized along with the importance of developing the analytical machinery to understand the relationship between the two. Unfortunately, a high level of mathematical skill and knowledge of a specialized set of analytical methods was required to understand the global behavior of simple individual-based systems. Even then, the application of this machinery is limited to a small number of special cases. Thus, up until recently, most scientists working seriously on such problems were physicists and mathematicians. Due to a lack of a common language, communication of these ideas to other fields was slow.

The recent development of cheap computational power has reduced the need for powerful analytical tools and has expanded the range of examples that can be studied. However, simulating large systems of interacting individuals on a computer is fraught with practical difficulties. Often, the dynamics of such models are as complex as the system they are trying to model. It is almost as difficult to analyse them as to conduct an experiment with the real system. The advantage of computer models is that the cost of experimentation is much lower than with real systems. Even with the possibility of very inexpensive experimentation, the value of the analysis of such models can be quite limited simply because 'outsiders' do not have time to unravel all the subtle assumptions underlying the model and the way the computations were carried out (which can introduce 'artifacts' in the model dynamics). Further, because of the fact that nonlinear systems (even with as few as three state variables) can be extremely sensitive to small perturbations in either initial conditions or model structure, a seemingly trivial inaccuracy in a complex model can lead to wildly different results. Finally, even cheap computational power can be overwhelmed with what Bellman (1957) (in the context of dynamic optimization problems) referred to as 'the curse of dimensionality'. This refers to the simple fact that as the number of agent types and their allowable actions grow, the number of trajectories that need to be explored to gain a qualitative understanding of the system explodes.

Aside from the brief discussion above, this chapter is not intended to give a thorough introduction to the huge field of agent-based modeling. There are several introductions on this topic that provide a wealth of examples, discuss the merits of the approach, and extend the discussion above in many directions. For more details specific to the social sciences, see, for example, Epstein and Axtell (1996) or Terna (1998). The aim of this chapter is to provide a framework to help increase the benefits that can be obtained from multi-agent approaches. The proposed framework provides a common language to relate traditional approaches such as ordinary differential equations, partial differential equations, or difference equations to multi-agent modeling. Using the framework to put multi-agent approaches into context can both minimize some of the problems discussed in the previous paragraph, and enhance the understanding that both traditional and multi-agent approaches can provide. This is examined in more detail in the following section.

2.2 TRADITIONAL APPROACHES AND THE ROLE OF MULTI-AGENT PERSPECTIVES

One of the dangers of multi-agent models is that they are developed simply because it is now possible to do so. One could argue that in many cases simpler aggregate models are adequate and that complex multi-agent models may even confuse the issues. Unless some tangible benefit can be derived from disaggregation in models in social economic systems, aggregated models should be used. The challenge is to know when to use a multi-agent approach and how to get the most out of it.

An example will help make this clear. Consider one of the most well-known multi-agent models: Axelrod's (1984) simulation of agents playing the repeated *Prisoner's Dilemma* game. In a recent review, the game theorist Ken Binmore (1998) points out that Axelrod's work may have contributed more if it had been more closely linked to game theory. First, the fact that the 'nice' strategy, Tit-For-Tat, could win the competition was already well known; it had been proven in the 1950s. Second, Axelrod ran finitely repeated games of *Prisoner's Dilemma*. In an infinitely repeated game, Tit-For-Tat is not the only survivor. Rather a mixture of strategies remains. Additional game-theoretic results also establish that it is possible to make the system converge to whatever stable equilibrium one chooses, by starting the system at the appropriate initial conditions. Thus, to obtain reliable results, initial conditions must be systematically explored. Binmore suggests that Axelrod did not explore a sufficient number of initial conditions. Subsequent work that did systematically explore initial conditions produced quite different results (Linster, 1992).

Binmore, nonetheless, points out that Axelrod's work did indeed make a major contribution to game theory and our understanding of the evolution of social norms. However, this contribution has nothing to do with the Tit-For-Tat strategy that has become the focus of this work. Rather, the contribution of the multi-agent approach he took was to focus the attention of researchers on the importance of evolutionary mechanisms in selecting a particular equilibrium strategy. The point here is that carefully relating multi-agent models to their more traditional counterparts can accomplish three things: 1) minimize systematic, or theoretical errors that reduce the significance of the results; 2) prevent wasted effort on rediscovering what is already known; 3) focus attention on what the multi-agent approach really adds. The goal of this chapter is to provide a framework, a common language to link multi-agent thinking to more traditional approaches, to facilitate comparison across different multi-agent models, and to provide a template for the development of multi-agent models. It is important to emphasize that the framework is not intended as a cookbook for multi-agent modeling, but rather as a template for multi-agent thinking. Finally, I hope that the framework will enhance the value of the models presented in this volume.

2.3 MULTI-AGENT MODELS IN SOCIAL-ECOLOGICAL SYSTEMS

Indeed the history of resource management science has been a slow realization that aggregated models and the analytical methods associated with them are in many cases inadequate. Early models of resource management emphasize the optimal control of very simple systems (Clark, 1990; Dasgupta, 1982; Dasgupta and Heal, 1974). The value of such models lies in the general insights they provide; powerful enough to completely change the way problems are viewed (Clark, 1973). They are, however, of little use in actually addressing real world resource management issues. The reason for this is simple: the scientific understanding necessary to define the problem sufficiently well to meaningfully apply optimal control techniques is either too costly or altogether impossible to achieve. Put another way, too many tacit assumptions must be made about the way markets work, how much information is available to decision makers and their ability to process it, and the goals of resource users which simply do not hold. This is where multi-agent modeling has much to offer.

Not only is the optimal control approach not practical, it is based on a paradigm of centralized control by agencies exogenous to the system of interest. The resource management research community has shifted from the exogenous control approach to approaches based on adaptive, experimental learning from within the system (Walters, 1986). This approach still emphasizes 'management' in the sense that the goal is, through active experimentation with the system, to progressively learn to manage a given resource base. There are many problems with this approach. One of the main difficulties is actually carrying out an experiment at the necessary scale. Stakeholders in resource systems seldom want to 'experiment' with their livelihoods.

A more comprehensive approach to human-resource interaction is to view it not as a problem of humans learning how to manage a system, but rather of learning how to live within it. A key distinction between these approaches is that less information may be required to live within a system than to 'manage' it. For example, the Tsembaga people of the New Guinea Highlands live their lives around a complex ritual cycle of pig festivals and ritualized warfare. In a detailed ethnography, Roy Rappaport (1968) proposed that the ritual cycle of the Tsembaga, essentially a set of behavioral rules, acts as a homeostatic mechanism which prevents degradation of their resource base. His theory was strongly criticized by the anthropological community, but a later mathematical model of the system has shown under what conditions Rappaport's hypothesis might hold (Anderies, 1998). The point here is that the Tsembaga do not have to understand in detail how their resource base functions for their ritual cycle to effectively prevent them from degrading it. Rather, individuals acting through a rich set of rules produces 'wise' ecosystem 'management'.

Ostrom (1990) also takes a rule-based approach in her work on resource governance. In the institutional analysis and design framework developed by her and others, it is essential to define the system using agents and rules. Namely, any general 'resource appropriator system' is defined by clusters of rules. Ostrom has cataloged seven key clusters of rules that govern the dynamics of the system. Once the rule sets are defined, the possible evolutionary trajectories of the system are determined. In the context of resource governance, the task is to determine the rule sets that generate dynamics that meet certain criteria such as sustainability, robustness, etc. The search for rule sets is difficult, and suffers from the problem of dimensionality previously discussed (cf. Ostrom, 1999).

The difficulty of conducting an exhaustive search for rule sets has given rise to the idea of complex adaptive systems. Here, the objective is to construct systems that evolve along desirable trajectories, adjusting rules as learning about system properties takes place. Given the level of complexity of most systems of interest, it is assumed that even though learning occurs, the information available about the system falls far short of being complete. The key is to build adaptive capacity – the ability to learn and change – into the system to achieve a good fit between ecosystems and institutions. Multi-agent approaches play an integral part in constructing and studying such systems. We revisit complex adaptive systems in Section 2.5.5. We now turn our attention to a formal rule-based approach to social ecological systems.

2.4 RULE-BASED MODELS OF RESOURCE APPROPRIATOR SYSTEMS

In the literature, the interaction of ecological and economic processes is often discussed and modeled. Developing such models entails writing down equations that describe these processes which are usually related to how the state variables change over time. In rule-based models, processes are *emergent* properties of, rather than defining characteristics of, the system. In the strictest sense, any system can be defined by a set of rules and a description of the initial set of objects (agents) upon which the rules operate. Once set in motion, the set of agents and rules in the system evolve over time, e.g. evolution of life forms on Earth given an initial distribution of matter and a rule set consisting of physical laws and the law of natural selection. On shorter time scales relevant to models of social ecological systems, we can assume that the set of agent types and available rules in the system do not change. This allows us to define a system based on a fixed set of agent types and a fixed set of rules. Most multi-agent models of social ecological systems are a subset of this class of models.

Given that any model can be written in terms of agents and rules, it follows that all models are multi-agent models. This is an important point to consider when trying to make sense out of the plethora of models in the literature. To illustrate this, let us first define a set of classes of rules upon which all models can be built. In the context of institutional analysis, Ostrom *et al.* (1994) and Ostrom (1995, 1999) have defined seven classes of rules:

1. *Boundary rules* affect the characteristics of the participants in the system. An example of a boundary rule is that only members of the local community above age 18 are allowed to participate in a fishery.
2. *Position rules* define the positions in the system along with the

capabilities and responsibilities of those in positions. Examples of positions are voters, monitors, fishers, regulators, scientists, etc.

3. *Authority rules* define the actions that the participants in the system may, must, or must not take. An example would be that fishers have the authority to fish in a certain location or not. Monitors have the authority to report fishers who fish in areas off limits to people in other positions, or they may have the authority to force the fishers to cease fishing.
4. *Scope rules* define outcomes that are allowed, mandated, or forbidden. An example of a scope rule is a limit on the total biomass of fish removed from a particular location during a fishing season. This rule is distinct from an authority rule because it does not directly specify what each fisher may, must, or must not do. It specifies the context in which they might exercise their authority rules, e.g. if the maximum annual fish harvest has not been exceeded, fishers may catch as many fish as they are able.
5. *Aggregation (or transformation) rules* affect the way individual actions are translated into final outcomes. An example of an aggregation rule is a production function in economics which translates the individual actions of laborers combined with capital equipment into number of cars produced. In a political system, the way in which individual votes are translated into collective action is an aggregation rule.
6. *Information rules* affect the nature of the information present in the system and to whom it is available.
7. *Payoff rules* determine the costs and benefits associated with actions and outcomes. An example of a payoff rule is a fine attached to a prohibited action.

Given the set of rules, the final step in specifying a model is defining the nature of the agents in the system. The definition of an agent, however, is not always clear. First, because multi-agent approaches are used in a range of fields from artificial intelligence to sociobiology, the term agent is used in a variety of ways. In the most general terms, an agent is any object whose state can change in response to other system variables. This definition allows for agents to be anything from charged particles to people. The important point is that the definition of an agent is arbitrary, and depends on the objective of the modeler.

Matters are further complicated by the fact that there is not a fundamental distinction between agents and rules. An agent can be defined by a collection of rules and smaller scale agents. Even at the atomic scale this still holds: atoms are composed of sub-atomic particles (agents), and rules for their interaction. In the end, any system then is defined by fundamental sub-atomic particles and a set of rules. Given that everyone agrees on the set of sub-atomic particles, then all models can be specified by a set of rules only. The need for agents is purely practical. For social-ecological systems the agents will, of course, be living organisms.

Ostrom *et al.* (1994) suggest that agents have four key attributes: 1) preferences; 2) information processing capabilities; 3) selection criteria; and 4) individual resources. That is, individuals must 1) know what they want to achieve; 2) apply a set of data processing capabilities to achieving their goals; 3) apply a decision process to integrate across preferences (e.g. maximize some functional of a utility measure); and 4) apply a set of physical capabilities to achieve their goals (time or effort constraints, financial constraints, etc). I would suggest that it may be possible to collapse these four points into two: 1) objectives (includes preferences and criteria for evaluating those preferences); 2) resources (because information processing capability is a resource).

Again, Ostrom *et al.* (1994) discuss these points in relation to human actors, but they can just as easily be applied to any actor. For example, in the systems studied by Ostrom and others, position rules are often used to describe some sort of power hierarchy, i.e. resource users and resource regulators. Hierarchies are common in non-human systems as well. For example, two positions in social animals such as lions and wolves are dominant and subdominant males. There are fairly clear authority rules associated with these two positions. A dominant male may breed with the female of his choice. Subdominant males may not breed with certain females, at least not without incurring a cost of being injured or killed by the dominant male. In this sense there is a clear payoff rule associated with subdominant males attempting to act beyond the authority rules associated with their position.

It is important to clarify what I mean here. I am not suggesting that animal and human objectives and resources are similar. They are vastly different. Humans have goals and objectives. We cannot say that animals have objectives in the same sense. Since we cannot communicate with them there is no way to know. However, if we are interested in describing behavior, we can treat animals *as though* they have objectives. A classic example is the description of stotting behavior. Stotting is a display performed by many species of *Bovidae*, *Antilocapridae*, and *Cervidae* involving an 18 inch vertical jump with legs stiff that requires considerable energy – energy that could otherwise be used to evade capture if an attack ensued. Zahavi, in Dawkins (1976), suggests that one explanation is that the gazelle has the 'objective' of saving time and energy (and staying alive) by 'communicating' to the potential predator a simple message: 'Hey, look at me, I am so fit I can jump up and down here, waste considerable energy, and still outrun you. So don't waste your time with a chase. We will both come out ahead if you go find a gazelle that doesn't stot as well as I do.' Although Caro (1986) reviews eleven different hypotheses for this behavior, Fitzgibbon and Fanshaw (1988) showed that in the case of Thompson's gazelles facing a 'coursing' predator such as an African wild dog, Zahavi's hypothesis is probably correct. Even though we don't know if the gazelle has an objective or communicates in the way humans conceptualize objectives or communication, its behavior can be described as though it stems from a specific objective and uses a specific resource – namely the ability to send a message (an information rule) by stotting.

In the next section, a formal model is proposed based on the use of agents and rules as discussed above. Several examples including some models presented in this volume are then discussed in the context of the formal model.

2.4.1 A Formal Model

In much of the resource management literature there is a strong distinction made between human and non-human actors. The systems people study are often referred to as ecological-economic systems, where it is assumed that there is a (non-human) resource to be managed (by humans). In the development of this formal model, I ignore the human/non-human distinction. Let the sets A and R contain descriptions of all possible agents and rules, respectively. These are very large sets indeed. Nonetheless, any particular model, m is defined by a pair of sets (a, r) as follows:

$$m = \{(a,r) : a \subseteq A, r \subseteq R\}. \tag{2.1}$$

In practice, a and r will be proper (in fact tiny) subsets of A and R. A geometrical analogy to this notation is that m represents a point in the space of all models, and its coordinates are (a, r) just as (1, 6) represents a point in the plane, the collection of all possible (x, y) pairs. For example, suppose that we are interested in a social-ecological system with two types of harvesters. Each harvester type harvests one of two animal species. These species, in turn, depend on a third species for food. In this case, $a \equiv \{h_1, h_2, s_1, s_2, s_3\}$ is expressed as a list of 'positions', where agents in positions h_1 and h_2 harvest species 1 and 2, respectively, and the three species, s_i for 1, 2, 3 have certain attributes. The set r would contain a list of rules that pertain to each position listed in the set a.

Once the sets a and r are defined, we would like to study how the system evolves over time. To do so, we must translate descriptions of agents in a and rules in r into a set of measures that convey information about the agents, and algorithms about how these measures change over time. Let X be a vector of (random) variables associated with the agent types in a. These variables will usually be based on the position rules. For our example above, X might contain information about the population of agents in each position at time t. In this case, we would write $X(t) = (H_1(t), H_2(t), S_1(t), S_2(t), S_3(t))$ where $H_1(t)$ is the number of harvesters who harvest species 1 at time t, and so on.

The rules in r can be broken down into two broad types. One group of rules most often specifies rates of change of variables over time or thresholds. The effect of these groups of rules can be expressed through parameters. For example, a boundary rule might affect the number of new boats (fishers) allowed to enter a fishery each year. A payoff rule might affect the rate of payment for monitors in an irrigation system. These effects can be expressed as a rate parameter, i.e. boats per year or dollars per monitor-hour. A scope rule might refer to the minimum allowable water level in a reservoir used for irrigation. This is also specified with a parameter. Let the vector p contain the values corresponding to the class of rules whose effect on the dynamics of the model can be expressed as rate or threshold parameters.

The second class of rules cannot be written down as parameters. These types of rules affect the way variables are aggregated, or interact with one another. For example, suppose we are interested in modeling a voting process in a collectively managed common-pool resource system. Suppose that voters are voting for one of two possible actions. The action which receives the majority of the votes will be chosen. The 'aggregation' rule for the voting model is as follows: the votes for each of two possible outcomes are enumerated. Whichever of these two numbers is larger determines the outcome. In the context of a model, these types of rules are expressed in the functional forms used to map individual or group data into outcomes. There is no way to write them in terms of parameters because they are discrete choice rules. Let this subset of rules in the set r be represented by the mapping F that maps state variables into outcomes that determine the evolution of the system over time. Some of the rules in F will be simple aggregation and transformation rules while others will be much more complex.

This process of model parameterization is schematized in Figure 2.1. The leftmost box depicts the model description in terms of agents and rules. Parameterization entails moving from the leftmost box to the center box whereby agents' attributes are represented by metrics x_i and rules are converted into parameters in the vector p and functional forms in the vector F. Let Δ_t be an operator that describes how X changes over time. If the model is in continuous time, then Δ_t would be the derivative, i.e. $\Delta_t X = dX/dt = (dx_1/dt, dx_2/dt, \ldots, dx_n/dt)$. If the model is discrete time, then $\Delta_t X = \Delta X/\Delta t = (\Delta x_1/\Delta t, \Delta x_2/\Delta t, \ldots, \Delta x_n/\Delta t)$, where $\Delta x_i = x_i(t+\Delta t) - x_i(t)$. The following (generalized) dynamical system can then be constructed based on r and a:

$$\Delta_t X(t) = F(X; p) \quad (2.2)$$

where $F = (F_1, F_2, \ldots, F_n)$ is the map that indicates how $X = (x_1, x_2, \ldots, x_n)$ changes in the period t to $t + \Delta t$ and $p = (p_1, p_2, \ldots, p_m)$ is a parameter vector as discussed above. Each element of the parameter vector p is a function of a subset of rules in r, that is

$$p_i = p_i(r_{I_{i,1}}, r_{I_{i,2}}, \ldots, r_{I_{i,n}}) \quad (2.3)$$

where I_i is a set of indexes specifying the subset of n rules in r that affect parameter p_i and $p_i(\cdot)$ is a function that maps the rules into a real number. For example if $I_4 = \{43, 89, 456\}$ then rules 43, 89, and 456 determine the value of parameter p_4 via $p_4(r_{43}, r_{89}, r_{456})$. In the case of our hypothetical model, the parameter p_4 might specify the allowable harvest of species 1 per harvesting season. This rate of harvest depends on r_{43} which might specify how many days per season harvesting is allowed to take place. Rule r_{89} might specify the type of harvesting equipment allowed. Finally, r_{56} might specify the total number of harvesters allowed to harvest at any given time. The first two rules are authority rules, and the last rule is a scope rule. This final step, model realization, is depicted as movement from the center box to the rightmost box in Figure 2.1.

Note, F is not confined to mathematical functions frequently associated with expressions like that in equation (2.2). It is much more general: it contains instructions about how to convert the state of the system and the rules specified in the vector p into a description of how the system changes over the time horizon specified in the operator Δt (from t to $t + \Delta t$). Equation (2.2) provides a general specification for all the models presented in this volume.

The reader might legitimately be asking what the value of thinking in terms of the framework is. Is such a framework overkill for what is intuitively obvious? On the other hand, consider making sense out of the vast literature of multi-agent modeling versus a more traditional field such as game theory. In the latter, the reader has a very well developed context in which to place a new model. The reader will be aware of many ideas which a new model will be extending, or where a new idea fits in the existing knowledge context. This makes the motives and objectives of the new model easy to place. Because it is not so well established as a field, this frequently isn't the case with multi-agent modeling. There are so many models with so many objectives, it is often difficult to discover why the author has taken a multi-agent approach, whether it is appropriate, and what is gained by doing so. The framework presented above is intended to help organize information about models presented in this volume and about the modeling process in general. A means of classifying, comparing, and contrasting these models, i.e. putting them in a knowledge context, is of substantial value in improving the understanding of the systems they attempt to represent. Finally, the framework can be used as a starting point for organizing one's thoughts for the development of a new multi-agent model.

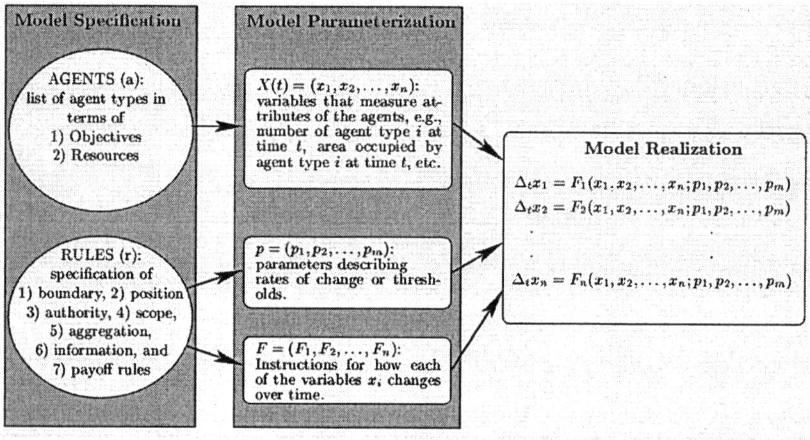

Figure 2.1: Graphical depiction of the framework. In the first box the model (m) is specified by agents and rules. This is typically the most difficult step in constructing a model, and most important in unraveling someone else's model. In the second box the model is 'parameterized'. Here the verbal description of agents is translated into numerical measures x_1, x_2,...,x_n of their attributes as shown in the upper box. There may be more than one metric attached to each agent. The rules must be translated into parameters p_1, p_2,..., p_m that specify rates (often related to authority rules, i.e. tons of resource harvest allowed per year) or thresholds (often related to scope rules, i.e. the water level in a reservoir may never go below a certain level), or instructions F_1, F_2,...,F_n describing how the measures x_1, x_2,...,x_n change over time.

2.5 EXAMPLES

In terms of the general model, the goal of modeling the relationship between humans and resources could be stated as follows: given the set a and a rule set r_u that contains rules beyond human control, we seek a rule set (not necessarily unique), r_c, such that the rule set $r = r_u \cup r_c$ generates dynamics consistent with a given welfare objective. Or, what changes in r most effect the model outcomes? What changes are not important?

The statement of the problem in this way may seem arcane, but it emphasizes the shift in perspective: understanding the relationship between human activity and the environment is more about people and institutions (the nature of humans and the rules by which they interact with one another and other agent types) than optimal resource use. It is in this context that the multi-agent models discussed in this volume have so much to contribute. In this section I discuss several models to illustrate the use of the framework.

2.5.1 A Baseline Example

We begin with the simple open access model of a fishery originally developed by Gordon (1954) and later extended in several directions, e.g. (Clark, 1990; Clark *et al.*, 1979). I discuss a particular extension of it in a subsequent example (Wilson *et al.*, 1999). Many readers will recognize it:

$$\frac{dx}{dt} = r_x x(1 - x / K) - qEx \tag{2.4}$$

$$\frac{dE}{dt} = r_E(TR - TC) \tag{2.5}$$

$$TR = P_q Ex \tag{2.6}$$

$$TC = CE \tag{2.7}$$

where x is the fish stock biomass, E is the fishing effort (vessel days per year), r_x is the intrinsic growth rate of the fish population, K is the carrying capacity of the fish population, and q is the 'catchability' of fish. Fishers enter and leave the fishery depending on the difference between total revenue TR and total cost TC. Total revenue is price P times harvest qEx. The cost of a unit of fishing effort C is assumed constant so that total cost is simply CE. The parameter r_E determines the rate of entry or exit for a given level of profitability.

Is this a multi-agent model? Although not what one would usually associate with a multi-agent model, it is. Stated in terms of the modeling framework, there are two agent types, a = {fish, humans}, and the following rules (not necessarily unique) apply:

1. Boundary: fish do not enter or leave the fishery. They are born, harvested, or die within the fishery. Fishers (anyone who wants to) enter the fishery if $TC > TR$ and leave it otherwise.
2. Position: There is one position for fish and they are all identical. There is

one position for humans: fishers. Like the fish, all humans are identical in their ability to fish and evaluate the financial state of the fishery.

3. Authority: each fisher has the authority to exert one unit of effort per unit time (i.e. fishing days per season). Fish are free to move around the fishery and breed with whomever they wish.
4. Scope: The total fish population may not exceed K.
5. Aggregation: The fishers and fish are 'well mixed': all members of the fish population interact with all others, and fishers interact with all members of the fish population. Thus the amount of fish caught is based on a 'mass action' rule given by the product of total effort (computed by multiplying per fisher effort by number of fishers) and total fish population. All other non-human agents that fish interact with are aggregated into K (intraspecific competition for resources), and r_x (intraspecific interactions leading to reproduction).
6. Information: All information needed for fish to successfully find food and reproduce at rate r_x is available. All information required for fishers to catch fish at a rate of q units of fish per unit of effort is available. All fishers have full information about the total number of fishers presently exploiting the fishery, the total number and price of fish being caught, and the cost per unit of fishing effort.
7. Payoff: fishers are entitled to all the revenue they receive from selling the fish they catch. Each fisher is responsible for their fishing costs.

Finally, the agent attributes (at a minimum) are as follows:

- Fish: Objective – as noted earlier, we cannot say what the objectives of a fish may be. However, if we believe in evolution by natural selection, fish should behave *as if* their goal is to maximize expected lifetime reproductive fitness. Resources: information processing and physical abilities required to survive to maturity, find a mate and reproduce.
- Fishers: Objective – maximize profit from catching fish. Resources: information processing and physical abilities to operate a fishing vessel, locate, catch, and sell fish.

This rule set allows for a very high level of aggregation and the use of a differential equation model to *represent* the system. It is important to note that there exist many rule sets that would generate the same representation.

All of the models one might term 'multi-agent' are generated by altering all or some of these rule clusters. For example, to include age structure in the fish population, the position rules would be altered to include several age classes. Each age class is a position and the rules for moving in and out of age classes are based on the passage of time. To include regulators, the position rules affecting humans would be modified to include two positions: fishers and regulators. In each case, the vector $X(t)$ would be modified accordingly.

2.5.2 The Effect of Position and Boundary Rules on Management

In a recent paper on scale misperceptions in a social-ecological system (Wilson *et al.*, 1999), the authors extend the basic model given by (2.4) and (2.5) and define three local fish populations (three different fish positions), fishers and regulators. The authors define a boundary rule via a 'transfer function' governing how fish move from one position to another. The model attempts to address whether scale misperception makes a difference in the effects of different management (i.e. regulator authority and scope) rules. Scale misperception is represented by the existence of a discrepancy between the position and boundary rules that affect fish populations and those assumed by the regulators. The authors distinguish between three different examples: 1) single population model (perfect mixing and random mating of all fish as in equation (2.4); 2) Panmictic with three local populations (random mating, but fish not well mixed); 3) Metapopulation (fish not well mixed, three populations reproductively isolated). These three descriptions can be characterized by three different sets of boundary rules (migration), and aggregation rules (mating). The question is whether scope rules based on boundary and aggregation rules for case 1) cause damage when applied to fisheries whose boundary and aggregation rules are actually in cases 2) or 3). In terms of the general model: how large is the impact on fishery dynamics when r_c are based on an incorrect understanding of r_u?

These first two examples have focused on describing models based on the seven rule clusters that describe interactions between agents. Often, however, the interest of the modeler lies not in the nature of rules but in the nature of agents. Standard neo-classical economic models usually assume that the objective of agents is to maximize some measure of utility, usually from the consumption of a stream of goods and services. Further, such agents, often referred to as *Homo economicus*, are usually assumed to have infinite information processing capabilities (i.e. infinite problem solving resources). There is a vast literature addressing the implications of relaxing these assumptions based on the early work of Simon (1947) on bounded rationality. Multi-agent models are not necessary, but can be particularly well suited for examining certain questions in this area. We can apply the framework to highlight exactly why.

2.5.3 Agent Types, Resources, and Information Rules

A recent example of a multi-agent approach developed by Jager *et al.* (2000) explores how assumptions about the behavior of two different types of agents, *Homo economicus* and what the authors term *Homo psychologicus*, affect the characteristics of a typical commons dilemma. The distinction between the agents is based on the idea from social psychology that people do not always optimize their outcomes but often engage in satisficing behavior. The authors define agents, called consumats, with a broader range of possible behaviors than *Homo economicus*. Unlike standard models of optimizing agents, consumats engage in one of four behaviors:

- Repetition (if the consumat is satisfied and certain). In this case the consumat is satisfied, is sure about their actions and simply repeats their actions from the previous time step.
- Deliberation (if the consumat is dissatisfied and certain). In this case, the consumat behaves as *Homo economicus* and solves a constrained optimization problem.
- Imitation (if the consumat is satisfied and uncertain). In this case the consumat is satisfied, but is not sure that his or her actions yield the best outcome. The consumat imitates the actions of other similar consumats from the previous time step.
- Social comparison (if the consumat is dissatisfied and uncertain). In this case the consumat wants to improve their satisfaction level, but is uncertain about the outcomes of their behavior. Consumats compare their own previous behavior with that of similar other consumats and choose whichever yields higher need satisfaction.

Given these assumptions the authors run several simulations that compare the impact of humans on the resource base, pollution and other model attributes in a simple ecological model that allows fishing and mining.

This model can be classified using the seven rule clusters and the description of agent objectives and resources. It is instructive to use the framework to compare this model to the simple model in (2.4) and (2.5). Both are models of multiple agents utilizing a common-pool resource. What are the key differences between the models, and why are they important? A careful comparison between the seven rule clusters for each of the models reveals that they are similar in many ways. The boundary, authority, position, scope, aggregation and payoff rules are quite similar. What is different between the two models is the set of information rules. In the simple model, fishers only have information about total fishing effort and are able to compute the total profit being made in the fishery. They make decisions based only on this information. In the model of Jager *et al.* (2000), the fishers have access to more information – they can observe what each fisher is doing, not only the total effort. This difference makes it impossible to use a differential equation model to represent the system.

Next, the capabilities and objectives of the agents are different between the two models. In Jager *et al.* (2000), the agents' objectives are defined more broadly in terms of psychological processes rather than in terms of the simple utility function of *Homo economicus*. Also the agents are able to engage in social comparison, i.e. they are able to process a wider range of information than in the simple model. In this case each agent must be modeled as an individual because they are all making comparisons with one another.

One final interesting issue is highlighted by the framework. By using a differential equation model to represent a multi-agent system, several different boundary and authority rules can be envisioned (recall that the rules given above are not unique). The boundary rule for the baseline example given by (2.4) and (2.5) allows fishers, each capable of exerting an effort e_i to enter the fishery. For the differential equation representation to make sense, we must assume a very large number of potential fishers, each with a very small e_i relative to E. A sensible alternative is to assume a boundary rule in which only one fisher is allowed in the system (sole owner). This single fisher has the authority to change effort E in arbitrarily small increments at will and does so via the rule given by (2.5). In this case the number of agents assumed in the system is not critical to the behavior of the model. It merely changes the interpretation. Is the same true for the model of Jager *et al.* (2000)?

In Jager *et al.* (2000) the authors use 16 agents in their simulations. In a case where agents are modeled explicitly, the number of agents may make a difference. If you have only a few agents, the influence of each agent's decisions on other agents is large. As the number of agents increases, the influence of individual agent's decisions on other agents will be weaker and may have less influence on resource use (Marco Janssen, personal communication). In this case, the model results may be a function of population size. As population grows, a simpler differential equation model might be a reasonable representation of the model dynamics. In some cases, it may be that only when the population is relatively small that the multi-agent approach is necessary or yields novel results.

The important point here is that careful application of the framework can help clarify attributes of the real system required to make a particular representation sensible. Are there rule sets for which the model with 16 agents adequately represents a situation with a large population? For example, we might think of the 16 agents as an approximation of a much larger population which is divided into 16 *groups*, each with several hundred people. Agents are well mixed within the groups but not well mixed across the groups. The groups might represent kin groups, clans, lobby groups, labor unions, etc. This represents a change in the position rules – instead of there being just one position for humans, there are 16. Consumats look at other groups and act accordingly. Indeed, Jager and Janssen (Chapter 6, this volume) have extended the consumat approach to help understand the outcomes of actual experiments on real agents in commons dilemmas. Because the original experiments involved eight agents, the multi-agent model can be easily matched to the real experimental set-up, and the value of the multi-agent approach is clear.

2.5.4 Farm Policy and Payoff Rules

In the previous example, we saw that the key reason for a multi-agent approach was the fact that agents observe each others' behavior as a part of their decision-making process. In this last example, I briefly discuss a quite different multi-agent model that concerns agricultural policy switchings (Balmann *et al.*, Chapter 8, this volume). In this model, the agents are not people, they are farms. There are 12 positions based on farm structure, i.e. dairy, pigs, mixed, etc. Each of the farms behaves autonomously to maximize expected farm income, and all use the same decision rules and expectations, i.e. they all have the same objectives and decision-making resources. The farms do not communicate with other farms, however they interact indirectly through markets. Most importantly, they may increase their size by purchasing land. Balmann *et al.* use this model to explore several questions including the effect different policy options (payoff rules) have on land-use as measured by frequency of farm sizes.

I leave it to the reader to examine the model attributes in terms of the other rule classes. I just want to emphasize one point here. Since the farms are identical in the way they behave, what is gained by a multi-agent approach in this case? Again, the answer lies in the nature of the information rules and the information processing ability of agents. The allocation of land to different farming activities could be treated using a traditional constrained optimization technique. However, this approach assumes unrealistic information processing abilities that real agents do not posses. Further, it assumes that all the information necessary to allocate land optimally resides in one location. As Hayek (1945) points out, such information never exists in one place. Rather, it is dispersed throughout the population. Thus, the solution to the optimal allocation problem (if it is possible to obtain) will not represent the actual situation.

As Balmann *et al.* point out, their model can be thought of as a recursive programming approach that is seeking the best allocation of land, livestock densities, and farming practices given that information is dispersed throughout the agents in the system. Information about the state of system is processed through markets, most importantly, the land market. At each period, farms must make decisions about their activities and whether to increase or decrease in size. As this process proceeds, more profitable farms grow, less profitable farms shrink. The system thereby moves toward increasingly profitable configuration. In this case, the multi-agent approach is being used as a solution method for a problem that traditional optimization techniques cannot address.

From the examples, one can see how the framework combined with the rule classes provides a taxonomy for models. The taxonomy not only provides a means to classify, compare, and contrast models, it can help relate certain classes of models to certain questions of interest. Further, using the framework forces one to think very clearly about the assumptions underlying the model. Assumptions are often taken for granted by experienced modelers and not spelled out explicitly in the presentation of the model. Often, a better understanding of underlying assumptions dramatically improves the ability of readers to understand and benefit from models developed by others.

2.5.5 Complex Adaptive Systems and Multi-Agent Modeling

In the examples discussed above, the range of possible rules for different models is enormous. Nonetheless, once a set of rules is selected, no matter how many, the evolution of the system through time is determined. In theory, given the initial state of the system and the set of rules, the state (or distribution of states if the model is stochastic) of the system at any future time is predictable. For a given social-ecological system, a careful search of the set of possible rules will yield a subset that produces dynamics consistent with a preset objective. However, as the set of allowable rules grows, such a search becomes practically impossible – like searching for a needle in a haystack. This difficulty is further compounded by imperfect knowledge of r_u. Thus, in the same way as it is impossible to find an optimal control solution for resource use, it is impossible to specify the rule set r_c to achieve a particular outcome.

A similar problem is faced by our immune system which has to cope with forever changing invaders (Janssen, 2001 and Chapter 3). The types of possible invaders far outnumber the types of proteins our immune systems can produce to counter them. It must rely on a different strategy. It cannot rely on central direction with a fixed set of rules to combat invaders. It must self-organize based on local information and adapt, or 'fit' to new invaders as they appear. Solutions to invasion problems emerge from the dispersed interactions of the individual agents that make up the immune system.

Analogous to the immune system combating invaders is a social system, made up of diverse individual agents, addressing environmental problems.

Systems made up of interacting individual agents such as immune systems, economies, ecological systems, and the brain share certain characteristics (cf. Levin, 1998):

- sustained individuality and diversity of components,
- localized interactions among those components,
- an autonomous process that selects from among those components based on some criterion,
- aggregation: i.e. global features emerge from local interactions.

These key features generate systems that evolve over time, learn, and can solve complex problems. Given these similar features, John Holland (1992) grouped these systems under the common name of complex adaptive systems.

The problems and issues humans face with regard to the environment emerge from local interactions between individual agents. Because of their emergent nature, these problems cannot be effectively addressed with approaches that focus on aggregate behavior. Instead we must craft rules (institutions) to govern interactions between individual agents so that problems do not emerge. Consider again the scale misperception model (Wilson *et al.*, 1999). The authors suggest that management rules based on aggregate behavior rather than individual interactions (although the authors do not model individual fish interactions, they model interactions between different classes of fish that the aggregate model ignores) can cause overfishing. In Tsembaga society the rules that are a part of the ritual cycle govern, to a large degree, interactions between individuals. They were not crafted in response to an aggregate behavior of their system. The rules that govern Tsembaga intrapersonal, pig–person, and person–garden relationships are such that environmental degradation does not *emerge* in their social ecological system.

It is impossible to work out a set of rules (institutions) that can successfully address a wide range of complex problems across a wide variety environments. The list would take too long to compile, and would be too large to store. The complex adaptive systems metaphor suggests that institutional structures should be composed of a set of rules that act as building blocks, and a set of rules that guide an adaptive rule construction/changing algorithm over time. Now the question reduces to the search for such algorithms. Does this help? Are there just as many algorithms as there are rules? As John Holland has recently put it, 'Are there "general purpose" adaptive algorithms that can discover good action sequences [rules] over a wide variety of environments, without prior tuning for each environment?' (Holland, 2001, p. 1). The discovery of such an algorithm in the context of common-pool resources would be a great step forward. In order to tackle this problem, multi-agent modeling approaches are *essential*. The continued development of multi-agent models of complex adaptive systems will help provide the necessary tools to help construct institutional structures able to cope with the complexity of social-ecological systems.

2.6 CONCLUDING REMARKS

The purpose of this chapter has been first to provide some context for the diverse set of modeling exercises presented in this volume, and second to set out a general rule-based framework which can be used to more carefully think about and compare different multi-agent models. It is important to emphasize that the framework is not a recipe for the models presented in this volume or any other multi-agent model, although it could be. Rather, it is intended as a tool the reader can use to increase the value of the models presented herein.

On the first point, I argue that the emergent nature of environmental/resource use problems in some cases necessitates the recognition of the importance of the diversity among agents and of local interactions. Further, the complexity of ecological systems precludes the discovery of a set r_c that will generate system configurations consistent with a given welfare objective. Instead, the set(s) r_c should contain a set of building block rules and a set of rules for adaptively constructing/changing rules using the building blocks. Multi-agent approaches are essential to address questions related to diversity, local interactions, and adaptive management. Many of the models presented in this volume grapple with these issues.

Regarding the second point, I outlined a general mathematical framework based on the rule set developed as a part of the Institutional Analysis and Development (IAD) framework (see Ostrom *et al.* (1994) for details). In the IAD framework these rule sets apply to humans. I have suggested that they can be applied to any and all agents in a system. The examples have been limited to animate agents, but inanimate agents could be described as well – i.e. molecules are agents which occupy well-defined positions with scope rules (the rules of physics such as light has a maximum speed, etc). This level of resolution is, of course, absurd for the types of issues addressed in this volume.

The framework serves two important purposes: it provides context for multi-agent models as a subset of a more general class of models, and it provides a mechanism by which multi-agent models can be compared. It is important to emphasize that there is nothing special about multi-agent models, all models are 'multi-agent'. What sets models presented in this volume apart from other types of models is the nature of the position and aggregation rules upon which they are based. In general, they have many more positions (diversity of agents) than in other 'process-based' or system-type models, more boundary rules (emphasis on space or movement between agent classes), and the aggregation rules emphasize local interactions, i.e. agents can only interact with their nearest neighbors in contrast to differential equation type models where it is implicitly assumed that every agent interacts (randomly) with every other agent.

The comparison across multi-agent models is also essential for the reasons discussed earlier. Namely, what level of resolution is necessary to address particular classes of problems? The aim is to use the simplest model possible, but no simpler. The tendency with multi-agent models can be to use excessive levels of complexity. As model complexity increases (more positions and boundary rules) model utility decreases. A careful study of appropriate rules for appropriate problems will be important for the continued development of useful multi-agent models.

3. Changing the rules of the game: lessons from immunology and linguistics for self-organization of institutions

Marco A. Janssen

3.1 INTRODUCTION

Multiple agents use multiple services from most ecosystems. Without rules about who is allowed to consume which ecosystem service when and at what rate, conflicts may arise. To avoid tragedies, we need rules that shape interactions between people. The question is how to formulate these rules of the game, the institutions, in order to sustain the long-term development of the social ecological system. Game theory has been used to provide insights about which rules, or which mix of rules, lead to the best performance. However, the set of rules in game theoretical analysis is fixed. The puzzle is how new rules are created, and how effective rules get selected and are remembered. This may contribute to a better understanding of how institutions evolve.

Self-organization of institutions has been prominent in the field of common-pool resources. The reason is that laboratory experiments and field surveys show that people are able to create effective institutions without governmental intervention (Ostrom *et al.*, 1994). These findings are in contrast with conventional economic theory, which assumes rational selfish individuals. The picture of empirical research shows the existence of social norms (Ostrom, 2000) and reciprocity (Fehr and Gächter, 1998; Gintis, 2000a). The hypothesis in institutional science is that the existence of social norms explains observed reciprocity. The question remains how formal rules (such as law) and informal rules (such as social norms) have emerged and how they may evolve over time.

Similar questions on self-organization have been addressed in other disciplines such as immunology and linguistics. Immune systems are of interest since they are able to produce effective responses to new microbiological invasions, and remember successful responses. Key questions in linguistics are how a language can evolve in a group of agents, and how more complex syntaxes evolve. These two disciplines study the origins of rules in quite different systems. Besides being interesting analogies, these fields have developed computational models of their systems of interest and this led to the hypothesis that a better understanding of the self-organization of institutions might be derived when we cross-fertilize insights and tools with these other disciplines.

The next section discusses in more detail the research on management of common-pool resources. In Section 3.3 the puzzle on the origins of rules is reviewed in relation to immune systems and language development. Sections 3.4 and 3.5 describe the relevant dynamics of immune systems and language development. Methodology to simulate the evolution of rules is discussed in Section 3.6, while Section 3.7 relates this to a general framework of modeling self-organization of institutions. Section 3.8 concludes.

3.2 COMMON-POOL RESOURCES

When people have free access to a resource they will consume ecosystem services to the point where private costs equal the benefits, whereas the whole group is affected by external effects. This phenomenon is known as the tragedy of the commons (Hardin, 1968) and is in line with the solid theoretical models which predict how people behave in non-cooperative games. Conventional theory predicts that players in a non-cooperative game will follow a Nash equilibrium.

Laboratory experiments have been performed to study this prediction. Even in the simplest experiments, without communication between the participants, anomalies were found that are not in line with the predictions (Ostrom *et al.*, 1994). In none of the reported experiments on common-pool resources was the predicted Nash equilibrium observed. Furthermore, the total consumption of the common-pool resource fluctuates in time (Ostrom *et al.*, 1994). In additional laboratory experiments the role of institutions was analysed. It was found that communication could have an important influence on the level of cooperation in a group of players (Ostrom *et al.*, 1994). Furthermore, this cooperation will increase when sanctioning of defectors is allowed, especially when the players themselves have determined the rules of sanctioning.

Careful analysis of a variety of common-pool resources in different parts of the world shows that there are some common characteristics among self-organized institutions of common-pool resources. These include the presence of boundary rules, and authority rules related to allocation, and active forms of monitoring and sanctioning (Ostrom *et al.*, 1994). Furthermore, traditional societies, which have established a sustainable interaction with their environment, use rituals, and taboos as mechanisms to practice and remember ecosystem management (Berkes and Folke, 1998).

According to conventional economic theory, a solution to the tragedy of the commons is to calculate the optimal level of resource extraction and to allocate individual permits of harvesting to the individuals of the group. Based on these insights private property, or governmental ownership and control, was recommended. This implicitly assumes that the government has a perfect understanding of the system and acts in the public interest. The empirical evidence gives a radically different perspective. Cooperative solutions can emerge spontaneously. Unfortunately it is not known precisely under which circumstances this may happen. However, it is clear that mutual trust among the participants is important to derive such a self-organizing process. This might be the reason that governmental interventions, as recommended by conventional theory, are often not effective. The government is not one of the original participants with whom a trustful relationship has evolved.

Common-pool resources have evolved and nowadays we experience multiple-use resources too (Steins and Edwards, 1999). A common-pool resource can be, for example, a lake where fishermen catch fish. In a multiple-use resource, other uses of the lake become important such as tourism, natural conservation and industry. The involvement of different stakeholder groups makes the management of resources more complex. Steins and Edwards (1999) advocate the use of platforms for resource use negotiation. Such participatory approaches are discussed in Chapters 12 and 13. From an analytical perspective the question appears on how multiple use may influence the possibilities to reach a collective action.

3.3 THE ORIGIN OF RULES

The general aim of this paper is to outline an agenda for the understanding of the evolution of rules, by recognizing that similar questions are addressed in immunology and linguistics. This section formulates a number of research questions for such an agenda.

Emergence of new rules relates to the difference between individual selection (individual rules) and group selection (cultural rules). Various scholars provide different explanations for cooperative behavior (Sober and Sloan Wilson, 1999). Some argue that cooperative behavior does not exist due to individual selection (selfish genes), while others argue that genetic selection operates on different levels which explains that groups that contain altruistic individuals can have a higher fitness. Institutions can be seen as an expression of cultural rules. We want to understand how such rules emerge in social-ecological systems.

Rules can be formal, like law, taxes and fences, but also informal such as norms, taboos and rituals. The aim is to understand general system characteristics. How do rules of interaction between agents, and between agents and the environment, emerge, selected and remembered? How does the expected self-organization of rules depend on the dynamics of the environment?

As an organizing framework the social-ecological system is approached from both an immune system perspective as well as a language development perspective. The purpose of using these rather different systems is to derive inspiration about how to analyse the evolution of rules, like lymphocytes evolve in an immune system, and words and grammar evolve in languages. Obviously, there are important differences. For example, social-ecological systems are not organisms such that the analogy of the immune system perspective is limited.

Janssen (2001) argues that there are a number of fascinating similarities between social-ecological systems and immune systems. Social-ecological systems can be viewed as distributed systems in which organisms (humans, animals, and plants) respond to local changes. Traditional ecosystem management, that has proved to be successful over long time periods, has developed mechanisms to detect anomalies, develop effective responses and remember successful policies. These characteristics of successful ecosystem management are similar to the functioning of biological immune systems. Nevertheless, the analogy can be useful to explore. A detailed discussion on the immune system perspective for the evolution of rules can be found in Janssen and Stow (2002).

Language consists of words and grammar. Language development is a dynamic process. Words and grammar change in time. The meaning of words can be different for each person. Sometimes the meaning of words overlap and sometimes we cannot find words to express our feelings. Notwithstanding the complexity of language, children are able to learn to speak without having experienced all possible utterances. This relates to the problem of learning in complex environments. When a stranger enters a group (s)he has to learn the formal and informal rules of this group. This is a complex task since the stranger would not be taught the rules explicitly. (S)he has to learn the institutional rules like a child learns to speak.

Before the discussion on self-organization of institutions is continued, the general system dynamics of immune systems and language developments are reviewed in more detail.

3.4 THE IMMUNE SYSTEM

The immune system protects the body from harmful pathogens. Pathogens include a large set of micro-organisms such as bacteria, parasites, viruses and fungi that constantly invade the body. These pathogens are the source of many diseases. Rapid recognition and elimination of harmful pathogens is therefore required for sustaining the health of the organism. Furthermore, the immune system is able to remember successful solutions to overcome harmful invaders. A brief sketch of how the immune system functions follows. It is primarily based on Hofmeyr (2001), who provides a clear description of the immune system from the perspective of systems dynamics.

The immune system consists of certain types of white blood cells, called lymphocytes, which cooperate to detect pathogens and assist in the elimination of those pathogens. Each of these lymphocytes is a kind of mobile independent detector which circulates around the body via the blood and lymph system. There are millions of these lymphocytes, forming a system of distributed detection without central control. The surfaces of lymphocytes are covered with various receptors, while the surfaces of the pathogens are covered with epitopes. Both the receptors and the epitopes have complicated three-dimensional structures. The more complementary the structure of the receptor and the epitope, the more likely it is that binding will occur. The strength of this is called affinity. Therefore each lymphocyte is specific since it can only bind a particular set of similar pathogens. Pathogens can have different types of epitopes, thus several different lymphocytes can be specific for a single kind of pathogen.

This functioning of receptors leads to local signals of recognition that mediates the immune response. Local recognition of a harmful pathogen occurs when the number of bounded receptors on a cell exceeds a certain threshold.

The detection and elimination of pathogens is a consequence of trillions of cells interacting through simple, localized rules. Detection of pathogens focuses on the identification of harmful 'non-self' pathogens. The immune system has evolved in such a way that false negatives and false positives are rare. A false negative is a non-detected non-self, and a false positive is an element corresponding to self but recognized as a non-self. Consequently, the immune system is very resilient to the failure of individual components and attacks on the immune system itself.

The immune system has a repertoire of various types of responses because different pathogens must be eliminated in different ways. New types of responses are generated continuously like a kind of random generator. When a particular harmful pathogen has been detected, the immune system selects an appropriate response triggered by chemical signals. So-called antibodies are generated which neutralize pathogens by binding pathogens and self cells.

The immune system is adaptive, because it adjusts the distribution of different kinds of proteins as a response to its recognition of specific classes of pathogens, and retains a memory of the successful responses to speed up future response to those and similar pathogens. This adaptation occurs during the first response when a new pathogen invades. This initial response is slow and the organism will experience an infection, but the immune system retains a memory of the kind of pathogen that caused the infection. If the body is infected again by the same kind of pathogen, the response of the immune system is faster, because it remembers the earlier response to this pathogen.

The memory of the immune system is related to the fact that it is unable to contain a sufficient diversity of proteins to respond to all possible pathogens. The immune system contains about 10^6 different proteins, while there is a potential of 10^{16} different foreign pathogens to be recognized (Hofmeyr, 2001). Therefore, the immune system needs to contain enough diversity to respond to new kinds of pathogens. One of the main mechanisms for producing the required diversity is a pseudo-random process involving the recombination of DNA. Furthermore, it needs to have a memory function sufficiently powerful to respond rapidly to pathogens that invade frequently. How an immune system is able to remember successful responses can be understood in the following way. Once the immune system experiences pathogens of a type it has not experienced before, it may take several weeks to eliminate the infection. During these few weeks the immune system is learning to recognize the new type of pathogen. This occurs due to generation of new designs of receptors and replication of cells that are able to recognize the pathogen. If the pathogen invades the body again, the response is more rapid and effective. However, this is dependent on the response being remembered. There is only a limited capacity for remembering successful responses. When the body is not infected for a long time, memory can be lost. A good example is immunity to malaria infection. When persons are not exposed to the malaria parasite for longer periods, immunity declines.

3.5 LANGUAGE DEVELOPMENT

Language is a dynamic system of information exchange between agents. Language emerges at the intersection of three processes which act on different time scales: *learning* (children learn a language in order to communicate with their environment), *cultural evolution* (languages change over generation when words enter or leave languages, meanings shift and phonological and syntactic rules adjust), and *biological evolution* (the mechanism to learn and process language adapt over evolutionary time) (Kirby and Hurford, 2001). A traditional approach of linguistics was to unravel the universal elements of languages. There are indeed universal tendencies in natural languages, such as the distinction between subject, object(s) and verb, but also languages exist which do not fit with these universal patterns (Steels, 2000). The reason for the hypothesis of universality lays in genetic information, such as a language organ inside the brain (Steels, 1997). Language, from this perspective, would largely be transmitted genetically. Children learn their mother tongue through parameter setting. Empirical evidence exists that lower animals like crickets, quails and song birds have stored their species-specific communication patterns in highly localized hard-wired neurological structures (MacWhinney, 1998). However, an important difference between lower animals and humans is that children learn language gradually. Besides biological characteristics of humans, which determine the ability to speak, the social environment determines the languages that can be learned. For example, a Chinese child who grows up in the Netherlands can learn to speak Dutch perfectly, and may even not be able to learn to speak Chinese very well if (s)he is not taught to speak Chinese during his/her youth. Another example is a simple test that was used during World War II to check whether somebody was really Dutch. If you are able to pronounce 'Scheveningen' in the right way, one is almost sure that you have learned to speak Dutch during your childhood because the pronunciation of 'sch' in Dutch is something very specific. Thus the right pronunciation of 'Scheveningen' is an indicator for the identity of Dutch persons.

A specific characteristic of the human language is the use of writing to transmit information. There are various writing systems. The distinction between writing systems is the division of *logographic* writing (like Chinese) and *phonographic* writing (like English) (Sampson, 1985; Robinson, 1995). Japanese is a special language since it is a mixture of both language systems by combining Chinese characters and Japanese and English letters. The phonographic writing system has its origins in Mesopotamia about 5000 years ago. The number of letters varies between different languages. The original Semetic alphabet probably consisted of 22 letters, the Romans had a 21-letter alphabet, while the English alphabet consists of 26 letters. The Chinese writing system is at least 3000 years old. The writing systems have to balance flexibility and efficiency. For example, the Chinese writing system is hard to learn, and its interpretation is very context dependent, but once one is able to use it, one can read and write much faster compared to the English language, which is however easier to learn, and less context dependent. One may expect that writing systems become more efficient and context dependent when they are not much influenced by other languages.

A more recent approach to languages is rooted in the study of complex adaptive systems. Language is seen as the result of local interactions of language users (Steels, 2000). Like species of organisms, languages evolve due to local interaction, selection, mutation, adaptation, etc. Agents produce and interpret sentences partly in an automatic way but occasionally novel situations or errors lead to adjustment of language (Steels, 2000). Languages are the result of a self-organizing process. Since agents benefit from being understood, agents benefit from having the same language as those with whom they communicate frequently. By imitation and teaching children to speak a certain language, local clusters evolve of the same language. This explains linguistic diversity (Nettle, 1999). Often expressions and pronunciation differ between two neighboring villages.

Language evolves as we are confronted with new situations, other languages and due to social processes. New technology such as computers leads to new terms like *e-mail*, *zip-drive* and *internet*, but also adapts the meaning of existing words like *apple*, *window* and *mouse*. Due to contact with other languages new words invade the native language. In the Netherlands language is very much influenced by English through English music and television programs. On the other hand, countries like France and Germany try to limit the invasion of other languages by dubbing foreign television programs into their own language. Social processes also play an important role. Younger generations want to derive their own identity and start to use their own expressions and words which may be adopted by older generations later on. Furthermore, languages differ slightly between social classes due to local adaptations and selection of specific terminology. This is very prominent in science, where specialists have trouble explaining their research to non-specialists without using jargon.

3.6 METHODOLOGY

The possible analogies with language development and immune systems provide a basis for studying system characteristics that are of interest for understanding observed phenomena in the management of common-pool resources. In this section the main formal approaches of studying immune systems and language developments are discussed.

Within the studies of artificial immune systems (Dasgupta, 1999; Hofmeyr and Forrest, 2000) and computational linguistics (Steels, 2000; Cangelosi and Parisi, 2001), three types of approaches are frequently used to capture the system dynamics: game theory, neural and immune networks, and evolutionary computation. Although other methods are available such as differential-equation models and cellular automata models, these three approaches can serve as starting points for the development of the stylized models of social-ecological systems.

3.6.1 Game Theory

Game theory is a powerful tool for the selection of strategies of interaction and is currently one of the formal approaches for studying institutions. Game theory is also used to study evolutionary processes in language. The players have the chance of using different rules, whether this is grammar or vowels, and they will benefit when other players understand their messages. Evolution of cooperation is in this case the evolution of communication. Results of these games show clusters of rules, or the dominance of one rule. De Boer (1997), for example, describes the use of adaptive imitation games to simulate the emergence of vowels. The question is how agents can come to share a system of vowels without having been given a pre-programmed set and without central supervision. Agents produce random sounds from their repertoire, which other agents will try to reproduce. The agents may have to learn new sounds, reinforce sounds from their own repertoire or may not recognize these sounds. Robotic simulation leads to a repertoire of shared sounds that emerges through self-organization. Furthermore, the kinds of vowel systems that emerge have similar characteristics as those of natural vowel systems (De Boer, 1997).

An example on the evolution of grammar is given in Nowak *et al.* (2001) who describe a game where players benefit when they understand each other. Since players with a higher payoff generate more offspring, Nowak *et al.* (2001) can analyse the conditions under which a universal grammar emerges, and when rules-based grammar can evolve compared to a list-based grammar. Emergence of universal grammar depends on the accuracy of the learning process and the number of candidate grammars. The condition for rule-based grammar depends on the number of sentences types related to accuracy of learning and candidate grammars. If a population can effectively communicate by remembering only a few sentence types, a list-based grammar will evolve, otherwise a rule-based grammar will evolve.

3.6.2 Networks

Neural networks are powerful learning tools in complex environments. In a neural network, knowledge is represented in the strengths of the connections. Learning must be a matter of finding the right connection strengths so that the right patterns of activation will be produced under the right circumstances (McClelland and Rumelhart, 1986). Due to these system characteristics, neural networks are used for learning and remembering complex patterns. An illustration is a language development project where neural networks were used to study how children learn the rules when to use 'die', 'der' or 'das' in the German language, where 'die', 'der' and 'das' are used for 'the' in different circumstances (MacWhinney, 1998).

Networks also play an important role in immunology. Immune systems can be described as networks of molecules which simulate the dynamics of lymphocytes (Jerne, 1973; Perelson, 1989). The immune network dynamically maintains the immune memory using feedback mechanisms. Thus if something has been learnt, it will be remembered if it continues to be reinforced by other parts of the network.

3.6.3 Evolutionary Computation

Evolutionary computation covers a wide range of algorithms that contain the basic mechanisms of mutation and selection. The most prominent algorithm is the genetic algorithm, which in addition to mutation and selection includes crossover (Mitchell, 1996). Evolutionary algorithms are powerful in generating robust solutions of problems in a noisy environment.

Evolutionary algorithms are often used to generate new combinations of 'genetic' information. For example, new vowels are created by a random combination of articulatory controls, or mutation of existing vowels (De Boer, 1997). Simulation experiments to study the effects of evolution on the genetic encoding for antibody molecules were based on genetic algorithms (Hightower *et al.*, 1995). They found that an optimal solution to the antigen recognition task requires a balance between the conflicting issues of coverage redundancy and coverage gaps.

3.7 MODELING THE SELF-ORGANIZATION OF INSTITUTIONS

This section describes different stages in the self-organization of institutions: coding, creation, selection and remembering of rules. The hypothesis is that different types of rules emerge under distinct socio-economic and ecological conditions. It is expected that the methods already discussed can be used to develop a kind of computational laboratory. By careful experiments with such a laboratory it is expected to be possible to derive better understanding under which conditions observed rule-systems can evolve.

3.7.1 Coding of Possible Rules

To understand the emergence of rules, we have to understand the decoding of rules. Just as DNA is the code by which living beings can be described, we need a genetic structure of rules. Interestingly, Crawford and Ostrom (1995) provide us with a useful starting point with the introduction of a grammar of institutions which provides a theoretical structure for analysis of the humanly constituted elements of institutions like rules, norms and shared strategies. There has been a discussion in institutional science whether institutions are rules, norms or strategies. Crawford and Ostrom (1995) propose a broader framework, which encompasses all three concepts. The grammar of institutions enables them to generate structural descriptions of institutional statements. The syntax of the grammar institutions contains five common components. Different compositions of these components lead to strategies, norms or rules. Crawford and Ostrom (1995) argue that such a framework can help to identify and classify institutional statements. Besides the grammar, Ostrom *et al.* (1994) distinguish seven types of rules within institutions (see also Chapter 2).

3.7.2 Creation of Rules

Given the space of possible rules, how do new rules emerge? The development of language is a process of probing, imitation and mutation of words and rules (Tada, 1997). This leads to local clusters of linguistic rules. Sometimes words from other languages try to invade the language. When this is perceived as not distorting meaning, the new word is tolerated. Another factor that triggers creation is the pressure that is experienced. For example, our immune system is constantly confronted with microbiological invasions, but is organized efficiently to recognize and eliminate harmful cells or molecules. One of the reasons for the success of immune systems is the continuous random generation of new responses, the lymphocytes.

Translated to institutions and rules, new rules emerge due to experimentation and by contacts with different types of institutions for diverse ecosystems. In the spirit of Jacob (1977), creation of new rules can be viewed as tinkering. In contrast to an engineer, a tinkerer does not know exactly what (s)he is going to produce but uses whatever (s)he finds around him or her. Once new rules are proposed, the selection process takes over.

3.7.3 Selection of Rules

Before proposed rules become effective, that is, before becoming a social norm or a law, a selection process tests the rules. In fact all rules are nested in another set of rules which defines how the first set of rules can be changed. Kiser and Ostrom (1982) distinguish three levels of rules. An operational rule which relates to day-to-day decisions, collective-choice rules which determine decisions that affect the operational rules of individuals. At the collective-choice level decisions are imposed on individuals. Those who are monitored and break rules can be sanctioned. Constitutional-choice rules define formal procedures on how to change collective-choice rules. An example is to have elections for a new parliament that can decide on new regulation that may affect the daily decisions of the citizens.

Besides formal constitutional mechanisms, informal mechanisms are also important, such as the spreading of trust in a social network. It is known that the level of trust is higher if the network has a higher density (Granovetter, 1985), and the connectiveness of a social network – especially links between clusters – influence the spread of information (Granovetter, 1973). Models of trust can be formalized by repeated games, where agents derive information about each other due to repeated confrontation (Janssen and Ostrom, 2002). An interesting example of a more advanced computational model on the evolution of trust and cooperation between strangers is developed by Macy and Skvoretz (1998). This model combines a social network in which agents play iterated prisoner dilemma games. Using a genetic algorithm, and decoding characteristics like greeting behavior and distrust of strangers, clusters of trust relationships emerge.

The selection process of rules can also be endogenous. Constitutions change over time, but this is a much slower process than the actual selection process of the operating rules between agents. This relates to the concept of resilience (Holling, 1986). These slow variables can be a source of collapse. Therefore it is important to test institutions, by violating the rules or by social debates.

3.7.4 Remembering Rules

Even though rules have been selected, they might not be useful in every situation. It might be efficient to limit the number of rules. However, in a system that is influenced by variability, a diversity of rules can be used to spread risks. This can be compared with the functioning of ecosystems, which are adapted to a certain amount of variability. Reduction of functional biodiversity can reduce the ability of ecosystems to recover from disturbances (Holling, 1986; Levin, 1999). Due to this functional diversity, ecosystems maintain a memory of useful responses in a large set of situations. This is related to institutional diversity. Although many different institutions exist, and even overlap in their functionality, a certain level of redundancy is essential for the resilience of social ecological systems. Low *et al.* (2002) suggest a number of conditions that make redundancy advantageous: when information transfer across subsystems is inefficient or slow, redundant local systems may be more efficient and in large, spatially heterogeneous systems, redundant local systems may work better. A possible reason for the importance of redundancy relates to the ability of local systems to have better access to local information and a more rapid response than higher system levels.

3.8 DISCUSSION

This contribution provides ingredients for an approach to study the self-organization of institutions. The main question is how rules are created, selected and remembered. Rules can be seen as an outcome of a continuous process of tinkering, experimentation, making errors, breaking rules, selection and competition. To understand how rules between people organize themselves these processes have to be investigated. During the last 20 years, computational models have been used to study immune systems and language systems in order to understand the key characteristics of these systems such as novelty, learning, selection and memory. These tools can provide a good starting-point for studying self-organization of institutions.

The computational models and methods of immune systems and language systems can be used to develop systems that are able to generate self-organization of institutions. These computational laboratories can be used to perform repetitive experiments under which social and ecological conditions for self-organization of institutions can emerge. Such experiments can provide a deeper understanding of institutional change. Besides better understanding of observed phenomena in the past, new situations can be explored.

ACKNOWLEDGEMENT

The author thanks Marty Anderies, Daniel Stow and Henry Wright for helpful comments and the European Union (IST-2000-26016) for financial support.

4. Futures, predictions and other foolishness

Roger Bradbury

4.1 INTRODUCTION

Science has got itself into a bit of bother with the problem of prediction (Bradbury, 1998b). It seems that, in this regard, science may have become the victim of its own success. Its success at predicting things, a feat of some utility, has led to a view that the goal of science is prediction. This is nowhere more evident than in the newish scientific area of complex systems, and is especially an issue in that part called 'complex adaptive systems'. In these areas, we can see the desire to be part of the science club – the yearning of the new member to be accepted – expressed as 'me too'. If science is about prediction, the new members say, then so is complex adaptive systems, and if good science means good prediction, then good prediction is what we will strive for in our new field.

Sadly, for these enthusiasts, it is not the case that science is primarily about prediction, nor is it the case that the bit of science called complex adaptive systems is at all about prediction. Contrariwise, it is the case that complex adaptive systems is an area of science where we can show that prediction is neither realistic nor possible.

Richard Levins, in his famous and prescient paper, 'Complex systems', perhaps put it most succinctly:

> I would assert that the goal of prediction is a subordinate one: a contingent objective which is legitimate only to the extent that it helps us decide among alternative hypotheses, or in situations where there are applied problems of importance, but not equivalent to the goal of science itself, which is explanation. (Levins, 1970, p. 78)

In this chapter, I want to explore this matter a little further. I want to tease out the role of prediction in science more fully, to argue why prediction is a particularly unproductive goal in the case of complex adaptive systems, and why science itself is a complex adaptive system. And finally I want to speculate what we should do instead of prediction if we wish to encourage the new science of complex adaptive systems to unfold and reach a full flowering.

But before I begin my analysis, I should offer a caveat, since much criticism of the ways of science is from outside science itself, while this comes, as it were, from the inside. I am not presenting, in any sense, a deconstruction of science, but rather bringing the most powerful analytical tool known, science, to bear on this scientific problem. I am not attracted to the phoney argument that we must step outside science in order to discuss it, and that the humanities or philosophy are the most appropriate venues for that discussion. I would perhaps go even further and argue that science has by far the most powerful set of intellectual tools for unpacking itself or any other sensible problem, and that recursion – using science on itself – provides no impediment to that process. I hope to show that the particular case of complex adaptive systems – where recursion is everywhere – provides strong evidence of science's muscularity.

4.2 IF EXPLANATION IS THE GOAL, WHY DID SCIENCE INVEST SO MUCH IN PREDICTION?

The answer to this question, as I advanced in an earlier essay (Bradbury, 1998a), is because it could, and people embraced science to the extent it could pull off the amazing trick of prediction for certain classes of systems.

Science has been terrific at knowing the future of those systems where past, present and future do, as a first approximation, lie along a line in some Cartesian geometry. Indeed since at least the time of Ptolemy, in the third century BC, we have known how to predict astronomical phenomena such as eclipses using just such geometries. Astronomical systems are the canonical examples of highly predictable systems that broadened in the succeeding millennia to embrace other dynamical physical systems, such as gases, machines and electricity. We can reach into the future with both precision and accuracy in such systems, given some knowledge about the state of the system in the present.

Science became good at predicting the future of such systems by using models. That is, science invented a fantastic trick: by creating a simpler caricature of the real system – one stripped of those bits extraneous to the matter at hand – in its own caricatured time and space, it was able to 'run the model' into the model's future (or past) and predict what would happen in the real world. Notice that I have not mentioned the *M-word* yet. Models do not have to be *mathematical*, although they often are. They can be mechanical, electrical, geometrical, pictorial or even narrative as well. All they have to do to qualify as models is to have the character of a caricature – being stripped down to the essential elements – and having their own world. Science found some models to be so powerful they were called physical laws.

I will return to the matter of the intricate relationship between models and prediction below.

Those systems whose future science successfully predicted with such models have in common the fact that they are all rather simple. They are made up of relatively few distinct entities – planets are all much the same when it comes to the way they orbit the sun – and their interactions tend to be limited: Newton's inverse square law of gravitational attraction ensures that, for all practical purposes, a planet is only affected by the very near or the very large, all the rest may be ignored.

But these are by no means all the interesting systems in the universe. They may, in fact, be the least interesting when compared to complex systems, and, in particular, to complex adaptive systems. Complex systems are characterized by, well, complexity. Where simple systems have few entities, complex systems have many; where simple systems have few interactions, complex systems have many. The weather, the oceans, and the earth beneath our feet are all examples of complex systems.

Complex adaptive systems, such as all living things and their parts – cells, say, or immune systems – and their assemblages – societies, economies, ecosystems and so on – take this complexity further in exhibiting the peculiar characteristic of learning, of evolving, of adapting. We speak more easily of the *behavior* of such systems to denote this flexibility, where we speak naturally of the *dynamics* of simple systems to denote their relative fixedness.

The science writer Laurence Marschall described it this way:

> Complexity is what financial markets, mammalian immune systems and ecological communities have in common. The ability to interact with the environment, to recognize patterns in the world and to apply acquired knowledge to the modification of future behavior is easily detected, yet the definition of complexity remains elusive. (Marschall, 1994, p. 45)

We know, from our everyday experience, something of the ways of these complex adaptive systems. The empirical evidence that we cannot predict them is all around us. This evidence runs up against the native optimism of science: perhaps, as science advances, we shall predict them, as we did the simpler systems of old.

We need to be able to temper such speculation with a realistic expectation of what science may be able to do with the notion of prediction in complex adaptive systems. We see below, as we dig a little deeper into the notion of prediction, that this leads us to some new ideas about its nature, the role of models and the nature of complex adaptive systems themselves, as well as a better understanding of what it is that science does.

4.3 THE NOTION OF PREDICTION

When we think about prediction, we mostly think about the techniques of prediction, whether through Tarot cards or weather maps. Or else we think about the prediction itself: that tomorrow will be fine or that a dark stranger will enter one's life. We tend not to think of prediction as an idea, a notion that we may be able, in some way, to say something sensible about things that have not yet happened, or about places that we have not yet seen.

I want to examine the notion of prediction at a very high and abstract level because, I argue, the problems of predicting the behavior of simple and complex systems are disparate. They can only be brought within the one intellectual framework at such a level. It also allows me to comment on the nature of this disparity without getting diverted, at this point, into a discussion about deterministic versus stochastic prediction or becoming bogged down in a discussion about issues of precision and accuracy in prediction. But it points to a fundamental problem about whether and which systems have a determined future.

Prediction embodies both the ideas of extrapolation – what happens beyond our purview – and interpolation – what happens where we have not yet looked within our purview. However it is in the sense of looking into the future, the sense of extrapolation, that prediction is mostly couched. It is in that sense that we explore the notion here.

The notion of prediction is a bundle of different ideas, each of which needs to be examined to see if together they make sense. In this bundle are ideas about objects and their boundaries, domains and environments, classes and instances.

When we predict, we are usually thinking about the future of some distinct thing. We do not as a rule mean the future of everything everywhere, but rather the future of some distinct bit of the universe. Moreover we are usually thinking, in some sense, about the immediate future, rather than all future time. Only some physicists have the hubris to imagine a 'Theory of everything', and only some economists have the wit to say 'In the long run, we are all dead'.

Thus we talk about the future of things like the weather, the stock market, the solar system or a pendulum. Each of these is some sort of bounded object – we can describe both what it is and what it is not. We can say that this is the weather, and that this, pointing to a rock, or to the stars, or to the stock market, is not.

Each of these things exists in some environment or domain, and it is the future of the thing within its environment that we are interested in. Thus we are interested in the weather of Australia or the stock market of Tokyo or our solar system or that pendulum there. This also introduces the idea of classes and instances into our notion of prediction. We accept that there is a broad class of weather, but that it is sensible to talk about Australia's weather, just as we accept that there are many stock markets, many solar systems, many pendulums, even as we are interested in one particular instance of any of these classes.

Of course the idea of classes of objects brings with it the idea of rational classifications of the world – the idea that we can see various similarities in the objects of the world and that we do not have to assume that the world is composed of singularities. The existence of such similarities is fundamental to the notion of prediction because it is a precursor to a rational empiricism, not to mention science itself. But it needs to be emphasized that it is not only science that bootstraps on this idea – all predictive approaches that use what we might call 'a logic' rely on the ideas of classes, objects and domains. Magics that see similarities between man and the cosmos use such ideas, as do neural nets. Science is only distinctive in that it uses scientific theory as well.

Taken together, the ideas of objects with boundaries, of domains or environments, of instances and classes, lead to the idea of models. We can create a model object that is like, to some degree of likeness, some object of interest. That is, there is a sense in which the model object and the object of interest are instances of some same class. Likewise, we can create a model environment that is like the modeled object's environment. Both are likewise instances of some same class. If we have chosen our model object carefully it will have the useful property of exhibiting dynamics. That is, it will have a model of time in its environment. And if we are even more careful, our model of time will tick over faster than real time.

In this way, we may be able to observe the state of our modeled object at some future modeled time. All scientific prediction relies on this simple philosophical-logical trick – if the model is like the object, that is, in some class with it, then the object's future is like the model's future. If we play the trick, we call the model's future a prediction about the object's future. It is the same trick whether we talk about big stuff like the future of the solar system or little stuff like the swing of a pendulum.

It is a matter of enormous moment that, for many systems, this trick works and works brilliantly. But it is also a matter of enormous moment that for some systems it does not work at all. How we distinguish those systems from each other is the matter I now turn to.

4.4 SIMPLE MODELS, GOOD MODELS

This sort of prediction depends on a whole lot of things being right, and getting them right is a creative act as much as a scientific one. This is why good predictive modeling is as much art as it is science, and why words like elegance and beauty can be used in modeling as in mathematics, itself a discipline somewhere between art and science. To build a good model we need to get the object and its environment right, as well as its class membership, and then we need to do the same for the model. We need to take particular care to ensure that the model and the object are each instances of some same class that will help our understanding.

Good models capture the essentials of the modeled object and its dynamics, but are usually simpler in ways that aid understanding. As I argued earlier, some physical systems make this process relatively easy. They are composed of relatively few different sorts of entities with strong clear interactions, and are often dominated by inverse square relationships, so that only the strongest and nearest entities really matter. Modeling such systems is relatively straightforward. We can create models that are so elegant that we call them physical laws, whether Newton's deterministic laws of motion or the more statistically based gas laws.

Such models are able to create powerful, detailed and precise predictions of the dynamics of the modeled objects. Indeed, such systems are highly determined and inherently predictable, and the modeling problem really is to find the model that best displays the predictions.

Models of complex adaptive systems are more problematic. They challenge the idea, developed since Newton's time, that good models equals simple models equals predictive models. Or put another way, they challenge our philosophical-logical trick. In the first place, complex adaptive systems are characterized by many different entities, most of which are more or less weakly interacting, and with interactions that are diffuse and non-linear. As systems they show openness, fuzziness, messiness, individuality, novelty, learning and adaptation. They blur the distinction between object and environment. Their dynamics show surprise and emergence. It is not clear in what sense they are determined, or if they are inherently predictable at all.

When we use the same ideas to model these systems as we use to model simple physical systems, we come up with different results. We find that 'getting things right' and 'capturing the essentials' are themselves more complex processes than for simple systems. We find that the character of the modeled complex adaptive system is embedded in the fine detail of the many entities and their interactions, not in the gross pattern of a few strong linkages. We find, in fact, that we need to build a model complex adaptive system in order to model a complex adaptive system.

So we have to 'set a thief to catch a thief'. These model systems are simpler, to be sure, than the systems they are modeling, but running them in 'model time' does not create dynamics that have the same character of prediction as does running a model of a simple physical system. The model systems show the characteristic adaptive and emergent behaviors of real complex adaptive systems that make them unsuitable for predictive purposes. We have thus created a classic recursion paradox reminiscent of *Gödel, Escher and Bach* (Hofstadter, 1979).

4.5 HOGEWEG AND HESPER'S BUMBLEBEES

Let me descend from these abstract, theoretical heights by discussing a concrete example – bumblebees. Paulien Hogeweg and Ben Hesper are theoretical biologists at the University of Utrecht. They have been working for many years on complex adaptive systems, in fact since well before they were called complex adaptive systems. They published an elegant paper (Hogeweg and Hesper, 1983) some years ago on the dynamics of bumblebee societies using a modeling technique that they then called 'individual-oriented modeling', but which we now call 'agent-based modeling'. The paper is of particular interest here, in that, being a pioneering work, it clearly describes how it departs from more traditional modeling and, in that light, comments on the significance of the results.

The natural history of bumblebees is fascinating from the point of view of emergent behavior. In the spring, new colonies of a few dozen individuals are created from eggs hatched by a queen. These eggs hatch into larvae which then pupate into adult worker bees that forage for the colony, and do other chores about the nest. The bees have a social dominance hierarchy led by the queen that leads to the emergence of two castes of workers – elite workers which interact frequently with the queen and who will lay eggs later in the season, and common workers who never lay eggs. The queen and the elite workers use their dominance to control the production of eggs. In the late summer or early fall, the queen may lose control of the nest and new queens rather than workers may be produced. This switch from worker to generative offspring can be a chaotic time in the nest with workers shifting their dominance levels dramatically, elite workers reverting to common workers, new elite workers emerging, and the queen sometimes being driven off the nest. The process is brought to a halt by the onset of winter in which the bees hibernate or die, sending out new queens in the next spring.

Ethologists are not able to predict, from a knowledge of the natural history, which of the initially identical workers will become elite workers, when or if a switch will occur, or what its consequences might be.

Hogeweg and Hesper built an agent-based model – itself a complex adaptive system – of a bumblebee colony, emphasizing the natural history of the bees, with the basic element of the model being an agent – a bee – that 'lived a life' and interacted with other bees. The model used a simple dominance rule for these interactions as well simple rules for the feeding, foraging and housekeeping behaviors seen in the bees. They show 'how individual behavior, based on simple rules and using little information, generates a social structure, which being the environment of the individuals, causes a more structured behavior of the individuals and individual variation in their behavior' (Hogeweg and Hesper, 1983). Their model shows emergent behavior strikingly similar to the real thing: elite and common workers emerge and progressively differentiate themselves during the life of the nest, the switch to production of new queens frequently occurs, and sometimes the queen is driven from the nest.

But this study is also important for what it does not show, for it does not show prediction. It shows instead a model that allows us to explore what we might call 'bumblebee space'. This space is quite unknown to us, and the model is like a simplified map, a sketch, of the space. The model colony is like a real bumblebee colony, but yet simpler than any real one. It is recognizably a bumblebee model, not a honeybee model, say, and yet it is not particularly more like one species of bumblebee than another. We may think of it as a new species of *Bombus*. Just as we learn something of all species in a genus by studying one species, we learn something of bumblebees in general by studying the model.

There is a sense in which we may say that the model explores the potential or possibilities of the space just as each species within the genus explores the potential or possibilities of the space.

However the model, like each of the real species, does not exhaust the potential of the space. Indeed, we may think of the space as only 'being built' or 'coming into being' because of the bees that give it its shape. Each is created from, or emerges from, the local interactions, and so is strongly contingent on its own local history. This sort of emergence is distinctly different from the emergence of ensemble properties of a gas, say, where we may generate a statistical dynamics that is a general solution to the 'gas space'. Our model is a purely local solution or exploration of the space.

There can be many models, of course, each exploring different parts of the space, but together they do not exhaust the space either. Because of the intense individuality of each model, it does not make sense, except at a very high and abstract level, to think of the ensemble properties of the collection of models. Thus it does not make sense to collect such models and produce statistical averages and so forth.

The bumblebees neatly expose the fallacy of the idea that there might be special sorts of (weak or attenuated) prediction for complex adaptive systems.

One version of such weak prediction might hold that we seek to predict stylized patterns in complex adaptive systems. This has the flavor of seeking the platonic ideal or genus of the system. The bumblebees show clearly that, while the model is obviously a member of the 'genus', it does not explore the genus-space exhaustively, nor is that space in any sense a fixed ideal, but always changing as a result of the creation of new members. The existence of the model changes the genus-space, but does not predict it.

Another version of weak prediction might hold that the statistical behavior of an ensemble of slightly (randomly?) different models might provide prediction. However, as I elaborate later, such ensembles spend most of their time in parts of genus-space that are never realized by any real member of the genus, and so they will only offer insights into behavior that will never be.

We can now see that, when used as models, complex adaptive systems create a new class of models that are explanatory but not predictive (Bradbury *et al.*, 1984). They are distinct, on the one hand, from the deterministically predictive models of the dynamics of individual entities, such as pendulums. They are also distinct, on the other hand, from the stochastically predictive models of ensembles or populations, such as gases.

This study also helps form the foundation for a different view of what science is, one that is grounded in complexity thinking rather than more traditional mechanistic thinking. We need to turn to this briefly in order to set the scene for a description of what might replace prediction in the science of complex systems.

4.6 WHAT SCIENCE IS

We can reach this new view of science by shedding some common misunderstandings of what science is, and allowing the real complexity of the science enterprise to reveal itself (Bradbury, 1999).

We need first to shed the idea that science is the truth. It is rather a body of knowledge about the world and a recipe for growing that body of knowledge and a bunch of people busying themselves with that knowledge and that recipe all mixed up together. In today's computer-literate world, we usually call such recipes 'algorithms' or 'programs', and the knowledge 'outputs'. But it is important to realize that science is neither just the algorithm nor the output, but the whole lot including the people. Science is a 3-tuple. It is also important to realize that we can push this analogy too far, that science is a special mix. Each part can change the other. The knowledge can change the recipe, the recipe can change the people.

I need to make it clear that this recipe is not what is often called 'the scientific method' – the approach used by most scientists since the seventeenth century. It is instead a recipe for discovering 'the scientific method', and all the other methods that achieve the same ends, which are to increase the body of scientific knowledge, to improve the recipe and to change the people.

We can best understand this by using complexity thinking – a product of science. Parenthetically, such recursivity – describing things in terms of themselves – is very much part of the way we are coming to understand such complexity (Dennett, 1995; Hofstadter, 1979). From this perspective, science is not some mechanical process – turn the handle and out pops science. Indeed, even the recipe part, the most likely candidate for being a mechanism, is not really mechanical, since it is constantly being changed on the fly. Instead science is a complex, interacting, adapting, and evolving system made up of living things and their ideas. What sets it apart from other such systems, such as art, religion or politics, is that the idea of belief is deliberately and persistently winnowed out of science, despite the persistent human tendency to keep putting it back in. Each of those other human enterprises has at its base a kernel of some fundamental and fundamentally unprovable, and hence illogical belief, whether it is belief in a god, an aesthetic or a manifesto, on which it then constructs the logic of itself. Even mathematics has this character, and is thus not part of the enterprise of science.

The second idea we need to shed is that science is relative. Now it might have seemed from the previous discussion that if science is not the truth, then it might not offer absolutes, and hence the best it might be would be some sort of relative description of the world, some sort of contingent or culturally bound knowledge. Thus at best all it might be is one view of the world among many, and competing with the many for its place in the sun. The 'science is relative' view says that science is but one way of knowing the world and that it is not superior to or more valuable than other ways of knowledge. This is a sad misreading of science, because it confuses, sometimes mischievously, a part of science, its body of knowledge, with the whole – the 3-tuple, the knowledge, the algorithm and the people.

The body of scientific knowledge is not, and cannot be, 'the truth', since it is always subject to revision. It is at best, our best guess at the truth, whatever that is. The knowledge is not absolute, but that is not the same as saying that science itself is relative – that it is no more than any other way of knowledge – and that anything goes. In fact, science is quite muscular and assertive in saying that some ideas about the world are just rubbish, and it says this independently of the sincerity or saintliness, or even the nuttiness, of those who cleave to these ideas.

No, what science is most like is a modern airport. It is always under construction, destined never to be completed, a place of great confusion and chaos, and yet a place where things get done, where progress is made every day. In this analogy, knowledge is constructed rather than discovered as a fully human activity, but constructed through the application of the most powerful recipe we know. Part of the power of the recipe resides in the way its constructive, creative part is always offset by its destructive, critical part. We can see the power in the edifice it has created, an edifice that can withstand the essentially trivial destructive forces from outside science as well as the formidable destructive forces from within. It is a feature of science that the external forces wax and wane, but the internal ones never abate. All of science, the whole edifice, is a work in draft, subject to revision and reconsideration.

Science is a construction, but a special one – a purposeful complex adaptive system.

The idea that science might be constructed rather than discovered is quite alarming to some scientists. But this need not be so. Their alarm is built on a misunderstanding, and a misplaced fear, that the deconstructionists of the humanities could have a field day deconstructing a constructed science. The idea that science may be threatened by such post-modernist nonsense has been beautifully put to rest by Alan Sokal (Sokal and Bricmont, 1998) in his delightful spoofs. Science's internal destructive forces, so essential for its proper working, are a far more potent source of criticism and hence renewal than any external fad can hope to be.

4.7 INSTEAD OF PREDICTION – A PALIMPSEST

I now want to bring these threads together – the idea that prediction is not appropriate for complex adaptive systems and the idea of science itself as a complex adaptive system – to see if we can imagine what might take the place of prediction in the analysis of these systems. We may not be able to imagine it as a fully fledged process, but we may be able to sketch its basic architecture. To do this, we need a few more ideas from the complexity rag-bag – we need a sense of the vastness of the universe of possibilities for a system, and some sense of how reachable those possibilities are. We need these to anchor our discussion of whether such systems are predictable even in principle.

We want to use our current notions of prediction to help us reach beyond them, just as Pascal used the traditional notion of prediction, the one we now call deterministic, to find the new idea of probability. Whether stochastic 'prediction' stretches or snaps the idea of prediction is a moot point, but it does show that we discover new ideas from old ones, as Wittgenstein pointed out:

> The wrong conception which I want to object to ... is the following, that we can hit upon something that we today cannot yet see, that we can discover something wholly new. That is a mistake. The truth of the matter is that we have already got everything, and we have got it actually **present**; we need not wait for anything. We make our moves in the realm of the grammar of our ordinary language, and this grammar is already there. Thus we have already got everything and need not wait for the future. (Sherman and Schultz, 1998, p. 60)

The idea of a vast, but not infinite, universe of possibilities for a complex adaptive system is crucial both for the way it actively excludes a stochastic approach to prediction and the way it offers some suggestions for a solution. Dennett (1995) provides a most lucid introduction to this universe where he explains the important differences between the merely vast and the infinite, and the sparse way in which this vast universe is explored by realized possibilities – the current or previous states of a complex adaptive system.

Stuart Kaufmann offers a very enlightening thought experiment in broad support of this idea:

> If we consider the species of organic molecules now on the earth, we can also consider those species of organic molecules that do not exist, but that are one reaction step away from molecules that currently do exist.
>
> ... since the number of possible molecules with several tens of thousands of atoms per molecule is vast, the Universe has not had sufficient time to make each such molecule at least once since the Big Bang. (Sherman and Schultz, 1998, p. 22)

Thus there are even vaster numbers of possibilities than have ever even been considered. Until a previous possibility has come into existence, an adjacent possibility could not be said to exist – that is, could not be found in any sort of step-wise way.

This idea of adjacent possibilities suggests a strategy of exploring the universe of possibilities by a series of stepping stones – the adjacent possibilities – starting from the realized possibility of the present state of the system. We know two useful things about the paths that these stepping stones create. Firstly, we know that only one path will be the one that actually unfolds, and, secondly, we know that, even though there is a vast number of paths created merely from the propagation of adjacencies, those paths are sparse indeed through the universe. We also can infer that two other strategies will fail. The 'futures' strategy, as practiced in scenario planning (Schwartz, 1991), where only a couple of supposedly canonical paths are explored, is unlikely ever to traverse interesting parts of the universe. The 'Monte Carlo' strategy, which attempts to randomly sample the whole universe, will spend most of its time exploring possibilities that can never be reached by the system.

We can envisage, as an alternative to these two approaches, a purposeful agent – a complex adaptive system – exploring the adjacent possibilities for a system. While one has not yet been built, I think we can sketch it in broad outline.

- It would be scientific. That is, it would conform to the 3-tuple I described above. It would not be capricious or arbitrary, but would be based in our scientific understanding of the world.
- It would be a model. I think we can safely assume it would be a model – a complex adaptive system to model a complex adaptive system – as we have discussed above.
- It would be historic. History matters because it gives the system a path to the present. It delivers the system to its present place in the universe of possibilities, and therefore it provides us with a starting point for exploring adjacent possibilities. And in a strong sense, it may be said to create those possibilities, since they can never be enumerated in any reasonable way. But history simultaneously freezes us out of vast regions of possibility which the system will never be able to explore, and which may be said to never come into existence, even as possibilities. In a sense, only the adjacent possibilities may be said to exist in the real computable world, with vast regions of impossible possibility beyond any complex adaptive system's computation.
- It would be visual. We nearly always make pictures and use the eye-brain channel to handle powerful complex new constructs such as this.
- It would be geometric. This just says that the pictures would be (scientifically) lawful, rather than fanciful, allowing it to be mapped to reality in a consistent and logical way.
- It could be Euclidean – the common geometry of landscapes. We are familiar with the idea of thinking of the system as traversing a landscape representing its environment. In fact, Murray Gell-Mann

uses a landscape metaphor in discussing exactly this problem of prediction in complex adaptive systems:

> We are all in a situation that resembles driving a fast vehicle at night over unknown terrain that is rough, full of gullies, with precipices not far off. Some kind of headlight, even a feeble and flickering one, may help to avoid some of the worst disasters. (Holland, 1998, p. 244)

But Gell-Mann, for once, misses an important point, as he paints a picture of a headlight searching ahead (predicting) out there in the night. It is that, for a complex adaptive system, there is no 'out there', since the system is in reciprocal interaction with its environment. In fact the boundary is open and we often choose as system that which is convenient. The landscape described here is, at once, system and environment.

Thus our landscape metaphor needs to be relaxed from the idea of the system (a point) traversing a landscape (its environment). In the metaphor we are building, the landscape comes into existence as we cast the light on it. And perhaps most importantly, there is no consistent single landscape to traverse 'out there'. If we approach it this way, we may see cliffs through the gloom, and if by another way, we may see valleys or plains.

- It should be dynamic. We should be able to fly through this landscape in real time.

This idea of the future as a 'contingent landscape' will change the way we think about complex adaptive systems. Like a chess playing program, we would look ahead through a cascade of adjacent possibilities, each one a model of a complex adaptive system, each one instantaneously brought to life and inheriting some of the properties of its adjacent progenitor, but each having a life of its own, and creating a partial image of the bit of landscape in its neighborhood. As we fly through this landscape it will unfold ahead of us, with the new terrain – the future – looming out of the dark, the whole becoming foggier as we move further ahead, as first tens, then thousands, then millions of instances of the model simultaneously draw their versions of the future on the palimpsest of the landscape. We will never see a single landscape, for it does not exist, nor will the landscape approached from this direction be joinable (in Euclidean space) with that approached from another direction – they are different.

4.8 LESSONS FOR ECOLOGY AND ECONOMICS

I conclude with some thoughts on what this means for ecology and economics, disciplines whose project is the understanding of particular complex adaptive systems.

The first has to do with the fitness of models. The best models of complex adaptive systems, in the sense of increasing our understanding of them, will be other complex adaptive systems. I argued earlier that model and object need to be both instances of some class. The 'tightness' of the class bears on the fitness of the model. The essential qualities of any complex adaptive system, such as its individuality, emergence, novelty, learning and so on, can really only be effectively captured by another complex adaptive system. It follows that the most powerful models of ecosystems and economies will be complex adaptive systems.

The second is that economies and ecosystems, as complex adaptive systems, are inherently unpredictable as a whole. Their futures are not determined. Their global behaviors emerge from their local interactions in complex, historically contingent and unpredictable ways. It may not even make sense to talk about prediction of these systems.

The third is a corollary of this: the extent to which models successfully predict bits of these systems is a measure of the extent that the bits may not be important to the understanding of the system. Thus the successful prediction of global variables like M3 in economies or carbon flow in ecosystems may be an indication that these variables are not really coercive in the behavior of the system. Hogeweg and Hesper would say they are not 'interesting' variables.

The last says that using models for exploration is not a *proxy* for prediction, it is *instead* of prediction. Its purpose is to enhance understanding, not to describe the future. We may even say that prediction is all we are able to do in simple systems, whereas the interesting problems, the scientific problems, await exploration.

Exploring is not only what we do with complex adaptive systems, it is what complex adaptive systems themselves do. Science itself, as I argued above, is a complex adaptive system, and it is this system that does the exploring of the worlds of models and reality. Having complex adaptive systems explore the behavior of other complex adaptive systems that model yet other complex adaptive systems is a piece of recursion that I will explore on another day.

5. Validation and verification of multi-agent systems

Steven M. Manson

5.1 INTRODUCTION

Use of multi-agent system models is moving from being an esoteric technique towards becoming mainstream research in a variety of disciplines. With general acceptance comes the responsibility of ensuring that multi-agent systems are subject to procedures commonly accepted for other kinds of modeling techniques. In particular, there is room for better validation and verification. These terms concern, respectively, the truthfulness of a model with respect to its problem domain and the correctness of its construction. In other words, verification means 'building the system right' and validation means 'building the right system'.

This chapter discusses validation and verification for multi-agent system models of socio-ecosystem dynamics. It begins by examining general criteria by which a multi-agent system model may be measured. The chapter continues with several sections on the challenges raised by the verification and validation of multi-agent system models. The chapter next explores issues of complexity, scale, and emergence, all of which are of particular importance to the validation and verification of multi-agent systems. The chapter concludes with a discussion of general themes and specific research directions.

5.2 GENERAL CRITERIA FOR VALIDATION AND VERIFICATION

Most of the multi-agent system models in this volume share roughly similar development paths and therefore can be considered via a common terminology. Most modeling efforts have underlying theory that is distilled into a conceptual model. The modeler then uses calibration data to instantiate the conceptual model in a software model, in particular, a multi-agent system model that represents a target system. The multi-agent system model is then subject to verification, which involves testing the software model to ensure the proper functioning of its underlying programming. The multi-agent system model is then run in order to create model outcomes. Validation comes in two varieties, structural validation, or how well the software model represents the conceptual model, and outcome-validation, how successfully model outcomes characterize the target system.

Just as verification and validation are two facets of a larger modeling enterprise, they are two ways of answering a general question asked of all models: how well does a model characterize the target system? There are a number of general criteria by which model success may be evaluated (Kwasnicki, 1999) and each bears on validation and verification of multi-agent systems:

- Correctness: model structure and outcomes must be similar to those of the target system.
- Consistency: the model must be internally consistent and match the conceptual framework in order to describe the target system.
- Universality: the model should be applicable to circumstances beyond those described by the calibration data.
- Simplicity: when choosing between two models, all other things being equal, the less complicated model is preferable.
- Novelty: a model should create new knowledge or outcomes.

Correctness bears directly on validation. How well does the model achieve its goals? These goals often focus on having the model reproduce behavior or phenomena found in the target system. In general, validation is achieved by comparing model structure and outcomes to pertinent characteristics of the target system. In order to accomplish this task, one has recourse to a number of techniques, explored below.

Consistency and *simplicity* are essential to understanding how well the model articulates the conceptual framework. In particular, how well do the various model components fit together to represent the ideas that underlie them? In this vein, a simpler model may be a better one. If the modeler can reduce the number of modules or lines of code without changing model function, then the model is arguably better since there are fewer opportunities for model artifacts. Similarly, the conceptual framework is more easily understood when it has fewer objects and linkages. There is also typically a declining marginal rate of return on making a model more complex in an attempt to better mirror reality due to error propagation and making the model less generalizable (Alonso, 1968).

Generalization, or *universality*, is a function of model construction. When a model is closely tailored to particular conditions, its ability to speak to other conditions is reduced. By definition, a model is not an exact representation of the target system. It follows that if a model is only applicable to the narrow confines of the data on which it is calibrated, the model may fail to represent changes in the target system. Since most social-ecosystems are dynamic, models of these systems must retain some universality.

Novelty poses a dilemma to modeling. In a general sense, if one has calibration data for a target system with which to create a model, then that phenomenon is known to the modeler and therefore not novel. In a more specific sense, however, generalizable models should be able to produce novel outcomes if they are fed data apart from that used for calibration. This is particularly so for a model that is well verified and validated, since its conceptual framework likely addresses situations beyond that represented by calibration data.

5.3 DATA

Verification and validation involves ensuring a model functions well and comparing its structure and outcomes to validation data. The nature of this data varies from observations of the target system to the products of another model or theory assumed to adequately characterize the target system. For social-ecological systems with significant human intervention, particularly valuable 'real' data are found through surveys, interviews, censuses, and broader scale remote sensing or geographic information sources. Validation data may also be taken from theory or the outcomes of other models. While these data are arguably not as realistic as other observations, many phenomena that are of interest to multi-agent system researchers are not readily measured. Modelers must look to all available sources of validation data.

For models meant to project future trends, validation data should not be used in model construction. There is a fundamental, data-driven relationship between validation and calibration. Studies of land-use and land-cover change, for example, are considered fortunate when they have data for one time, let alone data for two or three periods (Mertens and Lambin, 2000). There is an understandable desire to use all data for model construction and calibration that is, unfortunately, at cross-purposes with the necessity of holding back data for validation. In the end, some data should be held back from construction for use in validation.

If a model is less focused on projection and more towards shedding light on a theoretical question or explaining a phenomenon, then outcome-validation may not be as important as verification and structural validation. There is therefore less need to reserve validation data. The structure of the model may be the 'outcome' of interest since the model does not produce outcomes in the form of projections. The most common example of this distinction is the use of all available data for calibration of regression-based models (e.g., Mertens and Lambin, 2000). In this case, the model is less subject to outcome-validation and more so to verification and structural-validation with measures such as goodness-of-fit or expert opinion.

5.4 VERIFICATION

Multi-agent system models are flexible in their specification and design, which can lead to models that are difficult to manage. Verification reduces the problematic nature of flexibility by vetting model structure. Verification lies largely in forcing the model's underlying mathematical and computational components to fail by varying model configurations according to all foreseeable model inputs.

This 'breaking' of the model for verification purposes is similar to sensitivity testing, in which parameters are varied across repeated model runs in order to observe changes in simulation performance. Of particular interest is determining the spatial or temporal bounds of model applicability (Klepper, 1997). Sensitivity tests map incremental parameter changes against model outcomes to determine the influence of model structure on outcomes.

Two chapters in this volume, Deffuant *et al.* (Chapter 9) and Lynam (Chapter 10), offer prime examples of sensitivity analysis. As the latter notes, there can be too many model parameters to fully explore model sensitivity by incrementing variables. By using an appropriate sample design, however, key thresholds such as one standard deviation can be employed to give a good sense of where the model performs well. As the former work notes, it is also possible to break model results into clusters by use of a decision tree to classify results. The decision tree indicates the sensitivity of the model to parameters and allows the users to cluster parameter values.

5.5 IMPORTANCE OF DETAILS

Regardless of the exact means of verification, the overall process leads to careful examination of model objects and the linkages among them. Verification of multi-agent system models necessitates exploration of subtle subjects such as the effects of different randomization techniques. Ruxton and Saravia (1998), for instance, demonstrate how spatiotemporal ordering of events can create model artifacts. These sorts of errors are more likely in multi-agent systems given the potential for nonlinearities. The importance of space is also illustrated by several chapters in this volume (e.g., Bousquet *et al.*, Chapter 12; Hoffman *et al.*, Chapter 11; Janssen *et al.*, Chapter 7; Lynam, Chapter 10), where inclusion of spatial heterogeneity in a multi-agent system results in fundamentally different model behavior.

Even when attention is paid to order, it is possible to have slight, cumulative errors introduced into a model from seemingly well-established tools such as random number generators (Stroustrup, 1998). Error propagation and uncertainty are topics often left unconsidered in modeling (Robinson, 1994). Error propagation can be estimated by examining the kinds of operations performed on data (Alonso, 1968). Other approaches include treatment of error and uncertainty in geographic information systems within a Monte Carlo framework (Heuvelink and Burrough, 1993; Eastman, 1999) or studies of thematic error propagation (Veregin, 1995). Particularly promising is the use of Bayesian analysis to quantify a priori beliefs on the part of the researcher (Berger, 1986). Bayesian analysis proves useful in addressing the impact of error and uncertainty in both model parameters and model structure (Berk, 1994).

5.6 NORMALITY AND COMPLEXITY

Verification and validation of multi-agent system models is characterized by two general and sometimes conflicting themes. First, in order to use the bulk of statistical and mathematical methods, models should act in ways that support them. Model outcomes should change roughly in proportion to changes in inputs. Similarly, while a model cannot be expected to perform well with all sets of inputs, outcomes should gracefully degrade with incremental change in parameters away from calibration data. Proportionality and graceful degradation are key to statistical methods of verification. Measurement error, sampling regimes, observer bias, and data availability – all of these factors also feed into verification. A model should accommodate the imprecise nature of the world and the data with which it is characterized. The statistical characteristics of model operation are manifest in many assumptions of verification techniques.

The second theme is contrary to the first since multi-agent systems should be able to act as complex systems and experience large shifts in outcome in response to small changes in inputs. These shifts come from sources such as sensitivity to initial conditions or self-organized criticality (Manson, 2001). Interdependencies may create large, sudden shifts in model behavior due to small changes in parameter values. For many modelers, the potential for complex behavior is an important reason to use multi-agent systems.

Consider how complexity affects sensitivity analysis. In addition to the methods examined above, one can use an active nonlinear test (ANT) to discover model failures that result from nonlinearities (Miller, 1998). Since multi-agent system models are often meant to simulate complex, non-linear systems, they may not be amenable to traditional testing that modifies one parameter at a time. An ANT is an automated test that seeks out sets of strongly interacting parameters in a search for nonlinear relationships across variables that are not found by traditional sensitivity testing.

There is a key caveat, however, in considering the effects of complexity on verification and validation. Not all models will behave in a way that suggests that complexity or strong nonlinearities are present. The critical point is that sudden changes should be possible, but they must also be explained by the conceptual framework. It is not enough to look at a new, potentially surprising outcome and treat it as an outlier. Outcomes must be subject to verification and validation so they can be traced back to conceptual precursors and not poor model design.

5.7 SPACE AND TIME

Statistics is home to an array of techniques geared towards description and hypothesis testing, most of which is applicable to validation of multi-agent systems. In addition to standard statistical tests, the complexity of both ecological and multi-agent systems suggests the use of tests that measure spatiotemporal outcomes (Turner *et al.*, 1989). This need is evident in use of spatial statistical approaches to compare modeled outcomes to validation data in a variety of multi-agent system applications (e.g., Alberti and Waddell, 2000; Batty, 2000; Manson, 2000; Hoffman *et al.*, Chapter 11).

Hoffman *et al.* (Chapter 11) and Janssen *et al.* (Chapter 7) note the importance of having spatial components in social-ecological system models, which in turn necessitates use of spatial validation tools. In the former chapter, land-use and land-cover are key components of, and outcomes from, agent processes. It notes the importance of matching both spatial and temporal outcomes to data and experience garnered from the target system. This also holds true for examining aggregate changes over time, such as rates of deforestation and afforestation; when exploring the causality inherent in spatial inputs, such as the relationship between slope and agriculture; and, finally, more in-depth measures such as pattern and ecological landscape function.

A number of issues are raised by even seemingly simple validation metrics. By way of example, consider the simplest class of spatial validation techniques, which act on model outcomes in the form of mutually exclusive categorical map layers (e.g., Gilruth *et al.*, 1995). While useful, these error matrices are flawed in that the proportion of correctly classified cells is difficult to interpret since they may be classified correctly due to chance. The Kappa statistic accounts for this problem (Hudson and Ramm, 1987) but as noted by Pontius (2000), Kappa fails to recognize when model outcomes have accurately determined the relative quantities of each cell state. He therefore introduces a new Kappa measure that differentiates between location and quantity.

Pattern indices offer another example of how even seemingly simple validation tools can be problematic. These metrics include dominance, patch size variance, fractal dimension, nearest neighbor probabilities, contagion, and adjacency (Turner *et al.*, 1989; Giles and Trani, 1999). These pattern indices roughly translate to the statement 'if the state of a cell is known, by how much is uncertainty about neighboring cells reduced?'. Ecologists use fractal dimension and its various offshoots to understand landscape ecologies (Krummel *et al.*, 1987). Fractal measures are also used to validate model outcomes but these statistics must be applied with care due to the effects of resolution, scale heterogeneity, and the uncertain links between fractal metrics and ecological function (Li, 2000). These issues are emblematic of the problems of projecting biotic diversity in multi-agent system models, as noted by Abel *et al.* (Chapter 13).

5.8 STRUCTURE AND REALITY

On the boundary between calibration, validation, and verification is the issue of how well a software model maps on to the cognitive model that underlies it. A completely inductive 'black box' model may produce seemingly valid outcomes but it is not readily verifiable or structurally valid if causal mechanisms are not adequately represented. This becomes a problem when a structural change in the target system renders the model less able to produce valid outcomes.

Bousquet *et al.* (Chapter 12) illustrate how a multi-agent system, CORMAS, is calibrated and verified through involvement of stakeholders. Over repeated interactions between stakeholders and modelers, the objects and rules that make up the software model are more closely mapped on to the target system. Repeated examination of the fit between the model and target system makes model failures more apparent; verification and structural validation are therefore more easily achieved.

Deffuant *et al.* (Chapter 9) present a multi-agent system model of the diffusion of agri-environmental measures oriented towards understanding the role of uncertainty, the importance of information, and the decision-making process. In their model, the authors use real data for both structural and outcome-validation. Another chapter in this volume, Jager and Janssen (Chapter 6) details how experimental data can be used to both calibrate and validate a multi-agent system model. Here the focus is on the use of experimental laboratory data on common-pool dilemmas for validation of multi-agent models.

When used with both theory and real data, multi-agent system models serve as a locus of theory, experiment, and modeling. All three can posit different behavior. Such is the usefulness of multi-agent systems for understanding the dynamics of social-ecological systems, whether real, artificial, or theoretical. The most cogent result of validation may be illustrating the need for more validation data since this need can define how further experiments should be conducted. As a result, in addition to looking to theory for hypothesis construction, we can look to multi-agent systems. This holds especially true in terms of narrowing down the host of all possible options and parameters to a manageable number.

5.9 OUTCOMES AND THEORY

Multi-agent system models are also concerned with abstract concepts, which affect validation and verification. Agent-based computational economics, for instance, is concerned with the 'constructive grounding of economic theories in the thinking and interactions of autonomous agents' and 'the formulation and testing of conceptually integrated socioeconomic theories compatible with theory and data from many different relevant fields of social science' (Tesfatsion, 2001, p.283). In this case, modeling is driven by theory in order to affirm theory. Also noted, however, is the importance of validating models with 'analytical studies, econometric studies, field studies, and human-subject laboratory studies' (ibid).

Balmann *et al.* (Chapter 8) compare model results, in this case farm characteristics, to general characteristics of the target system. The model is calibrated with survey data and then model structure is compared to data from a real-world system of agriculture. In terms of outcome-validation, it does not involve comparing spatial patterns, as above, but it is concerned with comparing model statistics with idealized, yet realistic, characteristics. Real-world processes such as specialization, diversification, and stratification are recreated in the model outcomes. Similarly, model scenarios are created to match policy scenarios advocated in reality. These scenarios can then be examined from a number of stylized, theoretical perspectives to see if they are qualitatively reasonable.

In many respects, the data-theory dialectic lies at the heart of validation. While multi-agent system modelers are often interested in adding another dimension to our understanding of a target system, they are also interested in challenging the status quo. Two chapters in this volume use different theoretical models in a form of validation (Jager and Janssen, Chapter 6; Janssen, Chapter 3). Both demonstrate the importance of theory to validation when examining the evolution of institutions. They note there is a long-standing body of theory on common-pool resources, exemplified by Hardin's 'tragedy of the commons' (1968). A modeler could draw on this body of work in order to validate a multi-agent system model of common-pool interaction. As both chapters note, however, experimental evidence suggests that there are instances in which the theory is incorrect (also see Feeny *et al.*, 1990). In this case, its use as a source of validation data would be potentially incorrect. Both chapters also note how multi-agent system models highlight how individual agent behavior can be at odds with theory, particularly when there is explicit interaction between agents.

5.10 SCALE AND MULTI-AGENT SYSTEMS

The bulk of the statistical methods for validation and verification described above suffer from three scale-related problems that affect their use for multi-agent systems. First, if the resolution, or granularity, of a model outcome is divided into exclusive categories, then other outcomes may be separated from processes not directly mapping on to this resolution. Using watersheds to define spatial resolution, for instance, makes it difficult to examine phenomena not tied to hydrology. Second, if a given resolution is achieved through aggregating finer-grained phenomena or objects, then there is the potential to treat units at this level as just simplifications of those found at lower levels and not as entities that can have their own behavior (Kimble, 1951). This problem is especially pertinent to validating collective behavior in agent-based models.

A final scalar challenge lies in the fact that the validation statistics noted above are impacted by changing resolution (Lam and Quattrochi, 1992) and extent (Saura and Millan, 2001). Ecological fallacy occurs when the characteristics of an individual are incorrectly inferred from those of the population from which it is drawn (Robinson, 1950). The related modifiable areal unit problem (MAUP) is the sum of an aggregation effect, when larger spatial groupings of data create better correlations, and a zoning effect, whereby an area can be sub-divided into an almost infinite array of configurations that all share a common statistic, such as area, yet differ in others (Openshaw, 1977).

Ecological fallacy and MAUP can be statistically causal when covariation between variables is affected by the scale at which they are compared. In essence, scale can become an ill-chosen independent variable (Bian, 1997). Research in land-use and land-cover change, for instance, has demonstrated that changing resolution can affect the apparent magnitude and direction of relationships among causal factors of land conversion (Kummer and Sham, 1994). Various remedies for the statistical effects of scale have been proposed, including inductive search techniques (Openshaw *et al.*, 1987), fractals (Lam and Quattrochi, 1992), and geostatistics (Chou, 1993).

5.11 EMERGENCE

In addition to potential problems faced by statistical approaches because of scale, challenges are offered by the concept of emergent hierarchies. The term *emergent* refers to the fact that a system can have qualities that are not analytically tractable from the attributes of its internal components (Baas and Emmeche, 1997). Emergence is a function of synergism, whereby system-wide characteristics do not result from superposition but instead from interactions among components (Lansing and Kremer, 1993). An economy has emergent qualities such as volatility and investor 'herd behavior' that are commonly attributed to irrationality or imperfect markets but in fact are intrinsic to rational, local, interactions (Andreoni and Miller, 1995).

The notion of emergence is a central tenet of multi-agent system modeling. Emergence challenges accepted scale levels. Economics, for instance, studies stability and repeated patterns, while perspectives that employ emergence examine the evolution and change of levels (Lane, 1993). Similarly, institutions of importance to global change are seen as coming into being, or emerging, from the rational actions of individuals (Ostrom *et al.*, 1994), a theme echoed by Janssen (Chapter 3).

Emergent hierarchies illustrate the importance of blending quantitative and qualitative methods of validation for multi-agent systems. Consider the challenge of abstraction. Many attributes and outcomes of human interaction, such as trust or learning, are imputed or abstract. As noted earlier, validating such abstract outcomes may be difficult since they are not easily measured. Validation is even more difficult if outcomes are simultaneously abstract and emergent since they are hard to define in addition to being hard to measure.

At this point, the modeler may be left with only large-scale or qualitative measures of model performance. Balmann *et al.* (Chapter 8), for instance, draw on similarities between modeled outcomes and real, aggregated measures such as structural change or economic phenomena such as economies of scale and income disparities. Bousquet *et al.* (Chapter 12) address the challenges of abstract and emergent outcomes with expert and stakeholder interviews that provide a sense of how these outcomes are related to model structure. The process described by the authors could conceivably be applied to very rare events projected by a model. Since it may be impossible to draw upon reality to ascertain the validity of an outcome, one is left to interview experts to determine its validity.

5.12 CONCLUSION

By way of conclusion, Hoffman *et al.* (Chapter 11) refer to a validation scale that has four levels that range from a model being able to generate micro patterns that quantitatively match empirical patterns to a model possessing agents whose behavior qualitatively matches agents in the target system. The schema presents verification and validation as ranging from standard statistical techniques, through those that measure more nuanced characteristics of outcomes, to more qualitative techniques that require iterative, structured exploration by stakeholders. The key point of this schema and the discussion above is that no one technique is the best, or even adequate, when used by itself. Good verification and validation requires use of multiple, complementary methods.

The various challenges faced in model validation and verification highlights the need for more sophisticated methods. In creating these methods, multi-agent system modelers can consider two broad questions. First, what does the modeler learn when different model configurations enjoy varying levels of success across different forms of verification and validation? Is a significant difference in one metric between configurations due to a difference in approach or the metric itself? Second, does a model behave as expected when key components or their interdependencies are varied? This question goes to the heart of the modeling process in that a model is only as strong as its underlying theory. It is the role of both verification and validation to determine which components are important and why.

In the meantime, verification and validation of multi-agent systems will become easier with better communication of model design. The appropriateness of verification and validation procedures is difficult to establish when publications do not describe model design in sufficient detail to permit the reader full understanding. Given the necessarily restricted scope of most research papers, however, it can be difficult to fully elucidate model design, although a number of chapters in this volume do a good job of conveying model structure with mathematics and graphics. A growing tradition of publishing software code along with manuscripts exists within the agent-based modeling community.

Similarly, adoption of common languages would allow streamlined communication of model parameters and development of common validation and verification tools. The growth of SWARM, REPAST, CORMAS, and other frameworks is heartening, but there is much room for improvement, particularly in terms of improving the linking of these packages to other kinds of software, such as geographic information systems. As evidenced by several chapters in this volume, there is still a need to create custom modeling tools that in turn affects the extent to which multi-agent system are subject to validation and verification.

6. Using artificial agents to understand laboratory experiments of common-pool resources with real agents

Wander Jager and Marco A. Janssen

6.1 INTRODUCTION

Most natural resource systems used by multiple individuals can be classified as common-pool resources. Conventional economic theory predicts that when agents have free access to a common-pool resource they will consume ecosystem services to the point where private costs equal the benefits, whereas externalities are imposed on the rest of the community. This can lead to the well-known tragedy of the commons (Hardin, 1968). Many laboratory experiments have been performed to study this phenomenon. Even in the simplest case of these experiments, without communication between the participants, anomalies where found that are not in line with conventional economic theory, which is non-cooperative game theory (Ostrom *et al.*, 1994).

Conventional theory predicts that players in a non-cooperative game will follow a Nash equilibrium. In none of the reported experiments on common-pool resources was such a Nash equilibrium observed. Furthermore, the total consumption of the common-pool resource fluctuates in time (Ostrom *et al.*, 1994).

Many studies have focused on this phenomenon. Dudley (1993) analysed individual data from Indiana University on resource-use and classified most of the participants according to the strategy they used, such as non-cooperative and cooperative behavior. Deadman (1997) developed a simulation model based on artificial intelligence, which reproduced patterns similar as observed in the laboratory experiments. Casari and Plott (2000) report laboratory experiments that are consistent with their proposed analytical model that includes heterogeneity of social orientation among the agents.

The original aim of this chapter was to explain the phenomenon in question as originating from heterogeneity in Social Value Orientations (SVO's) of the participants, using the consumat approach, a multi-agent simulation approach from social psychology (Jager, 2000; Jager *et al.*, 2002) as a research tool. Indeed differences in SVO explained the observed aggregated data. However, when we got hold of individual data, our model was not able to explain individual patterns as observed in the various laboratory experiments. Therefore, we broaden this chapter, and use the common-pool resource experimental data for a discussion on the validation of multi-agent models by using laboratory data.

The chapter is built up as follows. In the next section the original common-pool resource experiments are discussed, as well as relevant other studies that tried to explain the observations. In Section 6.3 we discuss Social Value Orientations, then we provide a brief overview of theories on decision making and introduce the consumat approach. The consumat implementation of the common-pool resource problem is discussed in Section 6.6, and the experiments with the model in Section 6.7. Section 6.8 discusses the problems which have arisen during the simulation experiments and propose new additional laboratory experiments. The chapter closes with a discussion on the relation between laboratory research and multi-agent modeling.

6.2 COMMON-POOL RESOURCE EXPERIMENTS

Ostrom *et al.* (1994) developed a series of laboratory experiments in an effort to understand the degree to which predictions about individual and group behavior, derived from non-cooperative game theory are supported by empirical evidence. The baseline experiments as described in Ostrom *et al.* (1994) form the basis for the simulation experiments in this chapter. In these baseline experiments, eight subjects were presented with a situation in which they had to choose to invest tokens in two alternatives, or markets. Market one, a safe alternative, provides a constant rate of return on investments. Market two provides returns that vary in relation to the total group investment and the investment of the individual. Market two is the common-pool resource. Ostrom *et al.* (1994) performed experiments for two amounts of tokens, namely ten tokens and 25 tokens. The eight subjects had information about the functional relationship of the two markets, derived the aggregated level of token investments after each round, and were not allowed to communicate with other subjects.

Each individual i has a number of e tokens each round. An amount of x_i is invested in market two, which functions as a collective resource, and an amount of $e - x_i$ is invested in market one. The payoff function is:

$$u_i(\mathbf{x}) = \begin{cases} 0.05 \cdot e & \text{if } x_i = 0 \\ 0.05 \cdot (e - x_i) + (x_i / \sum x_i) \cdot F(\sum x_i) & \text{if } x_i > 0 \end{cases}$$

where

$$F(\sum x_i) = (23 \cdot \sum_{i=1}^{8} x_i - 0.25 \cdot (\sum_{i=1}^{8} x_i)^2) / 100$$

According to this formula, the payoff of someone investing all his ten tokens in market one ($x_i = 0$) is 0.05**e*, thus 0.5 tokens. Investing a part or all of the tokens in market two ($x_i > 0$) yields an outcome that depends on the investments of the other players. For example, when player *i* invests all his ten tokens in market two, whereas all the other players invest in market one, player *i* will receive 2.05 tokens, considerably more than what market one yielded. However, when all players invest ten tokens in market two, the yields are – 0.2 tokens, thus implying a loss. In the case when the players have 25 tokens at their disposal, they may experience even larger losses when they all invest a lot in market two.

If the players behave according to the non-cooperative game theory, they would derive the Nash equilibrium where each player maximizes payoff given the strategies chosen by the other players. The Nash equilibrium for both $e = 10$ and $e = 25$ is equal to eight tokens invested in market two.

If the group act as one agent, they should fully cooperate to maximize their outcomes. In this case their total investment in market two should be 36 tokens, which implies an investment level between four and five tokens per individual. In order for perfectly rational individuals to make themselves better off by achieving the optimal group returns, each individual would have to substantially cut back his or her investment. However, when all other players restrain their investments in market two, it would be very tempting for a player to invest more in market two as this would increase his outcomes significantly. Once individuals give in to this temptation, other individuals may follow, and the group may reach a Nash equilibrium, thereby lowering the group performance.

Looking at the outcomes of different group strategies, we see that a fully cooperative strategy yields the highest outcomes for the group as a whole (Table 6.1). The Nash equilibrium yields higher outcomes than investing exclusively in market one.

Table 6.1: Individual earnings given different group strategies. e can be ten or 25 (when e-25 dividing u(x) is divided by 2)

Earnings/subject	e = 10	e = 25
Group maximum ($\Sigma x_i = 36$)	$0.91	$0.83
Nash equilibrium ($x_i = 8$)	$0.66	$0.70
No investment in CPR ($x_i = 0$)	$0.50	$0.63

The performance of a group can be measured as the extra earnings by investing in market two as a percentage of the maximum extra returns. The Nash equilibrium level of investment leads to a level of 39 (=(0.66 - 0.50)/(0.91 - 0.50)) percent of this maximum.

Many laboratory experiments have been conducted on finitely repeated CPR dilemmas. The central question was to what degree the empirical data support the outcomes as hypothesized by non-cooperative game theory for finitely repeated, complete information games (Ostrom *et al.*, 1994). At the aggregate level the results seem to converge to a Nash equilibrium. But instead of monotonic convergence to the predicted equilibrium a different repetitive pattern was observed. The net yield drops toward zero and then rebounds as subjects reduce the level of investment in the common-pool resource. There is a difference in observed aggregate behavior in the two levels of endowment (ten and 25 tokens) (Figure 6.1). In the low-endowment setting, the aggregate results remain close to the predicted Nash equilibrium. In the high endowment setting, however, the aggregate behavior is far from the Nash equilibrium during the first part of the experiment, but begins to approach Nash in later rounds. More interestingly, at the individual decision level, no behavior is found which is consistent with the Nash equilibrium.

When individuals do not use a decision strategy that yields a Nash equilibrium (abbreviated as Nash strategy), the question becomes what strategy do they use in deciding how many tokens to invest in which market, and to what extend is this strategy biased. Dudley (1993) tested how a player should have behaved, given he had perfect foresight, to maximize his/her outcomes given the fixed behavior of seven other players. Dudley derived the fixed behavior from the other seven players from 27 experiments, including 216 subjects. The optimal decision behavior of the artificial players has been calculated for each of the experiments. The strategies behind this decision behavior were categorized as (1) non-cooperative Nash strategy, (2) the cooperative strategy, (3) the average strategy that aims at generating the same returns in both markets, and (4) the remaining non-classified strategies. In the ten-token experiment 24% of the subjects followed a Nash strategy, no player followed the cooperative strategy, 12% played the average strategy and 64%

of the subjects could not be classified. The results for the 25-token experiments were somewhat different. Here, 30% followed a Nash strategy, 3% a cooperative strategy, 66% played the average strategy, and only 1% could not be classified.

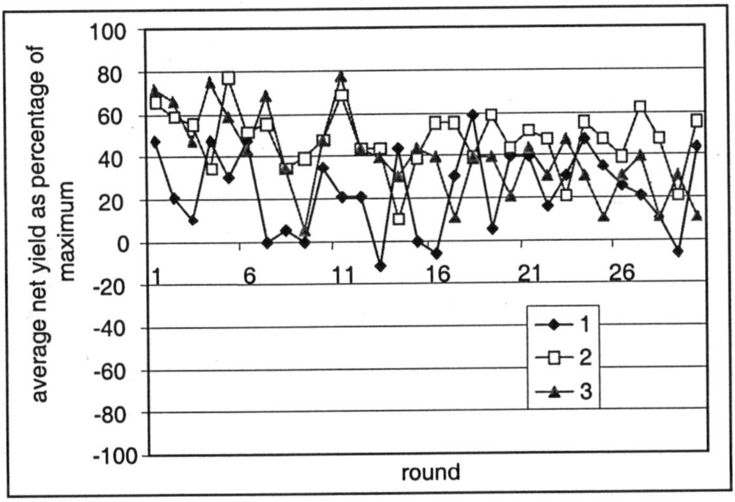

Figure 6.1a: Outcomes of three laboratory experiments with ten tokens. Nash equilibrium generates a yield of 39.5%

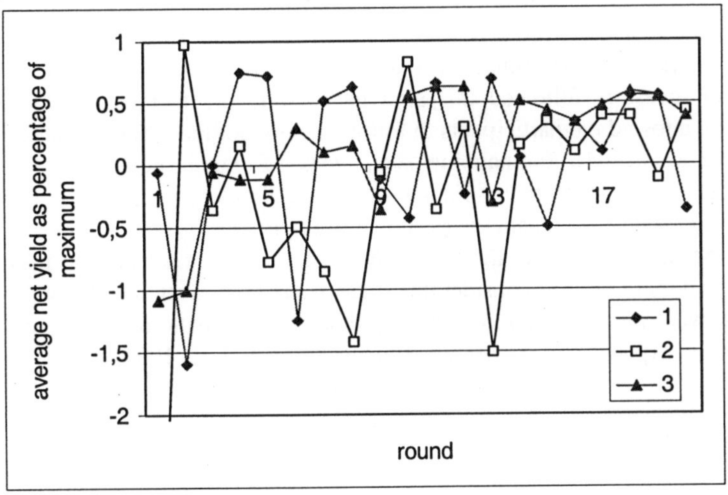

Figure 6.1b: Outcomes of three laboratory experiments with 25 tokens. Nash equilibrium generates a yield of 39.5%

The assumption of perfect foresight is of course not very realistic. Therefore Dudley performed experiments equipping the artificial agent with forecasts derived from real people. The forecasts were measured in 128 experiments with real people. Dudley found that in the case of ten-token experiments 29% of the subjects follow a Nash strategy, no subjects follow the coöperative strategy, 11% follow the average strategy, and 60% of the subjects could not be classified. In the case of 25-token experiments 13% followed a Nash strategy, 4% a cooperative strategy, 28% an average strategy and 34% could not be classified. The hypothesis that the reported forecasts of the subjects are unbiased could be rejected on the basis of these results. Dudley argued that there is strong evidence supporting that subjects use adaptive learning in their forecasts.

Deadman (1997, 1999) uses intelligent software agents to simulate the laboratory outcomes as reported by Ostrom *et al.* (1994). From the perspective of bounded rationality a limited set of rules of thumb have been formalized in software agents. These rules are based on questionnaires submitted by individual participants during the baseline common-pool resource experiments run by Ostrom *et al.* (1994). The outcomes of the questionnaires revealed that many participants followed a rule of thumb that stated: 'Invest more in market two whenever the rate of return is greater than $0.05 per token.' Whenever the per-token rate of return for a market two investment exceeded that of market one, participants increased their market two investment. When the rate of return fell below that of market one, participants invested more tokens in market one. In the ten-token endowment experiments, the authors found a tendency for participants to invest all their tokens in market two whenever the rate of return exceeded that of market one. Many investors followed this strategy, despite the fact that the full information allowed the participants to follow a more optimal (Nash) strategy (Ostrom *et al.*, 1994). Formalizing such rules of thumb in agents, and allowing agents to switch their strategy during the simulation, Deadman was able to replicate the cyclic patterns as shown in Figure 6.1.

However, if we look at the individual data of the players, we may conclude that there exists a considerable diversity in how much people invest in market two. Figure 6.2 shows some individual harvest patterns of subjects in a ten-token experiment (experiment 36 from Ostrom *et al.,* 1994).

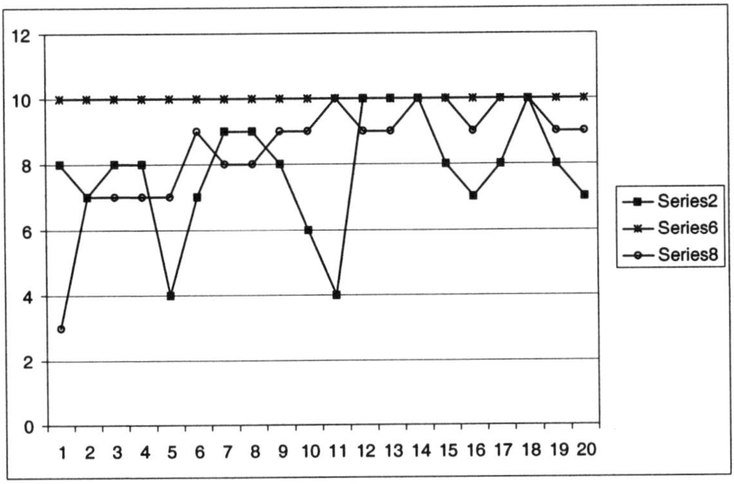

Figure 6.2: Some exemplary individual data from a ten token experiment (data by kind permission of J. Walker)

What can be observed here is that the subject represented by series 2 shows large variations in his/her investing. Series 6 on the contrary shows a subject that always invests ten tokens. Series 8 shows a subject that continuously changes his/her investment in market two, but the changes usually involve one token. Apparently the subjects behave quite differently in the same experiment, and hence heterogeneity amongst the subjects may play an important role in understanding the aggregated data.

Casari and Plott (2000) suggest that there is heterogeneity among the agents regarding how they want to interact with other agents. They distinguish altruistic, self-interested and spiteful agents in their analytical model. Spiteful agents derive utility from decreasing the earnings of others. They replicate the Ostrom *et al.* (1994) baseline experiments, and added additional experiments to test different sanctioning regimes. Their experimental data are consistent with their analytical model and they conclude that heterogeneity of social orientation explains why a Nash equilibrium is not reached. However, Casari and Plott (2000) did not test social orientations of the participants directly by surveys.

In sum, previous studies argue that heterogeneity of strategies among the participants is the main cause of the observed aggregate phenomena. Especially differences in the way people want to interact with each other are assumed to explain the cyclic behavior of the investments in the common-pool resource. In the next section we discuss in more depth Social Value Orientation (SVO) as a formalization of this interaction.

6.3 SOCIAL VALUE ORIENTATION (SVO) AND THE VALUATION OF OUTCOMES

In the research on social dilemmas, much attention has been given to how the Social Value Orientation (SVO) of persons affects their harvesting behavior. The SVO of a person is defined as the preferences one has for particular distribution of outcomes for oneself and others. Because in common-pool resources the choice behavior of people may depend on the preferences they have for a certain distribution of outcomes, this perspective may be relevant for understanding differences between people regarding their behavior in a common-pool resource.

The SVO can be measured by using a task where people have to make a choice between two distributions of outcomes for oneself and the others. People are confronted with a series of different choice dilemmas as formulated in, for example, *The Ring Measure of Social Values* (Liebrand, 1984). Graphically depicting the preferences for outcome preferences on a x-axis (own outcomes) and y-axis (other outcomes) results in a circumplex of SVO (Figure 6.3, see also Wiggins, 1980). In Figure 6.3 the most commonly researched prototypical SVOs are being depicted. The SVO of a person can be expressed by the angle of the vector, and the length of the vector can express the coherence of one's SVO. A person who makes choices that are perfectly consistent with his/her SVO is indicated with a vector that touches the outer circle. The less consistent a person is in his choices, the shorter the vector gets. In Figure 6.3 some vectors are denoted with grey arrows.

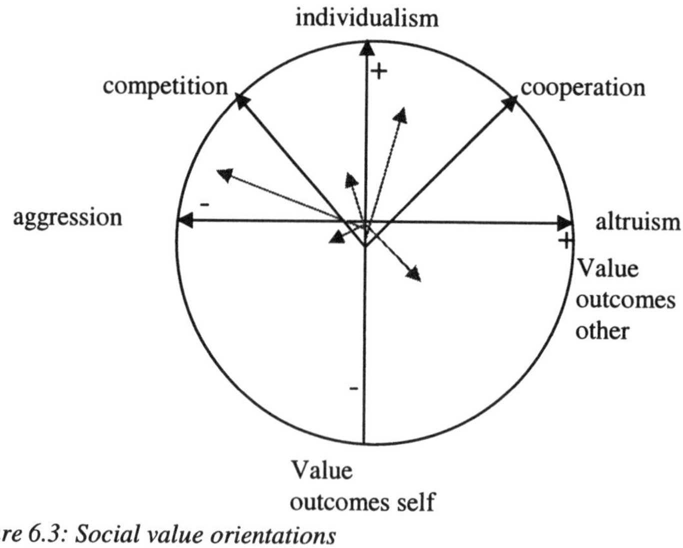

Figure 6.3: Social value orientations

For matters of simplicity we will not formalize different vector lengths in our model, but rather focus on the most common prototypical SVOs. Whereas in principle eight prototypical SVOs can be imagined, the three empirically most frequently occurring orientations have been the topic of theorizing and empirical study. These three orientations are (1) cooperation, aimed at maximizing the outcomes of self and others (2) individualism, aimed at maximizing one's own outcomes at the neglect of the others' outcomes, and (3) competition, aimed at maximizing one's own outcomes in comparison to others' outcomes. In one experiment Van Lange (1999) reports that the prosocial or cooperative is the most frequently observed orientation (57%), the individualistic orientation is observed less frequently (36%), and the competitive orientation the least frequently observed orientation (7%). Two other orientations that have been investigated are (4) altruism, aimed at maximizing the others outcomes at the neglect of own outcomes, and (5) aggression, which is aimed at minimizing the others' outcomes at the neglect of the own outcomes. An important conclusion from research on SVOs is that not all people are *a priori* inclined to value only their own outcomes, or to see the pursuit of self interest as rational (Van Lange *et al*, 1992, p.17). Including the outcomes of others in some way in the outcome matrix leads to a transformed outcome matrix, which may lead to other optimal solutions than choices on the basis of pure self-interest (Kelley and Thibaut, 1978; Kuhlman and Marshello, 1975; McClintock and Liebrand, 1988). The SVO people have is thus an important behavior determining factor in social dilemmas (Messick and McClintock, 1968; McClintock, 1978).

This SVO appears to be an important factor in describing heterogeneity between people in the management of a common-pool resource. However, the SVO of a person does not neatly describe the harvesting behavior of a person. For example, a cooperative person is likely to maximize the joint outcomes, but when the other players are systematically exploiting him, the chances are high that this person abandons a joint maximization strategy, in favor of a punishing strategy. Rather we conceive the SVO of a person as a factor determining his/her satisfaction level with a certain distribution of outcomes. This satisfaction level may affect the decision-making process of the person, which may continually change as a consequence of the own and other players' behavior. People may spend more or less cognitive effort in their decision making, and may use more or less information regarding the behavior of others in this process. First of all this introduces a heterogeneity *between* people, as one person may be inclined more towards extensive elaboration than another. On top of that, heterogeneity can also be observed *within* people, as they usually change their decision-making strategy when repeatedly making the same type of decision, as is the case in a common-pool resource game. In the next section we therefore provide a psychological perspective on human decision making.

6.4 DECISION MAKING

People make decisions all the time. During a day you decide for example what to eat for breakfast, what alternative route to follow to your work when the usual route unexpectedly happens to be closed, what to wear at that reception tonight and how to set up this laboratory experiment with 120 respondents. Because we make thousands of decisions each day, and our cognitive capacity is limited, people have developed very smart strategies to allocate their limited cognitive capacity over this multitude of decisions. Instead of maximizing the outcomes of behavior, as a prototypical *Homo economicus* would do, humans also optimize the decision-making costs (cognitive effort) that are associated with making a choice. As such, people use an abundance of decision rules or so-called heuristics in their daily lives. For example, people may habitually eat cereals for breakfast, follow the traffic stream (imitating) when confronted with the closed route, think of what other people of the same age will wear at that occasion (norm), and contemplate extensively on the advantages and disadvantages of different experimental set-ups. This allocation of resource works very efficiently, and allows us for example to deliberate about an experiment whilst (almost automatically) driving a car.

The critical question is of course how people decide on which decision strategy to employ in a given situation. The work of Simon (1955, 1959, 1976) on bounded rationality offers a perspective on why habits and complying with a norm may be a rational thing to do. The essential argument is that humans optimize the full process of decision making (*procedural rationality*), not only the outcomes (*substantive rationality*) (Simon, 1976). This holds that consumers may decide that a certain choice problem is not worth investing a lot of cognitive effort in (e.g., deciding on your breakfast), whereas another choice problem requires more cognitive attention (e.g., setting up an experiment). The less important a decision problem is, the less cognitive energy one is willing to invest in the decision, and, hence, the simpler the decision heuristic that will be employed.

Often people use their own previous experiences in a heuristic. For example, when you are satisfied with a certain type of cereal for breakfast, you may not waste any cognitive energy on deciding what to eat, but rather grab for the cereal in an automated way (which may be very convenient early in the morning). However, people may also employ the behavior and experiences of other people in their decision making. For example, when confronted with an unexpected road obstruction, the car driver may use the behavior of the other drivers as a clue regarding which direction to proceed. Instead of deliberating about all the possible alternative routes, the driver may assume that the predecessors have thought about it, and hence he/she may follow without giving too much thought on the issue. Also when thinking about which clothes to wear on that occasion, you use the experience of other people indirectly. Remembering the negative remarks people made on the too-casual outfit of a colleague on a previous similar occasion, you may decide to wear a suit, despite your personal preference to wear a more casual outfit. It appears that the cognitive strategies that people employ can be organized on two dimensions: (1) the amount of cognitive effort that is involved, and (2), the individual versus social focus of information gathering.

Regarding the first dimension, amount of cognitive effort, the basic idea is that people allocate their limited cognitive capacity over various decision problems they face so as to maximize their utility. When one is frequently being confronted with the same or similar decision tasks and the previous behavior yielded satisfactory outcomes, it is a good strategy to economize on cognitive effort by using simple heuristics or a habitual script in making the decision. This allows for allocating most of the cognitive capacity to decision problems that require more attention in order to find a satisfactory solution, such as non-routine decisions with important consequences. Because cognitive processing takes time, using simple decision heuristics will save time. This explains why people tend to use simpler decision heuristics when under time pressure (e.g., Smith *et al.*, 1982; Wallsten and Barton, 1982; Wright, 1974; Ben Zur and Breznitz, 1981). Also when the decision is less important (in terms of consequences) the decision maker may use a simpler heuristic instead of using all information available (e.g., Tversky, 1969, 1972). The simplest type of behavior in terms of cognitive effort refers to preconscious habits (Fiske and Taylor, 1991), i.e. behavior which bears a reflex-like character.

Regarding the second dimension, the individual versus social focus of information gathering, uncertainty is the key-factor that determines the focus of the information search process. When people are certain of themselves, they usually refer to their own previous experiences when making a deliberate or automated decision. When uncertain, people may use the experiences of other people to come to a decision in a cognitive efficient manner. Especially the behavior of other people with about similar abilities may provide a useful clue in the decision-making process. Simple imitation may be an economical way of allocating cognitive capacity to a decision. The *Social Learning Theory* (Bandura, 1977, 1986) states that seeing someone else's behavior being reinforced may affects one's own behavior. This imitating however requires more cognitive effort than a simple habit, because one should be *attentive* to the behavior of someone else, *understand* and *remember* that behavior, be able to *reproduce* that behavior, and experience *reinforcement* after performing the behavior yourself (Bandura, 1977). Following simple norms can also be considered as a social focused heuristic that requires relative little cognitive effort. However, according to the *Theory of Planned Behavior* (Ajzen, 1985, 1988, 1991), the *subjective norm*, may require more cognitive effort in making a decision. The subjective norm here refers to a person's perception of the opinion of others about him/her performing the relevant behavior. The subjective norm is proposed as a function of one's beliefs that referents think whether the person should or should not perform the behavior (called the injunctive norm), weighted by the motivation to comply with those referents. Social comparison (Festinger, 1954) is a key process here, involving the fact that people consciously compare their opinions and abilities with those of other people. These comparisons follow dimensions such as the possession of material goods, financial means, status, principles, attitudes and skills. With respect to opinions, people have a drive to roughly conform to others. With respect to abilities, people have a drive to be (somewhat) superior to others. Becoming aware of a subjective (social) norm would involve an assessment of relevant others and an appreciation of their behavioral intentions, which involves considerable more cognitive effort than simple imitation or obedience to a simple norm.

In organizing the various decision strategies that people employ, we find it instructive to use the two dimensions as graphically depicted in Figure 6.4.

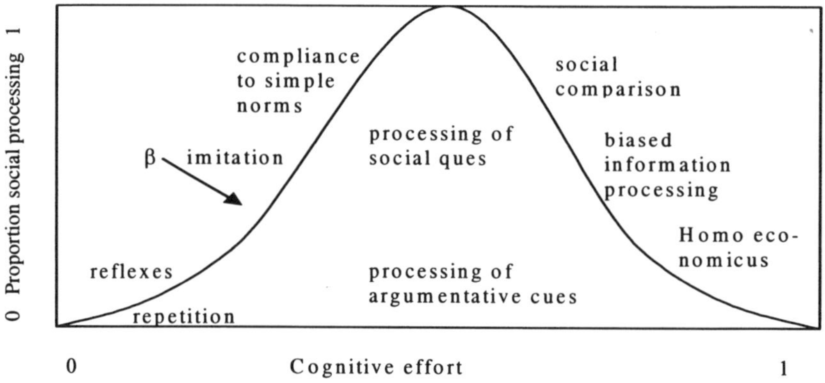

Figure 6.4: Different decision processes organized along the dimensions of cognitive effort and use of social information. β indicates the maximum level of social processing that allows for procedural optimality

Figure 6.4 shows that decision strategies that hardly require any cognitive effort (reflexes), or require very much cognitive effort (the prototypical *Homo economicus*) do not use social information. Strategies that require an intermediate cognitive effort may use both social and non-social information. Here, uncertainty is a key factor that determines the degree to which social information is being used in the decision-making process.

In organizing the decision strategies along these two dimensions, a perspective emerges regarding how people differ regarding their abilities and motivations to invest cognitive effort in a decision, and to what extent they use social information. Hence, this perspective contributes to the understanding of heterogeneity between people as regards their decision making. For example, some people may be more inclined towards using social information, and other people may have a larger cognitive ability, making it easier to invest cognitive effort in the decision-making process. On top of that, understanding how these abilities and motivations may change in a repeated decision-making situation provides a perspective on how people switch between decision strategies over time, and hence contributes to the understanding of heterogeneity within people. For example, when people become more uncertain, they will tend to use more social information in their decision making, and when people are not satisfied, they may be inclined to spend more cognitive effort in their decision-making process as to find a better behavioral opportunity.

To test hypotheses regarding the effects of heterogeneity in the decision-making process on collective outcomes we developed the consumat approach. This approach involves a multi-agent simulation model of decision-making processes. In the next section, we briefly elaborate on the consumat approach.

6.5 THE CONSUMAT APPROACH

The consumat approach is based on a comprehensive conceptual model of choice and decision-making behavior (Jager *et al.*, 1999; Jager, 2000). As such it tries to offer a more psychological based meta-theory of human decision making than the frequently used 'rational actor' approach. The consumat approach considers basic human needs and uncertainty as the driving factors behind the human decision-making process.

Based on this conceptual model, a multi-agent simulation model has been developed, in which the agents are called 'consumats'. The driving forces at the collective (macro-) and the individual (micro-) level determine the environmental setting for consumat behavior. This may be represented by a collective resource. The individual level refers to the consumats: they are equipped with needs which may be more or less satisfied, they are confronted with opportunities to consume, and they have various abilities to consume opportunities. Furthermore, consumats have a certain degree of uncertainty, depending on the difference between expected and actual outcomes of their behavior.

The various decision-making processes as organized along the two dimensions of Figure 6.4 are reduced to four decision rules: deliberation, social comparison, repetition and imitation. This simplification serves to keep the simulation model simple, and the results transparent for interpretation. Which of these decision rules consumats use at a given moment in time depends on their level of need satisfaction and degree of uncertainty. Consumats having a low level of need satisfaction and a low degree of uncertainty are assumed to *deliberate*, that is, to determine the consequences of all possible decisions given a fixed time-horizon in order to maximize their level of need satisfaction. Consumats having a low level of need satisfaction and a high degree of uncertainty are assumed to engage in *social comparison*. This implies comparison of own previous behavior with the previous behavior of consumats having roughly similar abilities, and selecting that behavior which yields a maximal level of need satisfaction. When consumats have a high level of need satisfaction, but also a high level of uncertainty, they will *imitate* the behavior of other similar consumats. Finally, consumats having a high level of need satisfaction and a low level of uncertainty simply *repeat* their previous behavior. When consumats engage in reasoned behavior (deliberation and social comparison) they will update the information in their mental map, which serves as a memory to store information on abilities, opportunities, and characteristics of other agents.

After the consumption of opportunities, a new level of need satisfaction will be derived, and changes will occur regarding consumats' abilities, opportunities and uncertainty. Moreover, the environment the consumats behave in, for example, a collective resource, will change as a consequence of their behavior, thereby affecting the behavior in subsequent time steps.

6.6 A CONSUMAT MODEL FOR COMMON-POOL RESOURCES

To study how heterogeneity in SVO and decision rules affects the behavior in a common-pool resource in a very controlled setting, we decided to formalize the consumat approach for the common-pool resource paradigm as sketched in the introduction. Three needs are formalized: a personal need, a social need and a need for exploration.

The personal need relates to subsistence, and is assumed to be equal to $u(x)$ the returns from investment. This personal need is equal for all agents ($N_{I,i} = u_i(x)$).

The social need relates to how an agent wants to relate to other people, and is a formalization of the SVO, and hence comprises individualistic, competitive and cooperative preferences for outcome distributions. The agents thus differ regarding their social need.

When an agent is purely individualistic then the social need satisfaction is equal to $u(x)$.

When an agent is competitive the social need satisfaction takes into account the relative returns compared to the average returns. The higher the relative returns, the higher the social need satisfaction.

$$N_{S,i} = 1 - \exp(-c \cdot u_i(\mathbf{x}) / \sum_{j=1}^{8} \frac{1}{8} \cdot u_j(\mathbf{x})) \tag{6.1}$$

The social need satisfaction of a cooperative agent is higher the closer the returns come to the cooperative optimum. This cooperative optimum is measured in difference with the cooperative amount of tokens, although the cooperative amount of returns could also have been used.

$$N_{S,i} = \exp(-c_2 * (\sum_{j=1}^{8} x_j - 36)^2) \tag{6.2}$$

Besides the personal and social need, we also formalized a need for exploration. The exploration need is a combination of understanding, creation and freedom needs. Whereas exploration is often described as specific, and may relate to, for example, the search for food, Berlyne (1966) also proposed a diversify type of exploration, motivated by the need to know. Large numbers of experiments by Berlyne and others led to further notions to be linked to exploration, such as the novelty of a situation (e.g., Hutt, 1970; see also Gibson, 1988). This type of exploration appears to describe how people learn to understand how a resource system (the resource and the other players) reacts to certain actions. Hence exploration serves to increase the understanding of the system. This exploration need is conceived to be less satisfied the more stable the outcomes are, because in such a situation nothing new is being learned about the system. The exploration need $N_{E,i}$ is being formalized as a standard deviation in outcomes over the last n (=5) rounds:

$$N_{E,i} = \sqrt{\frac{\sum_{j=t-n}^{t} \{u_i(\mathbf{x}(j)) - u_i(\bar{\mathbf{x}})\}^2}{n-1}} \tag{6.3}$$

where $u_i(\bar{x})$ is the average return for n rounds.

When consumats are dissatisfied, they deliberate about all possible courses of action. When all three needs are formalized, the investment opportunities are evaluated on expected outcomes for the self, the expected outcomes for the other and the contribution to the standard deviation in the outcomes. We assume that dissatisfaction is not absolute: it is more a rule to decide when to employ cognitive energy. Exploration is less important than subsistence, but dissatisfaction may lead to the same level of cognitive effort. Stated differently, if you don't have serious problems to think about, you'll think about less serious problems.

The total need satisfaction is defined as the weighted sum of needs, where $\Sigma_i \beta_i = 1.$:

$$N_i = \beta_1 \cdot N_{S,i} + \beta_2 \cdot N_{I,i} + \beta_3 \cdot N_{E,i} \tag{6.4}$$

Uncertainty, *U*, is defined as the standard deviation of the individual returns during the last five rounds.

$$U_i = \sqrt{\frac{\sum_{j=t-5}^{t} \{u_i(\mathbf{x}(j)) - u_i(\bar{\mathbf{x}})\}^2}{4}} \tag{6.5}$$

The agents are equipped with a memory, which is being used in calculating the expected returns from the decisions. In this calculation the agent makes use of an expected aggregate level of tokens. To estimate how many tokens other agents are expected to invest, a neural network is used. A neural network is an algorithm that resembles the way in which the brain works. It is composed of *nodes* representing physiological neurons, and *weights*, which are connections of differing strength between two nodes. Some of the neurons receive their input from the environment and some others give back their output to the environment. We use a single layer neural network, which is a method to describe changes in the weights based on physiological principles, and is described by the following equation:

$$EY_{t,i} = w_{t,i,0} + \sum_{k=1}^{ks} w_{t,i,k} \cdot y_{t,k} \tag{6.6}$$

where $EY_{t,i}$ is the level of tokens of the other seven agents, $y_{t,k}$ are the inputs of the neural network, the observed total investments during the previous k_s (=3) rounds. Finally, the inputs are weighted by $w_{t,i,k}$.

A neural network is trained when new information about the input values is used to update the weights (w). The widely used Widrow-Hoff delta rule is used to train the neural network during the simulation (Mehrotra *et al.*, 1997). This simple neural network simply weights the observed token investments of the other agents during the last few rounds to estimate the total investments in the next round.

$$\Delta w_{t,i,j} = \eta \cdot \delta_{t,i} \cdot \frac{\mathbf{y}_i}{\|\mathbf{y}_i\|} \tag{6.7}$$

where

$$\delta_{t,i} = Y_{t,i} - w_{t,i,0} - \sum_{k=1}^{ks} w_{t,i,k} \cdot y_{t,k} \tag{6.8}$$

where $\delta_{t,i}$ is the difference between the observed values and the expected value, and $Y_{t,i}$ is the observed token investments by the other agents. The delta rule updates the expectations according to the observed errors. The rate of updating depends on the value of η, which is suggested to lie between 0.1 and 1 (Gallant, 1993). We will assume η to be 0.5 reflecting relative adaptive agents.

In making an investment decision, the agent employs one of the four cognitive processes as described in the previous section, depending on the (combined) level of need satisfaction and uncertainty and the thresholds U_{max} determining when an agent is uncertain, and N_{min} determining when the agent is satisfied. When the agent is dissatisfied and certain, it will engage in deliberation. Deliberation has been formalized as calculating the level of investment that maximizes the agent's expected need satisfaction. We assume that if consumats deliberate they have full information and understanding of the problem, and are able to calculate the Nash equilibrium.

When the agent is satisfied and certain, it will engage in repetition, and hence invest the same quantity as in the previous round. When an agent is dissatisfied and uncertain, it will engage in social comparison. This implies comparing the average investment of the previous round with the own investment of the previous round, and choosing that investment with the highest expected level of need satisfaction. Finally, when the agent is satisfied and uncertain, it will engage in imitation. This implies copying the average investment of the previous round.

The threshold values for aspiration level (N_{min}) and uncertainty tolerance (U_{max}) have an empirical counterpart in the Intellect factor and the Emotional Stability factor of the Big five personality structure (Goldberg, 1990, see also Janssen and Jager, 2001).

In the next section we discuss the results of simulation experiments in which the agents are confronted with the same common-pool resource as used in empirical studies by Ostrom *et al.* (1994).

6.7 RESULTS

In this section we discuss the results for a series of experiments. Tables 6.2 and 6.3 show statistics of six original laboratory experiments as described in Ostrom *et al.* (1994). The average investments are among the Nash equilibrium of eight tokens, although the experiments with 25 tokens are systematically above eight tokens. The variability among individual investments is higher in the 25-token experiments, showing that there is more behavioral change in this condition. Furthermore, variability among rounds is generally larger in the 25-token experiments. Furthermore, the higher the average investment, the higher the interround variability. These statistics are used to compare the simulation experiments with the laboratory experiments.

Table 6.2: Experimental values of three experiments with ten tokens. The first column of numbers denotes the average token investment in the common-pool resource over 30 rounds. For each round the standard deviation of the eight individual investments is calculated. The second column contains the average standard deviation over 30 rounds and is a measure of variability within each round. The last column shows the total absolute changes in the total investments, which is a measure of investment variability among 30 rounds.

Experiment	$\bar{x}$	$\overline{stdev(x)}$	$cum(\lvert \bar{x} - \bar{x}_{t-1} \rvert)$
1	8.46	1.37	139
2	7.72	0.60	125
3	7.94	0.88	137

Table 6.3: Experimental values of three experiments with 25 tokens. The first column of numbers denotes the average token investment in the common-pool resource over 20 rounds. For each round the standard deviation of the eight individual investments are calculated. The second column contains the average standard deviation over 20 rounds and is a measure of variability within each round. The last column shows the total absolute changes in the total investments, which is a measure of investment variability among 20 rounds.

Experiment	$\bar{x}$	$\overline{stdev(x)}$	$cum(\mid \bar{x} - \bar{x}_{t-1} \mid)$
1	8.64	3.36	283
2	9.22	2.89	316
3	8.54	2.15	108

Experimenting with cognitive processing, needs and social value orientation

A series of simulation experiments has been performed, and by varying the characteristics of the agents we created different conditions. A first characteristic we varied was the cognitive processing the agent could employ. Two conditions were created, namely the *Homo economicus* (HE), which engages exclusively in deliberation, thereby representing the rational agent from standard economic theory, and the *Homo psychologicus* (HP), which could employ all four decision strategies. For the *HE* conditions the values of N_{min} and U_{max} are put on such values that the agents only deliberate. In the HP condition, the values of N_{min} and U_{max} are 0.5 and 0.1 respectively, so that all four cognitive processes can be used.

A second characteristic we varied in the experimental design refers to the combination of needs the agents have. Four conditions were created, respectively; (1) agents with only an personal need (noted with P), (2) agents having an personal need and a social need (PS), (3) agents having a personal need and an exploration need (PE), and (4) agents having a personal need, a social need and an exploration need (PES).

The third characteristic we varied was the SVO of the eight agents. In the conditions where the social need is formalized a fixed number of the agents are either cooperative, individualistic or competitive. We analyse all the combinations of these three SVOs.

The initial expectations the agents have regarding the total investments are set in line with the equilibrium outcomes of the SVO of the agent. In case of a Nash equilibrium (competitive or individualistic SVO) this expectation is set at 64 tokens, and in case of a cooperative equilibrium (cooperative SVO) this expectation is set at 36 tokens.

Table 6.4 summarizes the statistics of the 16 conditions. In ten conditions we observe that the theoretical Nash equilibrium (x = 8) is derived, namely in the conditions where only the personal need is taken into account (HE-P, HP-P), or when the social need is weighted but where the agents are all competitive or individualistic.

Inclusion of the need for exploration leads to fluctuations in the token investments, which are more extreme in the case of the *Homo economicus*, since the *Homo psychologicus* can be satisfied with a lower variability.

Table 6.4: Summary of 16 conditions: Simulated values ten tokens, all four cognitive strategies and three social value orientations. The average standard deviation on between agent variation is not depicted since all agents have the same characteristics.

Experiment	$\bar{x}$	$cum(\lvert \bar{x} - \bar{x}_{t-1} \rvert)$
HE-P	8	0
HP-P	8	0
HE-PS-coop	8.03	448
HE-PS-ind	8	0
HE-PS-comp	8	0
HP-PS-coop	6	0
HP-PS-ind	8	0
HP-PS-comp	8	0
HE-PE	7.93	440
HP-PE	6.3	344
HE-PES-coop	8	424
HE-PES-ind	8	0
HE-PES-comp	8	0
HP-PES-coop	6	216
HP-PES-ind	8	0
HP-PES-comp	8	0

When we assess all combinations of SVO we have to simplify the presentation to keep an overview of how different experimental variations affect the results. Therefore we have developed an indicator *I* that describes the difference between the simulation results and the empirical results or theoretical expectation.

$$I = \frac{a_1}{3}(x - x_C)^2 + \frac{a_2}{3}(stdev - stdev_C)^2 + \frac{a_3}{3}(cum - cum_C)^2 \qquad (6.9)$$

This indicator I is constituted on three relevant outcomes (see Table 6.2 and 6.3). First, the difference between the simulated average investment behavior x and the empirical/theoretical average investment x_C is calculated in $(x - x_C)^2$. The empirical results are the statistics generated by the 24 real subjects in the three reported experiments. The theoretical expectation reflects the Nash equilibrium of eight tokens for these experiments. The second component of the indicator, $(stdev - stdev_C)^2$, reflects the difference between the variability in each round for the simulation and the empirical/theoretical investment. The theoretical (Nash) expectation for $stdev_C$ is 0, as all subjects are expected always to invest eight tokens. For the empirical results we calculate $stdev_C$ using the average indicator values of Tables 6.2 and 6.3. The third component of the indicator $(cum - cum_C)^2$, reflects the difference between the variability over the rounds for the simulation and the empirical/theoretical investment. Again, the theoretical (Nash) expectation of cum_C is zero, as all subjects are expected always to invest eight tokens. For the empirical results we calculate cum_C using the average indicator values of Tables 6.2 and 6.3. In this indicator I we defined the values of a_1, a_2 and a_3 in such a way that the indicator remained between 0 and 1.

In the following we present the results for a series of experiments. Graphically we present the value of I as a bar. Each bar represents the difference between the statistics of a simulation run and the theoretical expectation (upper, Nash equilibrium) or empirical observations (lower). In the left figures the artificial agents make decisions in line with the *Homo economicus* (HE), while in the figures on the right, the artificial agents make decisions in line with the *Homo psychologicus* (HP). The axes represent the number of individuals and the number of competitors. The number of cooperative agents is equal to eight minus the agents with individualistic or competitive SVO. In Figure 6.5 we present the results for agents having only a need for personal returns and the need for identity.

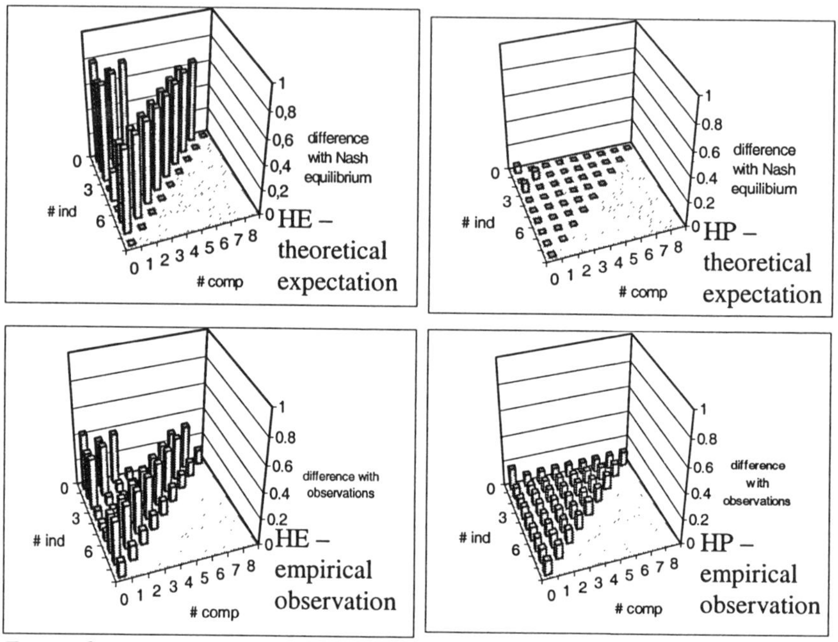

Figure 6.5: Results of all combinations of SVO of experiments with ten tokens with artificial agents without the need for exploration

It can be observed that in the conditions without cooperative agents (on the diagonal line), the theoretical outcomes can be reproduced in both the HE and the HP conditions. When agents act according to the HP, the results in most combinations do not differ much from the theoretical outcome. For the HE we observe that the inclusion of cooperatives causes a large difference with the theoretical outcomes. Only when the number of cooperatives is about four can we see a valley in the results, indicating that the results are closer to the theoretical outcomes. This valley is caused by oscillations that are less extreme than when fewer cooperators take part. Compared with the observed statistics, the HP clearly performs better than HE. It can be seen that more cooperative agents lead towards a better match with the empirical data.

Figure 6.6 shows the results for the same experiments, only now including the need for exploration. The 'valley' in the HP-empirical data figure shows that for the HP the observed statistics can be reproduced closely when about four cooperative agents (close to the proportion of cooperatives as identified by Van Lange, 1999) have been formalized. However, for most combinations of SVO the results are still more close to the theoretical expectations than to the empirical data.

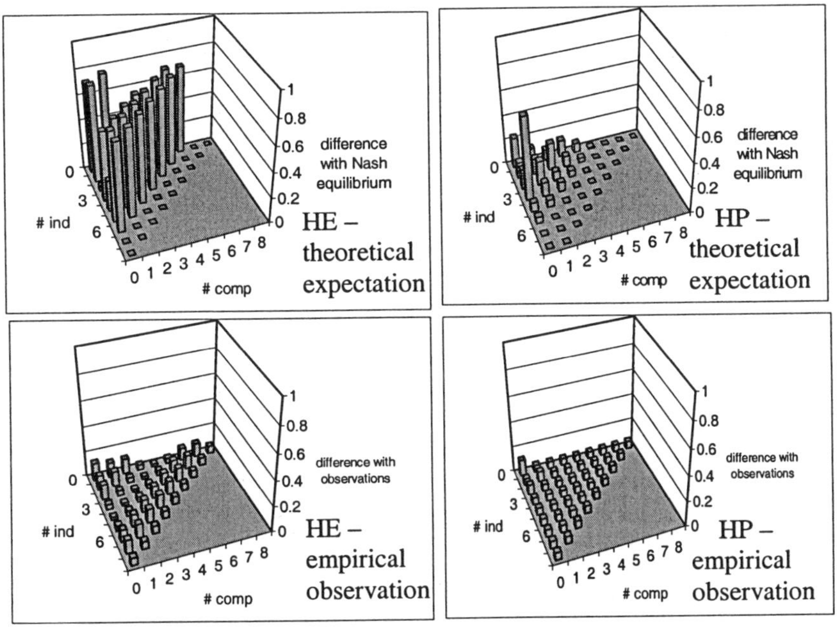

Figure 6.6: Results of all combinations of SVO of experiments with artificial agents with ten tokens, with equal weighting of needs

Experimenting with heterogeneity of needs

In the next series of experiments the agents differ with respect to the weighting of their needs, thereby adding an extra source of heterogeneity between the agents. The values of β_i are drawn from a random distribution. Still the sum of β_i's is equal to one. The procedure of generating the β's is as follows. For each agent the value of β is randomly drawn between 0 and 1, but then divided by the sum of β_1, β_2 and β_3. For each combination of SVO, we performed 1000 runs and calculated the average value of the indicators.

The results of these experiments are shown in Figure 6.7. Whereas in the previous experiments the agents performed better in matching the theoretical expectations than the empirical data, here we observe the opposite. However, this match on empirical data is less close than for the HP with about four cooperatives in the previous experiment. Remarkable is that the HP now have the best performance when there are only competitive agents. This is mainly due to its good fit on the second component of *I* that measures the difference between the variability in each round. The theoretical outcomes are now better reproduced when agents have an individualistic SVO, which reduces the variability of the decisions.

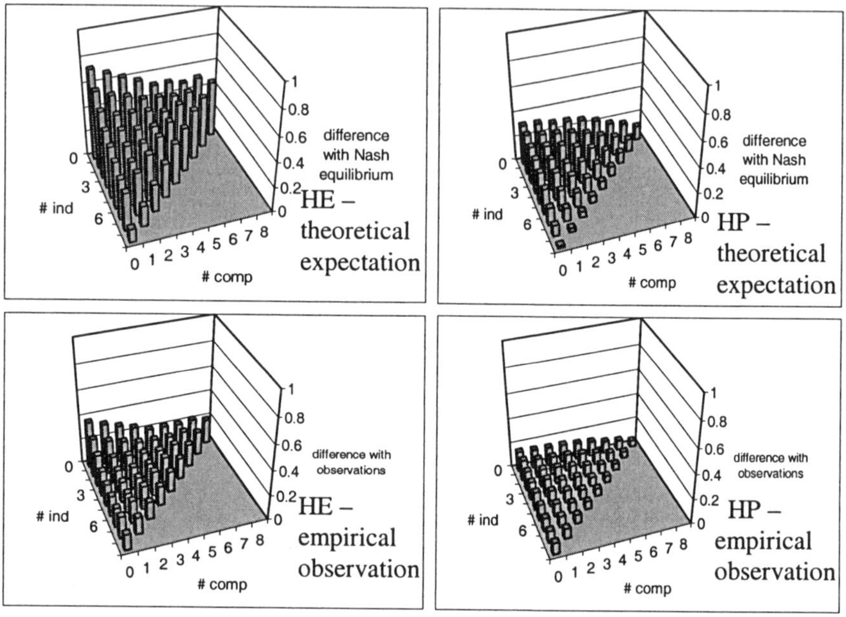

Figure 6.7: Results of all combinations of SVO of experiments with artificial agents with ten tokens, with heterogeneous weighting of needs

The same experiments with heterogeneous weighting of the needs as performed for the ten-token case are repeated for the 25-token case (Figure 6.8). For the HP, the results match the closest to the empirical data when there are only competitors. When there are less competitors, the results match quite well for as long as there is a balance between the number of individualists and cooperatives, as can be seen in the 'valley'. The theoretical prediction can be replicated the best when agents are individualistic and agents perform like the HE.

These computational experiments show that we are not able to replicate the observations perfectly, but that with different assumptions different types of agent formulations are more suitable to approximate the statistics of the observations. Remarkable is the influence of heterogeneity among the weighting of the needs. If we do not assume heterogeneity, groups that include cooperative agents are better able to approximate the statistics of the observations, while inclusion of heterogeneity leads to the result that no cooperators should be in the group in the attempt to replicate the statistics of the observations. What is clear is that all three needs are of importance to understand the observations, and that agents conforming to the *Homo psychologicus* have a better performance than the *Homo economicus* in approximating the empirical data.

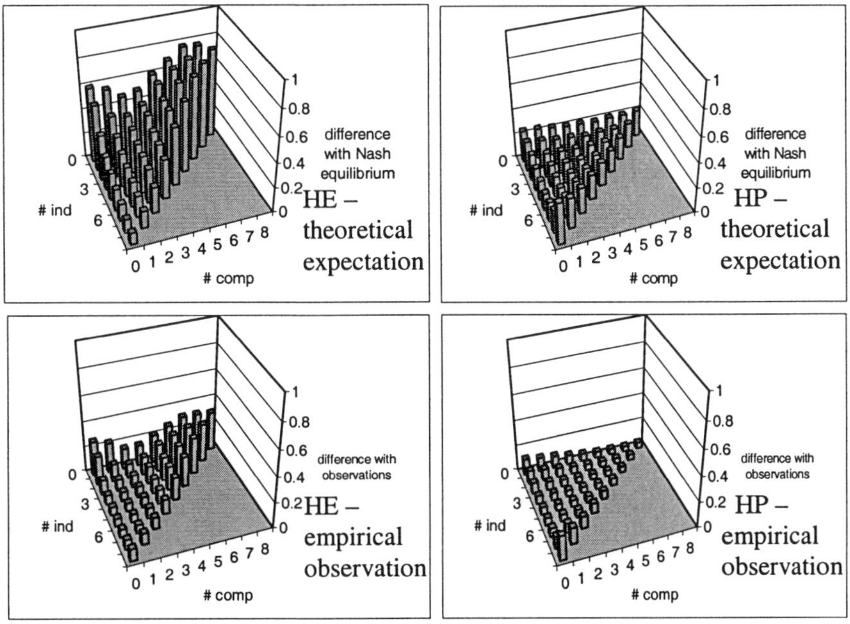

Figure 6.8: Results of all combinations of SVO of experiments with artificial agents with 25 tokens, with heterogeneous weighting of needs

6.8 DISCUSSION AND CONCLUSIONS

In this series of experiments we demonstrated how agents can be equipped with decision rules that are based on psychological theory, and experimented with different settings of these rules as to mimic the individual behavior of real people acting in a resource dilemma. We showed that mimicking the behavior of real people on an individual level requires more psychological realism in the agents than mimicking the aggregate outcomes as in previous experiments. However, we realize that the reproduction of statistics that fit with a limited set of empirical observations does not provide sufficient proof that the simulation model captures the most relevant dynamics that guide the behavior of the subjects in the Ostrom *et al.* (1994) experiments. To do so would require more empirical data. Hence we argue that more experiments are required to unravel the decision-making process of real people. These experiments should address the factors underlying the heterogeneity of decision making. In this chapter, we identified several factors that may contribute to this heterogeneity in decision making. We distinguished the different needs that play a role in the decision-making process, the relative importance of these needs, the SVO of the decision maker, the cognitive process that people employ when making a decision, the personality characteristics that determine the tendency to use certain cognitive processes more often than others (Intellect, Emotional Stability) and the time-horizon that is taken into account when making a decision. To test all these factors simultaneously in experimental research would yield an enormous task, both for the experimenter and the subjects. This task is especially difficult because we are dealing with complex behavioral dynamics, as different people having different (changing) values on the different factors are interacting for 20 time-steps. Therefore it would be practical if we could test beforehand the relevance of factors and develop hypothesizes concerning the effects of varying these factors. We argue that simulation research provides a tool capable of doing so. By performing many experiments and conducting sensitivity analysis one may identify the conditions under which certain behavioral dynamics are more likely to happen, and which factors play a crucial role. Following that, hypothesis and a research design can be formulated for testing these specific effects in empirical experiments.

We claim that it would be most efficient to combine simulation research and empirical research as described above to harvest synergetic benefits. For example, our simulation experiments suggest that the cognitive processes a person is most likely to use may be very important in his/her harvesting behavior. Someone having a low aspiration level is more likely to develop a habit, whereas someone that has a low uncertainty tolerance is more likely to engage in imitation and social comparison. Field experiments could be focused on the question how personality characteristics (aspiration level and uncertainty tolerance) of people are related to their cognitive processing and behavior. The data obtained in this empirical research can subsequently be used to formalize the relation between personality and cognitive processing more validly in agent rules. However, until now experimental and simulation research are rather distinct, despite the fact that they more and more deal with similar research questions. Only three studies are known to the authors that use multi-agent models to formulate hypotheses which are tested with real agents (Duffy, 2001; Pingle and Tesfatsion, 2001; Tobias, in preparation). We think that both multi-agent modeling as well as experimental research can benefit from more interaction between both fields. Whereas both fields appear to be concentrated around the research methodology that is being used, a focus on the research question would benefit this interaction. As a start we discuss five hypotheses based on our computational model, and formulate tests for the experimental research.

Hypotheses:

1. Groups of people having a high aspiration level and a high tolerance for uncertainty are more likely to engage in deliberation. As a consequence they are less sensitive to the imitation effect. This would result in fewer fluctuations in investment decision between agents and between rounds.
2. Groups of people that systematically differ as regards their distribution of SVO also systematically differ as regards their investment levels. We expect that homogeneous groups of cooperatives invest less in market two than relative homogeneous groups of individualists and competitors. Experiments would require a (unobtrusive) pre-measurement of SVO, which will be used to allocate subjects later to experimental settings.
3. Displaying only aggregate outcomes will inhibit the social need in the subjects' decision making, and hence will moderate the SVO effect in comparison to conditions where individual outcome levels are displayed.
4. Giving the subjects a fee for participating in the experiment on the basis of their returns would increase the weighing of the personal need, whereas providing a standard fee for participation would inhibit the importance of the personal need in the decision-making process.
5. Allowing the subjects to explore issues that are not directly relevant for the experiment (e.g., particular information on the other players) would decrease the influence of the exploration need on the harvesting behavior, and hence the results will show less variance.

To be able to compare simulation results involving cognitive processes with experimental data, it would be practical to measure the cognitive process in an experimental setting. The approach of Hine and Gifford (1996) demonstrates that it is possible and necessary to ask people about their decision-making strategies. However, a more structural approach of measuring cognitive processing is proposed by registering the quantity and type of information subjects retrieve from a matrix board (on screen) before making an investment decision. The decision-making process can be tracked by registering the quantity and type of information subjects retrieve from a matrix board (on screen) before making a harvesting decision. The information that can be retrieved relates to the state of the resource and the harvesting behavior of other people (aggregated or individual). Measuring information retrieval in real time allows for discrimination between habitual harvesting and deliberate stable harvesting. A possible avenue for further research on cognitive processes is to use MRI-scan data obtained during the decision-making task of the subjects. First experiments in which MRI scans are conducted in a cooperation game are conducted by McCabe *et al.* (2001) and show a relation between pre-frontal brain activity and promoting cooperative behavior in the other player.

Several other factors that have been identified as potentially influential in simulation research can be measured in empirical experiments. Moreover, many experiments can be performed in which subjects are being grouped together according to their scores on relevant factors. For example, to test hypothesis 2 we would have to obtain SVO data before assembling subject groups. The more clearly experimental research shows how a certain factor affects the behavior of real people, the better it is to formalize this factor into a valid way in a simulation model. This would give more insight in the dynamics behind the investment behavior and give rise to refining the hypothesis on relevant behavioral processes. We are convinced that the combination of different research tools to address the same research question is a promising way to get a better understanding of the very basic behavioral dynamics that determine our use of collective resources.

ACKNOWLEDGEMENTS

We thank Jim Walker for providing the laboratory data and giving feedback on an earlier manuscript. Furthermore we thank participants of seminars at Indiana University and the University of Amsterdam for providing constructive and helpful feedback.

7. Implications of spatial heterogeneity of grazing pressure on the resilience of rangelands

Marco A. Janssen, J. Marty Anderies, Mark Stafford Smith and Brian H. Walker

7.1 INTRODUCTION

Due to the scarcity of reliable surface water supplies, grazing pressure has been low for most of recent Australian evolutionary history. Fires set by Aboriginal people or lightning were relatively frequent in many ecosystems. These systems were also adapted to highly variable rainfall in both time and space (Friedel *et al.*, 1990). Following the establishment of water points after European settlement, wild (native and feral) and commercial animals combined to exert high grazing pressure on rangeland vegetation. This increased pressure has reduced grass cover, and thus the build up of fuel. This, in turn, has reduced the frequency of fire in many rangeland systems. Sheep and cattle are primarily grazers, rather than browsers, so they do not suppress shrubs. These changes have combined to cause woody plants to increase (Noble, 1997), and grass production per unit of rainfall to fall.

Rangeland managers must decide when to reduce grazing pressure in order to let the system recover, and if and when to use fire to suppress shrubs. The manager is confronted with many uncertainties: rainfall patterns, prices, imperfect knowledge of the ecological systems, and possible change in regulations. An important property of an ecological system such as a rangeland is the possibility that the system may occupy one of multiple states (Walker *et al.*, 1981). Depending on the management of the system, the rangeland can flip from a productive and sustainable state to an unproductive state. Ideally rangeland managers maintain the resilience of the system, that is, they will minimize the probability that the system flips to another (less productive) state due to a surprise (for example, a period of low rainfall).

Abrupt shifts among a multiplicity of very different stable domains are plausible in various different regional ecosystems. The likelihood of such shifts is determined by the resilience of these states as measured by properties of their stability domains. The costs of such shifts depend on the length of time spent in such states and their relative productivity. Since these features were first described for ecosystems (Holling, 1973), it has become evident that alternating states arise in a wide variety of ecosystems, such as lakes, marine fisheries, benthic systems, wetlands, forests, savannas, and rangelands. Although dynamics on different scales are explicitly included in the resilience analysis of ecosystems, spatial heterogeneity is usually neglected in the formal models which describe ecosystems with multiple stable states (though see Van de Koppel *et al.*, 1997).

Spatially explicit predator–prey models can show a reduction of oscillations compared with traditional predator–prey models (Jansen and de Roos, 2000). Given this ability of space to fundamentally change the dynamics of ecological systems, the question is, whether space can affect the resilience of ecosystems. The aim of this chapter is to explore the consequences of spatial heterogeneity for the resilience of rangelands. We show that including space can make the system even more vulnerable (i.e. it can be destabilizing), in contrast to studies in which space tends to be a stabilizing force (Grenney *et al.*, 1973; Pacala and Levin, 1997; Durrett and Levin, 1998).

In this chapter we analyse the implications of non-uniform grazing. The model is based on the stylized integrated model of rangeland ecosystem and management interactions from Anderies *et al.* (2002). With this model, Anderies *et al.* assessed the conditions under which the system flips from a healthy state to an unproductive shrub state. The model in Anderies *et al.* (2002) did not address space explicitly. In fact, a mean-field model was used for the rangeland system. The mean-field assumption can be a good approximation when the environment is homogenous and the population is well mixed and interacts over long distances. This assumption might be reasonable for many applications in physics and chemistry, but it often does not hold in ecology (Dieckmann *et al.*, 2000).

Using an agent-based model, we use the same framework as the Anderies *et al.* model but simulate individual sheep on a spatial lattice, and analyse the consequences for different assumptions related to the behavior of sheep, such as herding and the location of waterpoints.

The chapter is developed as follows. First, the importance of heterogeneity in rangeland ecology and rangeland management is discussed. Then, we briefly discuss the key equations of the Anderies *et al.* model. In the sections thereafter, the agent-based model is introduced and simulation results are discussed, which are analysed in more detail by a two-region model. The last section concludes.

7.2 SPATIAL HETEROGENEITY IN RANGELANDS

Spatial heterogeneity is a fundamental characteristic of rangelands due to their low level of productivity, which results in management units (like paddocks) having to be large relative to the underlying landscape heterogeneity. As a consequence, unlike high productivity pastures where different landscape units can be readily fenced off and managed differently, rangeland managers must focus on managing the heterogeneity *within* a management unit. This has long been recognized in management (e.g. Lange *et al.* (1984) report management investment in the 1890s to allow for this) as well as in range ecology (Coughenour, 1991), where space has been considered mainly in relation to the way soils and plants function within a landscape and in relation to animal behavior.

Work on landscape function has focused on resource re-distribution (e.g. Noy-Meir, 1975, further developed by Pickup, 1985, and a series of papers from Tongway and Ludwig, 1990 to Ludwig *et al.*, 2000, and, to a lesser extent, the interactive effects of plant competition (e.g. Archer *et al.*, 1988), usually at the landscape scale of ~1–100 ha. By contrast, animal behavior concerns have generally addressed a larger scale of ~1–100 km^2. The most notable spatially explicit models in this case were Goodall (1967) and Noble (1975), built on by the work of Senft *et al.* (1983), Stafford Smith (1984, 1988), Pickup and Chewings (1988), Stafford Smith and Foran (1990), and Coughenour (1992). Most of these models were aimed at specific implications of spatial grazing behavior on land degradation and paddock design, with a limited ability to link this to animal production. Further development of improved models of pasture and animal production has proceeded with point models that neglect most spatial processes (e.g. SPUR – Hanson *et al.*, 1988; GRASP – McKeon *et al.*, 1990). Although some effort has been made to place these in a spatial context (e.g. Pierson *et al.*, 1999; Milne *et al.*, 1999; Reynolds and Wu, 1999), none of these models have been explicitly used to explore the question of whether spatial heterogeneity and spatial processes per se confer more or less resilience on rangelands, and whether there are then lessons for management in different systems. Indeed most of the models were not formulated with this goal in mind, so that their simplistic use to test such questions would be doubtful.

This has not prevented pastoralists and extension agencies from continuing to develop management procedures that ideally should be based on the answers to these questions. For example, it continues to be contentious in northern Australia whether one should seek to fence to separate vegetation based on low and high productivity soils (so that they can be managed separately) or deliberately fence a diversity of vegetation in together (so the animals can actively select what is 'best for them' at different times) (see Ash and Stafford Smith, 1996). There is a considerable literature of recommendations based on perceived ideal waterpoint distributions for even grazing, which is in conflict with the desire to leave some areas un-grazed to preserve biodiversity that is incompatible with grazing disturbance (James *et al.*, 1999). Thus the issue remains in need of considerable clarification.

Based on the simple graphical models of Noy-Meir (1975) and the understanding arising from the many studies mentioned above, it is possible to assert that the answer to the question of whether spatial processes enhance or reduce resilience in rangelands is unlikely to be simple. For example, building on general observations made during the Responses of Savannas to Stress and Disturbance (RSSD) program (Frost *et al.*, 1986), Ash and Stafford Smith (1996) made explicit predictions about circumstances when one or another outcome might be expected, in the context of the management response that would then be sensible (see Table 7.1). In addition, Cridland and Stafford Smith (1993) have argued that the effects of spatial grazing behavior is driven mainly by differences between vegetation units in small rangeland paddocks, whilst in large paddocks distance to water becomes a more dominating driver; the definitions of 'small' and 'large' depend on the grazing animal, being distinguished by maximum distances to water that are smaller or larger than about 3–4 km for sheep and 5–7 km for cattle.

This chapter now sets out a preliminary model to explore these types of predictions in a theoretical sense, with the intent of sharpening up the quantification of conditions under which different responses might be expected. There are a number of different types of degradation responses which can occur, including loss of perennial forage species, increase in woody competitors, loss of local soil quality, and changes to landscape redistribution processes, any of which can reduce net productivity. This chapter focuses on changes in the grass/tree balance.

Table 7.1: Relationships between forage production and quality at the landscape scale and likely implications for production and management: predictions from Ash and Stafford Smith (1996) for a paddock which has two types of vegetation in it, where the vegetation types may grow at the same time or different times, where they may be more or less resilient to grazing pressures which are higher than the average in the paddock (in this paper, more or less likely to experience woody weed encroachment), and where one is preferred by the grazing animal over the other but both are edible. Column 3 indicates the predicted net result. This table neglects the effects of distance from water, which is an additional spatial factor within each vegetation type if the paddock is large enough. In this paper we analyse only conditions related to the last prediction.

Forage growth and quality between landscape elements	Preferred vegetation units	Impact on animal production and vegetation	Management implications
Asynchronous	Resilient	Enhanced animal production, vegetation stable through time	Paddock sizes can be large to take advantage of forage diversity with less risk of localized degradation
Asynchronous	Susceptible	Enhanced animal production but preferred areas at significant risk of degradation	Maintain large paddocks but strategic use of spelling fire, location of water and supplement points to reduce pressure on most preferred units. Scenario with the biggest trade-off between production and sustainability.
Synchronous	Resilient	Little production advantage in trying to exploit diversity, vegetation stable through time	More intensive stock management e.g. supplementation to achieve production targets
Synchronous	Susceptible	Little production advantage in trying to exploit diversity, preferred areas at risk of degradation	Smaller paddock size, preferred vegetation units in particular to be fenced off.

7.3 A STYLIZED MODEL OF RANGELANDS

The model describes the interactions between perennial grass, shrubs, fire and commercial stock in a stylized way, based conceptually on the functioning of semi-arid woodlands and shrublands in western New South Wales (cf. Ludwig *et al.*, 1997). Here a brief overview is given. A detailed description and analysis of the model can be found in Anderies *et al.* (2002). Four state variables are defined as outlined below. The state variables are scaled in order to derive dimensionless equivalent forms. For example, instead of measuring grass in kilograms per hectare, it is measured as a proportion of the maximum grass biomass.

It is assumed that the pastoralist will adjust sheep density in order to maximize income. Decisions concerning sheep density are also assumed to be related to ecological factors, in this case, the amount of shrubs on the land. Furthermore, the pastoralist can suppress the occurrence of fire to a certain degree and thus control its effect on the competition between grass and shrubs.

The grass plant consists of two parts: the crown and the shoots. By crown we mean the root system and growing points. The shoots are the above-ground grass portion of the plant. The biomass of grass shoots is denoted by s, and basically follows a traditional logistic function. The crown promotes growth of the shoots according to the tiller potential ca_c independent of grass biomass, and through its interaction with above ground biomass via the term cs. Competition between woody shrubs and grass reduces the grass growth. This is captured by the term $\alpha_{us}w^{\beta}$, where α_{ws} is a competition coefficient, and where β (>1) leads to a growth reduction effect of woody shrubs that does not kick in until shrubs reach a relatively high density. Grass is removed by grazing pressure via the term $\gamma_g s$. Finally, grass biomass can be consumed by fire I, which has a general response function of form $f()$.

$$\frac{ds}{dt} = c \cdot (\alpha_c + s) \cdot (1 - s - \alpha_{ws} \cdot w^{\beta}) - \gamma_g \cdot s - I \cdot f(s; a_s, b_s) \tag{7.1}$$

The response curve is formulated in general terms in equation (7.2), containing a variable k and parameters a and b. If $b = 1$, $f()$ is a monotonically increasing function bounded above by 1; if $b > 1$, the function is sigmoidal. The parameter a controls the location of the point where f is half its maximum value, and b controls the steepness of the increasing portion. The larger the value of b, the more rapid is the switching.

$$f(k; a, b) = \frac{k^b}{a^b + k^b} \tag{7.2}$$

The crown biomass c grows at rate $r_c s$ and dies at a rate 1. The grass growth is dependent on the presence of the crown.

$$\frac{dc}{dt} = r_c \cdot s - c \tag{7.3}$$

In the context of the consumption response of sheep, b =1 to capture the fact that at high grass biomass, sheep are limited by their ability to process grass, not grass biomass (constant offtake) while at low biomasses, search times become limiting and consumption drops off. The case is similar with fire.

The fire consumption index is governed by the following dynamics. A fire will break out when the grass biomass s grows a little beyond a_x. The term δ_I denotes the rate at which the fire begins to die out. The parameter r_I represents the rate of increase of the fire consumption index once sufficient fuel is present.

$$\frac{dI}{dt} = I \cdot r_I \cdot (f(s; a_I, b_I) - \delta_I) \tag{7.4}$$

Woody shrubs are simply defined as a logistic growth function, where r_w represents the intrinsic growth rate of shrubs. Furthermore, fire can consume woody shrubs as denoted by the last term of the equation:

$$\frac{dw}{dt} = r_w \cdot w \cdot (1 - w) - \gamma_{Iw} \cdot w \cdot f(I; a_w, b_w) \tag{7.5}$$

Table 7.2: Parameter values of the rangeland model

Initial values	Ecosystem parameters	Response function parameters
$c_0 = 0.3252$	$r_c = 3$	$a_s = 0.1$
$s_0 = 0.1262$	$r_I = 60$	$a_I = 0.5$
$w_0 = 0.4647$	$r_w = 0.1$	$a_w = 1$
$I_0 = 1\text{E-}9$	$\delta_I = 0.1$	$b_s = 1$
	$a_c = 0.1$	$b_I = 3$
	$\alpha_{us} = 0.5$	$b_w = 8$
	$\beta = 3$	
	$\gamma_{Iw} = 1$	

7.4 THE IMPLICATIONS OF NON-UNIFORM GRAZING

In this section we discuss the implications of non-uniform grazing of sheep on the biomass for grass and shrubs at the paddock level. Properties are typically 40000 ha, made up of about 20 paddocks. The stylized model is implemented in a spatial multi-agent simulation language Cormas (Bousquet, *et al.*, 1998). The paddock is split up into 100 cells (10 x 10) of about 20 ha each. Each cell contains equations (1) to (5). The starting conditions of the model correspond to a point in the stable cycle. The simulated periods are 100 years.

If a uniform grazing pressure is assumed (γ_g = 0.25) and a time step of 1/50 year, then the results are identical to the differential equation model (Figure 7.1). The model follows a stable cycle of about 9–10 years. The grass biomass and the shrub biomass grow until enough fuel is available to start a natural fire. The fire consumes the grass biomass almost completely, and reduces the shrub biomass significantly.

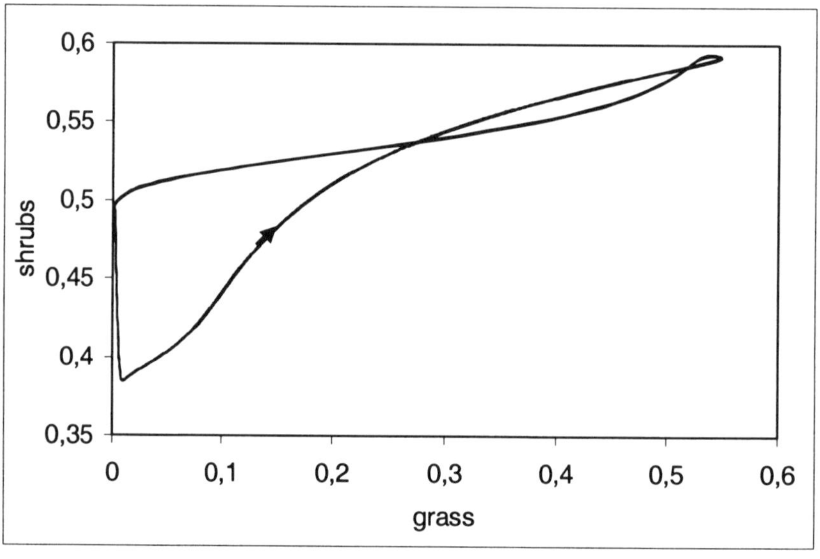

Figure 7.1: The trajectory that grass and shrubs follow due to periodic fires. This figure is based on the assumption of a uniform grazing pressure

The next step is to introduce sheep as individuals. A 2000 ha paddock might have around 250–300 sheep. The model is equipped with 300 agents representing the sheep. Initially, the sheep are randomly distributed over the paddock. When sheep move randomly in weekly time steps, the moving average of shrub biomass is similar to the experiment with uniform grazing, but the cyclic behavior over time vanishes (Figure 7.2). The levels of grass and shrubs continue to cycle within the same domain as for uniform grazing. For very short time steps and very small grazing units, the model simulations converge to the results of the mean-field differential equation model. However, with weekly time steps and 300 sheep, the results are not equivalent to a cloud of grass-eating, well mixed particles moving continuously over the landscape as is assumed by differential equation models. Therefore, the aggregated behavior of the systems includes some random behavior in a limited domain.

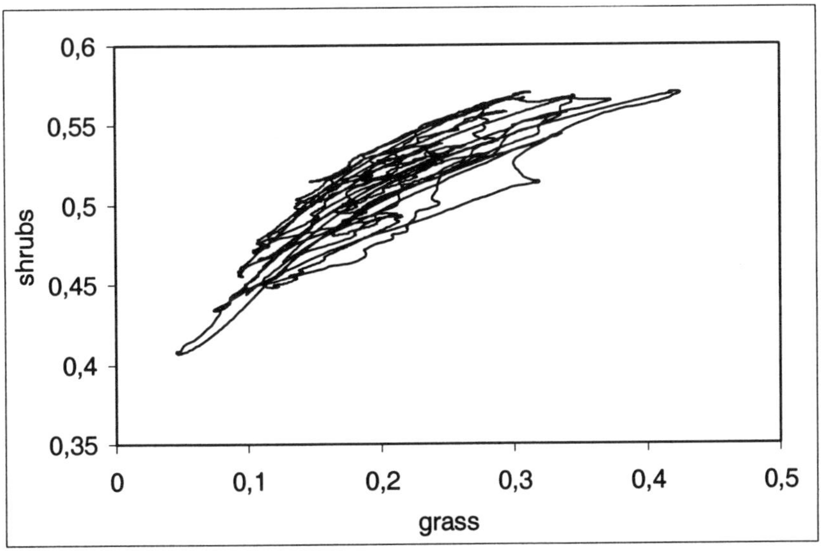

Figure 7.2: When sheep move randomly, over the paddock on a weekly time step, the resulting aggregated dynamics follows this erratic pattern

Of course, most grazing animals do not behave entirely randomly. Sheep select which part of paddock to graze in at various scales (Stafford Smith, 1988). First, they are obliged to return to the waterpoint at least once a day in hot weather (most of the year), so at least once a day they may choose to move to a different grazing location with minimal marginal investment of energy (that is, they are forced to make a major movement for water anyway, and moving to a different location is a marginal additional investment on top of this). The options that are available for this depend on the physical layout of the paddock, and in the absence of other stimuli sheep often move out into the wind. More locally, sheep also make minute-by-minute choices as they move during grazing, and these choices can also add up to significant movements at a broader scale. Here we examine the effects of various types of behavior which are observed in the field:

1. The sheep return individually to approximately the same area of the paddock each day but choose the best local cell within that area (with all parts of the paddock equally accessible);
2. As for 1. but the sheep prefer to be near others so that their behavior shows some contagion or flocking behavior;
3. Areas closer to the waterpoint are in fact accessed more easily and often, so we assume these are more attractive (for energetic reasons) and that activity is concentrated concentrically around the water;
4. We assume that the sheep know the state of the vegetation throughout the paddock and that preference for areas with greater forage overwhelms the effects of distance to water (i.e. vegetation preferences dominate movement choices.

1. Instead of moving randomly, we now assume that every time step, the sheep moves to the best of the nine cells centered on its current position. Given the weekly time step, this is akin to assuming that individual sheep prefer to return to the same location each day, but slowly adjust their preference. This process is simulated by comparing the state of the nine cells below and around the sheep. Interestingly, the resulting system behavior is similar to the case of uniform grazing (Figure 7.3). In fact, assuming that sheep will move to the cell with the relative highest level of grass biomass leads to a very efficient grazing of the paddock. Due to the existence of 300 independent sheep on 100 cells, the resulting grazing becomes similar to uniform grazing, especially since all cells start with the same conditions and have the same parameter values. This result shows that even in multi-agent models, the results might follow the results of the mean-field approximation. The interesting question is then, for which behavioral rules can the mean field results be replicated, and for which not?

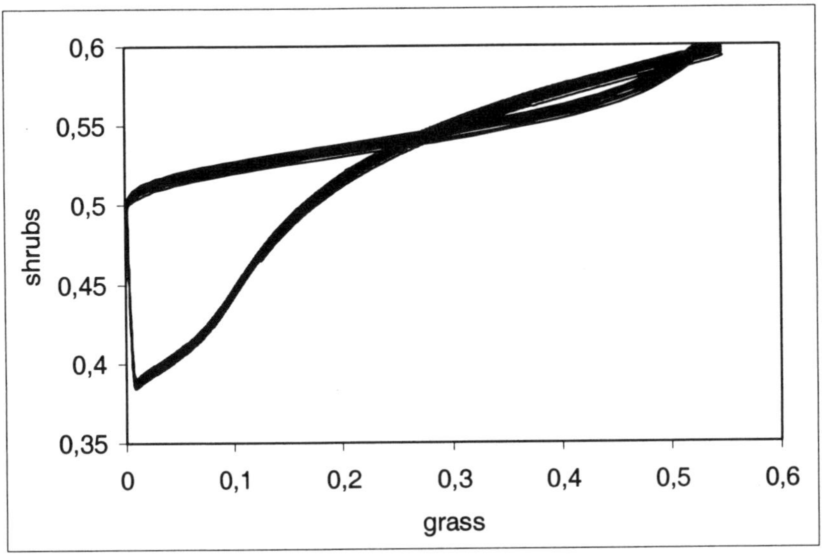

Figure 7.3: When sheep move to the neighboring cell with the relative highest grass biomass, the resulting pattern follows a similar figure as uniform grazing

2. To include more realistic sheep behavior in line with observation, we next include the impact of contagion. Sheep have a greater tendency to occur in larger flocks when conditions are good (Dudzinski *et al.*, 1978; Stafford Smith, 1984). This is implemented by defining an indicator of attractiveness of the neighboring cells. This indicator takes into account the amount of grass biomass and the relative distance to the waterpoints and the number of sheep in the neighboring cells:

$$rp = \beta \cdot s \cdot d + (1-\beta) \cdot p$$

where (7.6)

$$\beta = e^{-\alpha \cdot s}$$

and where *rp* is the relative preference, $d \in [0,1]$, the relative location of waterpoint ($d = 1$ is the waterpoint, $d < 1$ means not at the waterpoint. The lower d the further away from the waterpoint). p is the percentage of neighboring cells where other sheep are located. Including herding (with $\alpha = 0.04$) leads to a higher concentration of grazing pressure, and therefore increasing degradation of the paddock for a given grazing pressure (Figure 7.4). Flocks of sheep move around on the paddock while overgrazing local cells, which in the end leads to an irreversible effect at the paddock level. The system begins with shrub density cycling between approximately 0.43 and 0.55 between fire events; at the same time, grass varies between 0.08, just after a fire, up to 0.45, just before a fire. As the system becomes degraded, less grass (fuel) builds up between fire events (i.e. the width of the 'pyramid' decreases), fires are less intense and thus less effective at suppressing shrubs. Shrub density increases, adding further competitive stress on grass, i.e. the trajectory moves upward over time. Finally, the system can no longer carry a fire, the cyclic behavior ceases and the system converges to a shrub-dominated state and constant grass biomass. This process shows the flip from a fire-dominated to a shrub-dominated system.

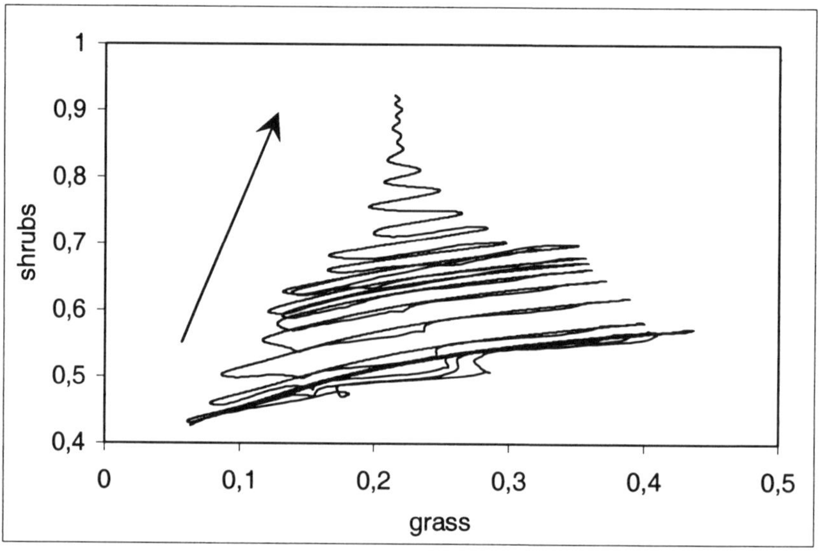

Figure 7.4: As Figure 7.3, but now including the assumption that sheep prefer also to be near each other. The herding effect has a destabilizing effect. The arrow depicts the direction of the system in time.

3. Now consider the effect of distance to water on grazing. Assume that the waterpoint is located in the middle of the paddock. Since the sheep have to go to the waterpoint every day, they have a preference for grazing near the waterpoint. Therefore we include an indicator related to the distance of the waterpoint. Two situations are modeled: The attractiveness of the point most distant from the waterpoint is reduced by 10% (modest effect) and 33% (strong effect), respectively (these correspond to different assumptions about the effects of scale on animal behavior; see Cridland and Stafford Smith,

1993). Initially the sheep are distributed near the waterpoint. The stronger the attraction of the waterpoint, for example due to high temperatures or salty forage and a consequent frequent need to drink, the faster the system will flip into a shrub-dominated case (Figure 7.5 and Figure 7.6). In this case where sheep are dependent on the waterpoints, the result is selective pressure around the waterpoint, and a degradation of these cells. The sheep have to move further away, putting pressure on a further set of cells. This eventually leads to a shrub-dominated property.

4. The final type of experiment focuses on the distribution of grass (now ignoring the effects of distance to water). If the carrying capacities of different cells are assumed to vary (by 10% (modest heterogeneity) and 50% (strong heterogeneity)), the system flips into a shrub-dominated state (Figure 7.7 and Figure 7.8). This is caused by the fact that the most favorite spots are grazed totally in the initial years, leading to lower potential biomass for the same amount of sheep.

To conclude, spatial heterogeneity in conjunction with specific assumptions about the movement of sheep can increase the risk of overgrazing. This is caused by localized pressure on cells which become overgrazed and which flip into an unproductive state. Although these model runs are merely illustrations of the consequences of spatial heterogeneity for specific parameter values, they clearly show the importance of including this heterogeneity in the analysis of management strategies. The results in this paper are in line with Jansen and de Roos (2000) who studied predator-prey systems and concluded that when the movement of the predator is homogenous, the system behaves like the differential equation variant, while when the predator moves diffusively, leading to local clusters, the dynamics differ.

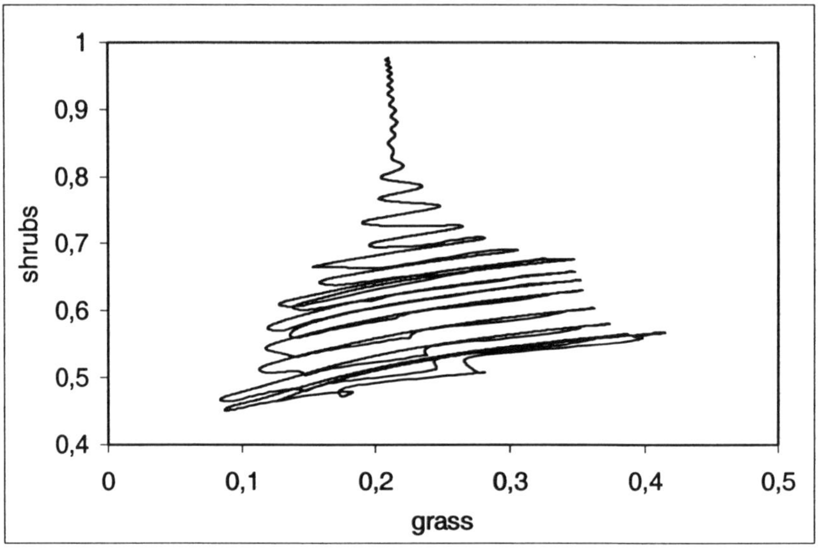

Figure 7.5: As Figure 7.4, but now including a modest effect of the availability of a waterpoint in the center of the paddock

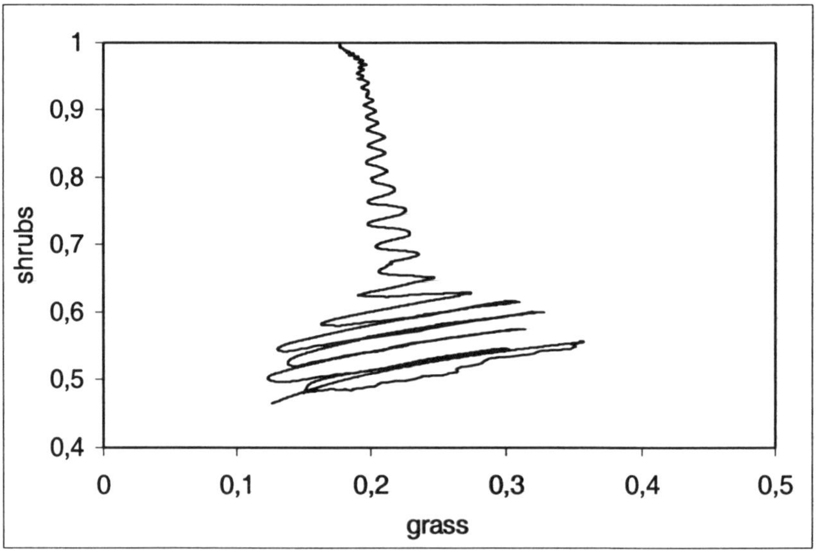

Figure 7.6: As Figure 7.4, but now including a strong effect of the availability of a waterpoint in the center of the paddock

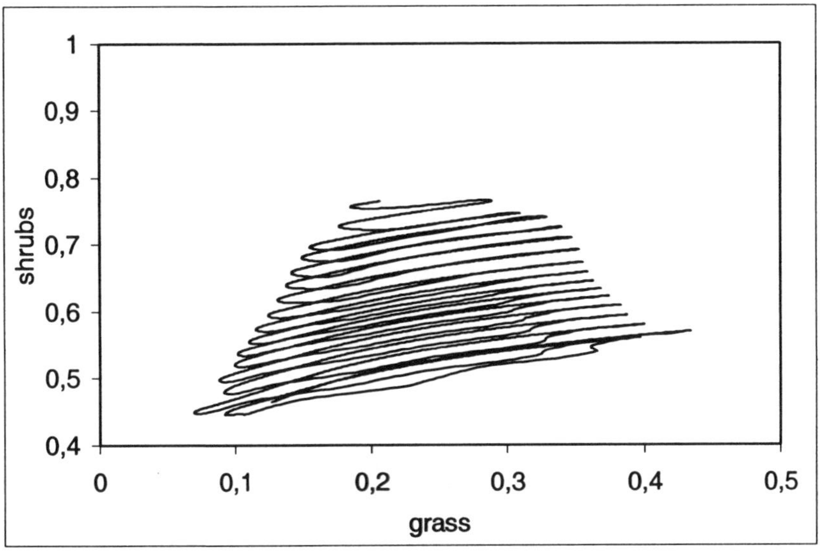

*Figure 7.7: As Figure 7.4, **but now assuming** modest heterogeneity of the local carrying capacity of grass biomass*

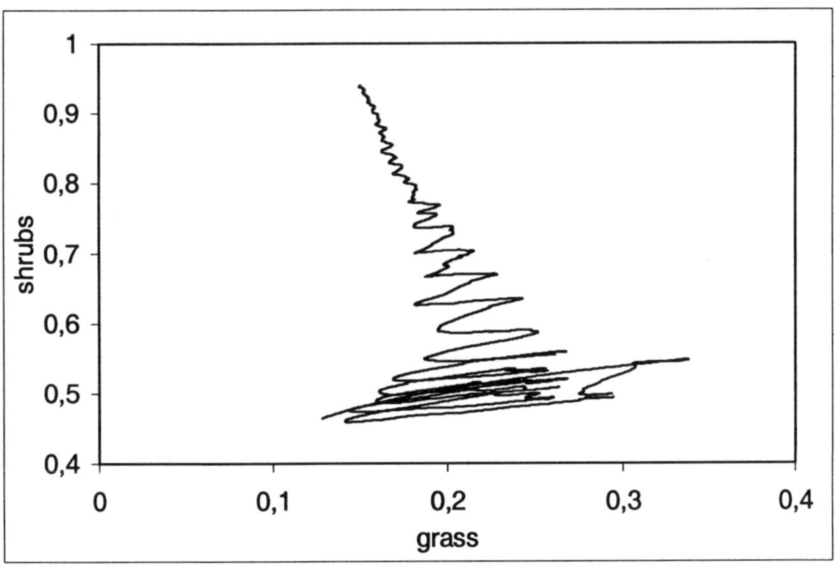

*Figure 7.8: As Figure 7.4, **but now assuming** strong heterogeneity of the local carrying capacity of grass biomass*

When herding behavior makes a difference

One of the key results of including space in ecology is its stabilizing role (Grenney *et al.*, 1973; Pacala and Levin, 1997; Durrett and Levin, 1998; Bascompte and Solé, 1998; Jansen and de Roos, 2000). However, the results in the previous section suggest that spatial factors can also play a destabilizing role in rangeland systems, as has long been suggested on empirical grounds (e.g. Ash and Stafford Smith, 1996; Table 7.1). An important difference between our rangeland model and the studies on predator–prey systems, is our presumption of the existence of multiple stable states (cf. Westoby *et al.*, 1989). We conclude our analysis by exploring the rangeland system in more detail through a simplified model. In line with Jansen and de Roos (2000) we developed a two-region model of the rangeland system, where livestock are assumed to migrate between two regions of a paddock. In order to use the simplest model possible, we exclude fire and shrub dynamics and focus only on the interaction of grass and grazing dynamics. The behavior of the sheep is modeled with simple rules to mimic flocking behavior.

The resulting model can be described as:

$$\frac{dc_j}{dt} = r_c \cdot s_j - s_j \tag{7.7}$$

$$\frac{ds_j}{dt} = c_j \cdot (a_c + s_j) \cdot (1 - s_j) - \gamma_{g_j} \cdot f(s_j, a_s, b_s) \tag{7.8}$$

where the index j refers to the jth region of the paddock, here j = 1,2. The sheep behavior is based on searching for food, and flocking behavior. Define $g() = 1 - f()$ where f is as defined as above (equation (7.2)). Then,

$$\begin{aligned}\frac{d\gamma_{g_1}}{dt} &= r_m \cdot [\gamma_{g_2} \cdot g(s_2, a_m, b_m) - \gamma_{g_1} \cdot g(s_1, a_m, b_m)] \\ &\quad + r_f \cdot [\gamma_{g_2} \cdot \max(\gamma_{g_1} - \gamma_{g_2}, 0) - \gamma_{g_1} \cdot \max(\gamma_{g_2} - \gamma_{g_1}, 0)]\end{aligned} \tag{7.9}$$

$$\gamma_{g_2} = \gamma_{g_t} - \gamma_{g_1} \tag{7.10}$$

The term $g(s_2, a_m, b_m)$ captures the idea that as grass density falls below the threshold a_m (with sharpness b_m) in region 2, the number of sheep there (γ_{g2}) will decline (at a maximum rate r_m) and transfer to region 1 – i.e. in migration to region 1. This decision does not depend on the density of grass in region 1, because sheep rely on local information to make their decisions. With just two spatial blocks, they are too large for a sheep to see the other region, the sheep must go there to investigate. In the previous model, the paddock was split up into 100 blocks so that the sheep could obtain knowledge of nearby patches during their daily movements.

The term $1 - f(s_1, a_m, b_m)$ captures out-migration from region 1. The term r_f $[\gamma_{g2}(\max(\gamma_{g1}-\gamma_{g2},0) - \gamma_{g1}(\max(\gamma_{g2}-\gamma_{g1},0))$ captures flocking behavior. If γ_{g1} is larger than γ_{g2}, sheep in region 2 will flock towards the larger population and migrate to region 1 and vice versa. The 'propensity to flock' is given by r_f. If we keep the total grazing pressure constant, the 'conservation' law makes a differential equation for γ_{g2} unnecessary, i.e. it is given by expression (7.10).

With this simple model, a stable limit cycle for sheep migrating back and forth between the two blocks of a paddock has been constructed. The requirements are: a fairly sharp threshold for when to move, i.e. $b_m = 8$ or more, and that sheep move very fast relative to ecological dynamics – which is reasonable, i.e. $r_m = 30$ or more. The model is rather sensitive to the threshold value a_m. If the sheep eat too much, they severely degrade the paddock block they are in before they move. They then severely degrade the other block and return before the one they left initially has recovered. Then the whole system is degraded. If they move after eating just a small amount of grass, this acts like diffusion and the model converges to the non-spatial version – i.e. $\gamma_{g1} = \gamma_{g2}$. This is compatible with some of the proposals in Table 7.1, depending on the resilience of the vegetation types.

In order to compare the resilience of the spatial and non-spatial systems, we attempted to compare the grazing pressure at which the equilibrium becomes unstable. In the non-spatial case, the equilibrium becomes unstable when $\gamma_g \approx 0.58$. For a two-patch model made up of identical copes of the non-spatial model (with no migration, i.e. $\gamma_{g1} = \gamma_{g2}$ always) then the equilibrium would become unstable when $\gamma_{gt} \approx 1.16$. The question is, then, at what grazing pressure does the model with spatial migration become unstable?

The system has two stable attractors. If the initial conditions are identical in both regions, the system remains at this stable point. The slightest perturbation will move the system to a periodic orbit (Figure 7.9). These results show how migration can reduce the resilience of the system. A periodic solution is shown in Figure 7.9 with $a_m = 0.2$, b_m , $r_m = 30$ and $r_f = 0.1$, super imposed on the isoclines for the non-spatial system. The average root and shoot biomass are less than they would be with uniform grazing. Using numerical experiments, we find that the periodic solution becomes unstable at $\gamma_{gtot} = 1.02$. Thus the spatial model is less resilient – it can support about 13.5% less grazing pressure for these assumptions.

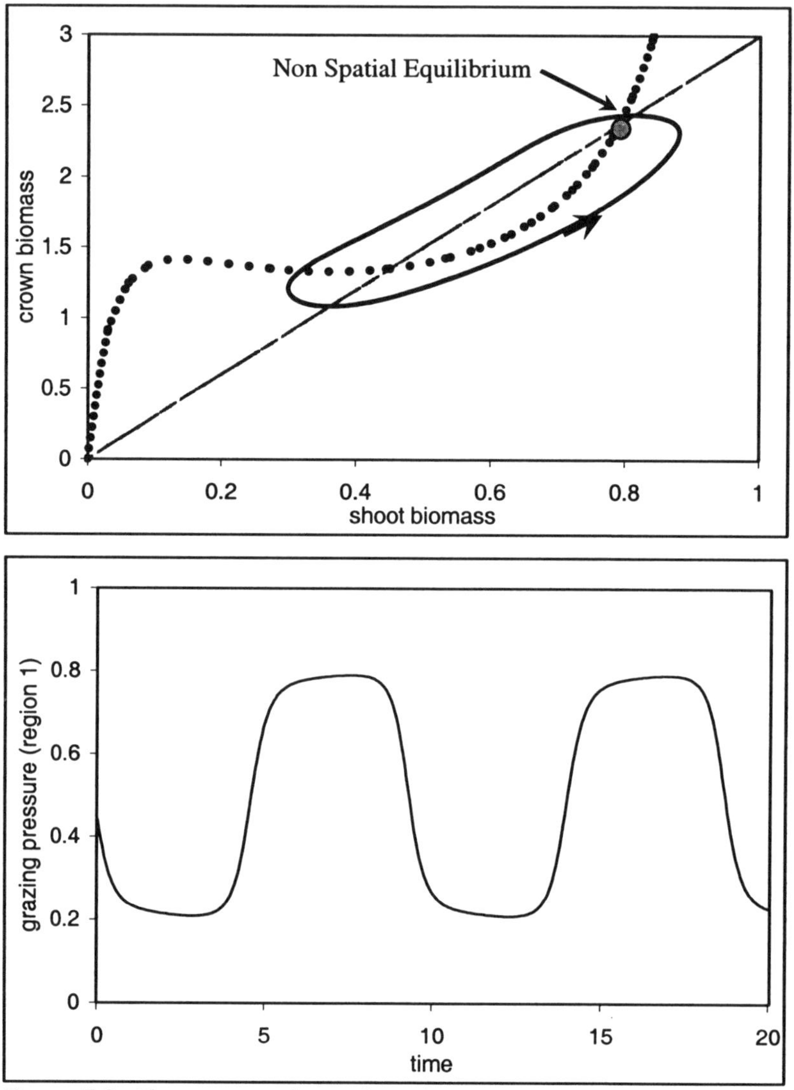

Figure 7.9: Shoot-crown phase plane for two-compartment model (above), and grazing pressure in region 1 over time (below)

To understand how the tendency to flock may reduce the resilience of the system, we performed an additional experiment. By comparing the dynamics of the systems for different levels of flocking tendency r_f, we determined that the key mechanism that reduces resilience is the degradation of the crown portion of the plant. Specifically, increasing r_f from 0.1 to 1 does not reduce the average shoot biomass, it just shortens the period of the cycles. However, it does reduce the average crown biomass. The relatively fast process of the analysis thus seems to show up the fact that flocking behavior is more detrimental to the relatively slower variable of crown biomass (note that, as modeled, grass dynamics are three times faster than crown dynamics and flocking dynamics are ten times faster again than grass dynamics).

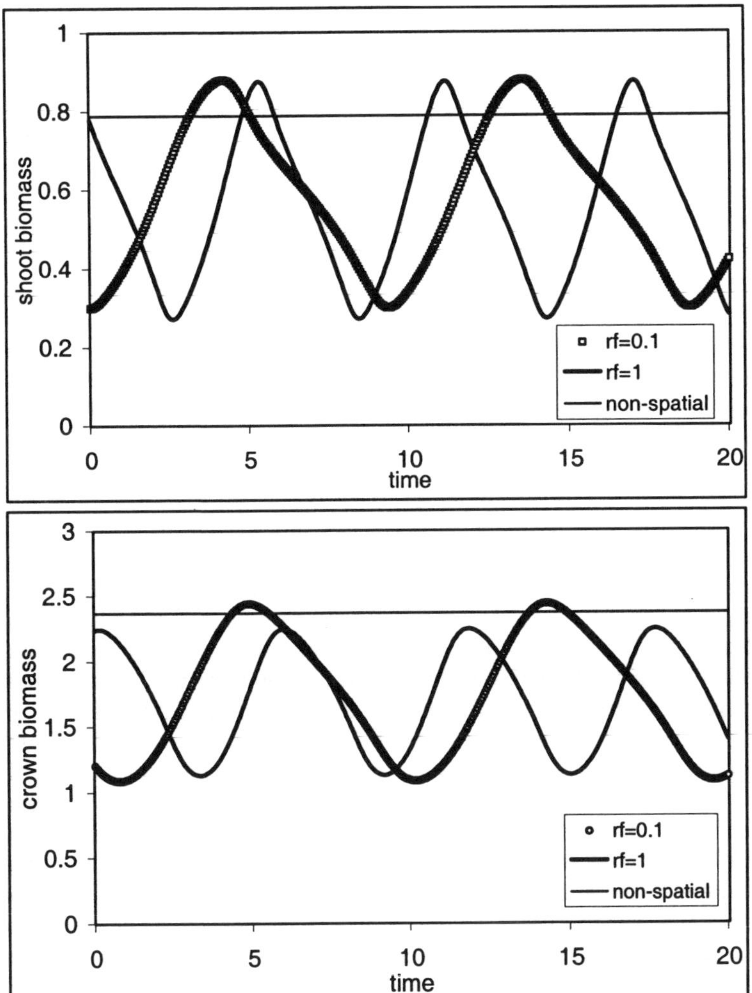

Figure 7.10: Shoot and crown biomass over time for different tendencies to flock. The horizontal line is the non-spatial case, the dotted curve is for r_f = 0.1, and the solid curve for r_f =1

7.5 DISCUSSION

In this chapter we have explored the consequences of a spatially explicit analysis of the resilience of rangelands. When sheep do not graze a paddock in a uniform way, but cluster due to herding behavior or physical attractants like waterpoints and preferred areas of vegetation, the results of our analysis show that this particular system will flip to a degraded state at a lower grazing pressure than in the same analysis without spatial explicit grazing.

Thus, in this case, inclusion of spatial heterogeneity makes the rangeland system more likely to end up in a degraded state for a larger set of parameter values. This means that a more realistic description of the rangeland system will support less optimistic stocking rates if resilience of the rangeland is too maintained.

This finding is compatible with some of the suggestions made in Table 7.1 on the basis of empirical observations. The practical experience of range management in Australia and other rangelands of the world show that ecosystem change has occurred suddenly and unexpectedly in some circumstances. The changes have been 'unexpected' compared with what might have been predicted from the experiences of higher productivity, smaller paddocks, or from the resilience of small grazing trials (see review in Ash and Stafford Smith, 1996), almost all of which suggest that higher rather than lower stocking rates should be attainable in rangelands. Indeed the practical experiences of the more conservative pastoral managers often suggest that lower stocking rates are attainable in rangelands than scientific evidence drawn from small plot trials would imply. This analysis begins to identify the potential role that spatial factors may have to play in this process of surprise.

However, Table 7.1 also provides some hints of the conditions under which this finding might not be general. An extensive literature has appeared in recent years suggesting that there are circumstances where rangelands may be said to be very much more resilient than the small plot evidence would suggest (see review in Illius and O'Connor, 1999). Whilst some of this literature relates to the purpose of production and therefore the significance of any change of ecosystem state, other elements probably relate to a genuinely buffering effect of spatial factors, particularly the selective behavior of herbivores, as noted by Ash and Stafford Smith (1996). In this regard, the detailed conclusions of this chapter are tentative since not all spatial oriented processes are included (e.g. seed dispersal, wildlife movements, heterogeneity of soil conditions). Even for those factors we have considered, not all possible combinations have been included, but this must remain a task for the future. What can be said confidently from the simulation experiments is that spatially explicit rangeland management is likely to be very important in some circumstances and should not be ignored in future modeling of these systems.

Understanding the implications of spatial processes is vital in the large open paddocks of rangelands; however, it is likely that these processes are also important in a perhaps more subtle way in other environments. Dealing with spatial heterogeneity has become an important part of range management in recent years (e.g. Coughenour, 1991), of equal significance as managing for temporal variability. The challenge in fact is to ascertain what spatial and temporal factors in what combinations create significant impacts, as well as when they do not; the comprehensive analysis of this question is the challenge for another study.

PART II

Applications

8. Adjustment costs of agri-environmental policy switchings: an agent-based analysis of the German region Hohenlohe

Alfons Balmann, Kathrin Happe, Konrad Kellermann and Anne Kleingarn

8.1 INTRODUCTION

A limitation of livestock density, i.e. tying animal production to farmed land, has become an issue in recent discussions on the future Common Agricultural Policy (CAP) of the European Union. Whereas it has been part of some voluntary management agreement programs for quite some time, the limit of two livestock units (LU)[1] per ha has recently – in the light of mad cow disease (BSE) and foot and mouth disease (FMD) in 2001 – been brought forward also on the national and European levels (cf. ZMP, 2001). The rationale behind this is that from environmental, food-safety, and political points of view it would be desirable to use in livestock production only fodder produced on-farm. A number of politicians (among them the German Federal Minister of Consumer Protection, Food and Agriculture) propose to reduce direct transfer payments to those farms that require a higher fodder input for livestock production than what can be supplied by the farms' own crops. Furthermore, it has been proposed to subsidize grassland and arable farming equally. Some policy makers argue that this would create incentives for farmers to reorganize production towards less-intensive production methods and towards a higher share of fodder produced on-farm. In order to receive the payments, farms with a high livestock density are expected to develop diversification strategies, either by reducing the number of livestock or by renting additional land. Both strategies would be in accordance with the intention of the policy makers. However, an inverse reaction to the policy change could equally be possible: If rental and sales prices of land are high and if reducing the number of livestock is also not attractive, farms may no longer be interested in farming land at all. At the expense of giving up all transfer payments a further specialization in pig or turkey production – which is not subsidized by direct payments under the current CAP[2] – may be more attractive. Eventually, such farms may even reduce acreage.

The relative attractiveness of the different strategies depends on land prices and, reciprocally, land prices may be affected by the farms' choices. For regions with a low livestock density, one can assume that intensive livestock farms will increase their acreage. Since in such regions relatively few farms are affected by the policy change, the adjustment has little impact on the land market. On the other hand, in regions with a high livestock density, the policy impacts on the land market may be more complex. If many farms try to increase their acreage, rental prices are likely to increase. Hence, some farms may increase their acreage, some may reduce the number of animals, and some may even specialize in intensive livestock production and reduce land to a minimum.

The intention of this chapter is to analyse the effects of such policy switching at a regional level for a region with a large number of intensive livestock farms. The region of 'Hohenlohe' located in the southwestern German State of Baden-Württemberg displays such characteristics (cf. MLR 2000, various years) and was chosen for the study. Apart from intensive hog finishing, and turkey production, there are also a number of dairy, farrowing, and arable farms present in the region. Farm sizes are relatively small and below the average in West Germany. The dominating organizational form is that of a family farm.

To tackle the problem of how farms interact within the region, the study is based on agent-based simulations. This means, the adjustment process is simulated with a spatial and dynamic model that considers approximately 2600 heterogeneous, individually behaving farms. These farms are spatially distributed in a region of about 75000 ha of agricultural land. Since data for 2600 individual farms are not available, the model was fitted to the agricultural sector in Hohenlohe on the basis of accounting data from a number of real farms operating in the region. These selected farms can be considered to be typical for the region and to cover the spectrum of farms in Hohenlohe. The initial farm structure in the model is determined by assigning a certain frequency to each of the selected typical farms. These frequencies are chosen in order to minimize quadratic differences between the model and the real region with regard to several key characteristics (cf. Balmann *et al.*, 1998 a,b).

In the remainder of this chapter we present the model's general structure, and discuss certain characteristics like the implementation of the land market and of technical progress in more detail. We furthermore inform about the region and how the model is calibrated. Thereafter, we present alternative policy scenarios and the simulation results with respect to structural change, efficiency, and farmers' incomes. Finally, we draw conclusions with regard to policy impacts and the applicability of the modeling approach.

8.2 THE MODEL[3]

8.2.1 General Aspects

Imagine taking a bird's-eye view of the map of an idealized agricultural region in which land is divided into plots of a fixed size like on a chessboard (Figure 8.1). The region consists of about 30 000 quadratic plots of 2.5 ha. each, i.e. the region's size is about 75 000 ha. To avoid border effects, the region is assumed to form a torus, i.e. plots at the border are assumed to be immediate neighbors of plots at the opposite side of the grid. Land is heterogeneous with respect to location and quality. Two qualities are considered: land for arable farming and grassland. Initially, a total number of 2606 family farms operate in this region. The plots on which farmsteads are located are surrounded by a highlighted border. Each plot managed by a particular farm has the same shade as the interior of the farmstead.

Each farm acts autonomously in order to maximize its expected individual household income. It can engage in 13 different agricultural production activities (e.g. dairy, cattle, suckler cows, hogs, sows, crops, sugar beets, pasture land) and it can invest in 28 different objects (Table 8.1). These investment alternatives allow for some economies of size, i.e. with increasing size, labor can be used more effectively and average acquisition costs per unit decrease. For instance – and in accordance with a number of empirical studies (Kirschke *et al.*, 1998; Helmcke; 1996, Peter, 1993) –; it is considered that in crop production, economies of scale exist up to a size of some 250 ha.[5] In addition to the different production and investment activities, the farms can use their labor and capital for off-farm employment, but they can equally hire additional labor and or take up loans. Additional land can be rented on a pure rental market and plots can be let on the rental market. Farms can give up farming and, in principal, new farms can be founded.[6]

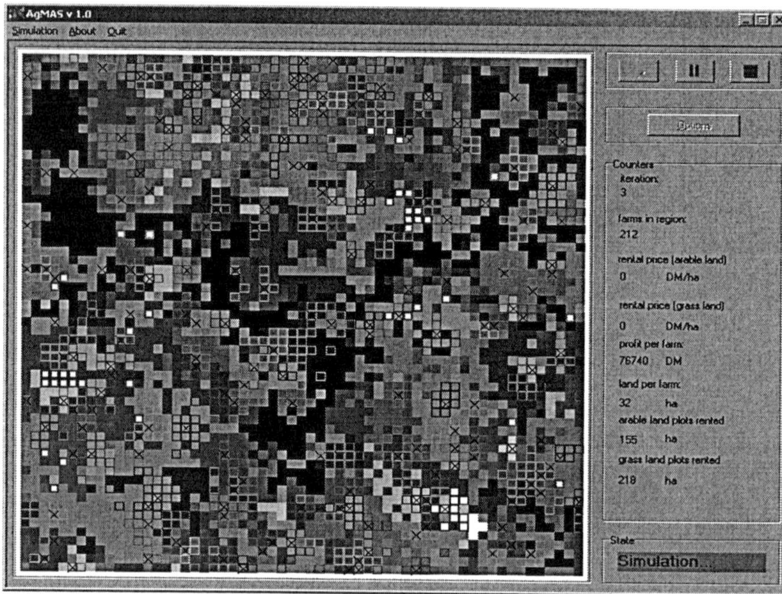

Figure 8.1: Graphical user interface of the simulation software[4]

Table 8.1: Types, capacities and useful lifetime of investment objects

Type	Capacity in places (animals) or acreage	useful lifetime in years
Sows	252, 170, 128, 64, 40	20
Hog finishing	1000, 600, 400, 200, 100	20
Bull fattening	200, 100, 40	25
Suckler cows	40, 10	25
Turkeys	15000, 10000, 5000	20
Dairy	480, 240, 120, 60, 30	25
Machinery (max. ha to be cultivated with machinery endowment)	150, 85, 55, 30, 15	12

The farms have to be understood as agents. Although all farms act autonomously and can evolve heterogeneously in many variables (cf. Table 8.2) they all follow the same decision rules and expectations. All decision-making routines are based on adaptive expectations. Production and investment activities are optimized by mixed-integer linear programming. If a farm invests, this has an impact on the farm's production capacities for the useful lifetime of the asset (Table 8.1), i.e. the investment costs are considered to be sunk. The same holds for the capital stock that depends on previous investments as well as on previously earned profits. A farm closes down if it is either illiquid or if the farm's expected profit does not cover the opportunity costs of the factors owned by the farm-household.

Table 8.2: Data structure (excerpt) and abbreviations[1]

Farms	$\boldsymbol{B}$ $(t = 1,\dots,K)$	**Investment alternatives**	$\boldsymbol{I}$ $(u = 1,\dots,R)$
farm number	FN_t	acquisition costs	C_u
farm age	FA_t	useful lifetime	N_u
on-farm labor units	FL_t	capacity	$\boldsymbol{b}_u$
capital withdrawal	WD_t	capacity of type k for factor j	b_{uj} $(j=1,\dots,m)$
arable land	FA_t	labor substitution	LS_u
grassland	FG_t	residual value	RV_u
equity capital	EC_t	cost factor affecting product i	$CF_{u,i}$
liquidity	L_t		
debts (long-term)	DL_t	**Production alternatives**	$\boldsymbol{P}$ $(i = 1,\dots,M)$
interest payments	IC	units produced	X_i
depreciation	D	price	P_i
rent expenditure	RC_t	variable production costs	vc_i
transportation costs	TC_t	factor demands	$\boldsymbol{r}_i$ $(j=1,\dots,m)$
overhead costs	OV_t	price change	dp_i
maintenance costs	MC_t		
total income	Y_t	**Plot**	$\boldsymbol{FL(y,z)}$[1)]
management factor	MF_t	number of the tenant	FN
total land input	F	state of plot (idle, rented, owned)	S
number of own plots	$F_{Own,t}$	distance to the farmyard	DC
number of rented plots	$F_{Rent,\,t}$	rent paid for plot	R
average number of adjacent plots	$F_{Neigh,\,t}$		
assets	$\boldsymbol{A}$ $(c = 1,\dots,S)$		
asset of investment type u	$A_{c,u}$		
age of asset u	AA_u		
production	$\boldsymbol{x}$ $(i = 1,\dots,M)$		
price expectation	$\boldsymbol{p^e}$ $(i = 1,\dots,M)$		
factor capacities	$\boldsymbol{b}$ $(j = 1,\dots,N)$		

1) *y* and *z* are co-ordinates on the grid.

For any period and farm the objective function and the constraints result in the following simplified maximization problem (for a list of abbreviations see Table 8.2):

$$\max Y^e\left(\mathbf{x'p^e}, \mathbf{A}, \mathbf{I}, \mathbf{r}, \mathbf{FL}, L, DL, IC, F, \ldots\right)$$

$$\begin{aligned}
\text{s.t.} \quad & Y^e = \mathbf{x'p^e} - IC - D - RC - TC - OV - MC && \text{with } x_i \geq 0, \\
& b_j \geq \mathbf{x'r} && \text{for } j = 1, \ldots, N, \\
& b_1 = FL \cdot ph && ph = \text{hours per person and year}, \\
& b_2 = L, && \\
& b_3 = F \cdot \text{land per plot}, && \\
& b_j = \sum_c A_{c,u} \cdot b_{u,j} && \text{for } j = 1, \ldots, N \\
& \mathbf{A} = \mathbf{A}(\mathbf{A}_{t\text{-}1}, \mathbf{p^e}, \mathbf{I}, \mathbf{r}, \mathbf{FL}, L, DL, b_{3,\ldots}). &&
\end{aligned}$$

Summarizing, on the micro level, each farm's decision making is defined in a way that can be called myopic or bounded rational. Although the farms are rather smart with regard to the applied optimization techniques, the farmers' cognitive abilities are limited. For instance, farms are not able to communicate directly with their neighbors and hence they are also not able to use machinery jointly or to merge. Moreover, they are not able to behave strategically. On the aggregate level, the model can be understood as a complex distributed recursive programming approach that is simulated for a number of periods.[7] In every period, all farms have to decide simultaneously on investments, renting land, and production activities. The farms' decision making routines are ordered and embedded in the program flow as shown in Figure 8.2.

The farms are linked via product markets and several local markets for inputs: arable land and grassland, milk quotas, and manure. Particularly relevant are the land markets, since a farm can increase its acreage only if there either is idle land which is momentarily not farmed by another farm or if other farms reduce their acreage and release land to the land market. The farms are also affected by developments on product markets. Product prices may change according to regional production activities. As already mentioned, it is assumed that the acquisition costs of assets are totally sunk after investing, i.e. the opportunity costs of assets are zero. Furthermore, it is assumed that each farm is handed over randomly to the next generation every 25th period. Then, for the decision whether or not to continue farming, the opportunity costs of farm-household labor are considered to be 15% higher. If the farm continues, opportunity costs for off-farm use of labor remain at the lower level, since this is understood as an investment into agricultural training which reduces the chance of a profitable off-farm employment.

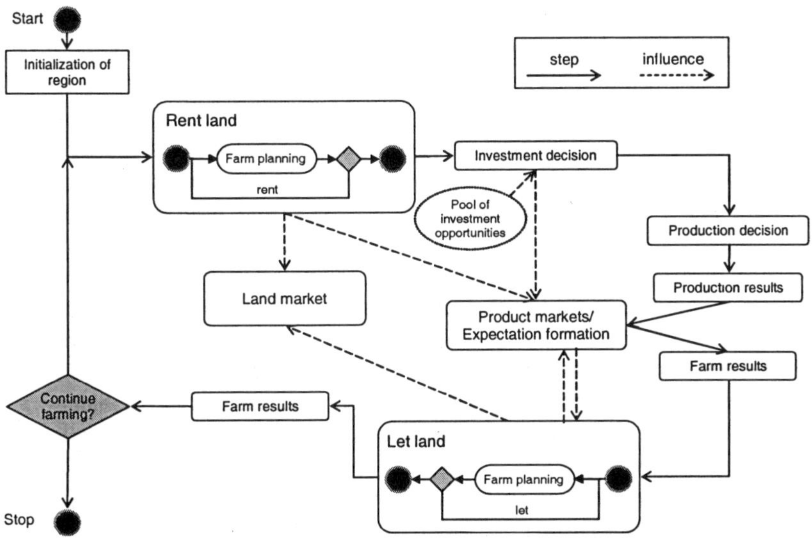

Figure 8.2: Decision sequence for each farm agent

8.2.2 Special Implementation Issues[8]

8.2.2.1 Land

The land rental market has been implemented as a central iterative auction for arable land and grassland. At the start of a bidding round, each farm searches for one plot of arable land and a plot of grassland that it wants to rent.

Then farms make bids for each of these 'wanted plots' according to their shadow prices for land, taking distance-dependent transportation costs and the number of adjacent plots into account. The number of adjacent plots and the farm's bid are positively related, since the more adjacent plots there are, the more a farm can exploit economies of size. The bidding process alternates between arable and grassland starting with arable land since about 75% of the land in the region is arable land. In each bidding round, the bids of all farms are compared and the highest bidder receives the plot it wishes to rent. Thereafter, this farm and all other farms which were interested in the same plot compute new bids so that they can be compared again with all other bids. The process is repeated and continued until no land is left to rent or the bids are zero. To reflect other costs associated with renting land (e.g. taxes, fees) the actual rent paid does not equal the bid offered for the plot, but about 75% of the original bid.

At the end of each period farms have the possibility to finish unprofitable rental contracts in order to increase their yield expectation for the next period. In the following periods, rental prices of land which are kept by the farms are adjusted towards the average price level of newly rented plots. This is done to consider dynamic developments on the land market as well as to smooth the fluctuation of rents between periods.

In recent studies the cost saving effects associated with larger field sizes were underlined (cf. Breitschuh, 2000). Larger field sizes allow for a more efficient use of existing as well as larger machinery. This is particularly an issue in regions with a small field structure. According to Breitschuh, the costs per ha associated with crop production activities decrease by about 30% if the field size is increased from 2.5 ha to 20 ha. This is introduced into the model in the following way: Assume that for farm B_t with each new investment I_u in larger crop production technology the variable production costs vc_i of crop production activities i (crops, sugar beet, rape seed, maize silage, intensive grassland, extensive grassland, protein plants) decline by a factor $CF_{u,i}$ which is defined as

$$CF_{u,i} = \left[1 - \frac{0.15}{1 + 100/(F_{own,t} + F_{rent,t})}\right] \cdot \left[1 - \frac{0.45}{1 + 100/(10 \cdot F_{neigh,t} + 1)}\right] \quad (8.1)$$

Accordingly, farms with initially little and scattered land can realize large cost savings if they increase their acreage considerably. The potential costs effect is much lower if a farm's acreage is already large and if the plots are in the neighborhood.

A critical examination of the renting process reveals that there are still quite a few shortcomings. First of all, no direct bilateral negotiations between farms are implemented, but land is rented to and from a central auctioneer.[9] This can be justified on the one hand by computational reasons. On the other hand, in reality land is increasingly rented from land owners who are not farmers. Hence, negotiation between active farms would not reflect the true situation either. Moreover, the central auction does not mean that all farms compete for the same land. Rather, competition for land is only local, since it is only farms located closely together that bid for the same land.[10] A second critical aspect which is more problematic in our respect is the fact that only one plot at a time can be rented. Because a farm's bids for land do not necessarily decrease with increasing acreage, it would be more reasonable to give the farms the opportunity to bid for a group of plots. However this requires additional computing power and the development of a rather complex algorithm for renting land, especially if combinations between arable and grassland have to be considered. Therefore, this extension has not been implemented at the present stage of the model. And, thirdly, in reality the renting of land depends not only on economic factors but often it is very much determined by personal factors, which are difficult to reflect in a normative economic model.

8.2.2.2 Technical change

In the agricultural sector technical improvements are mostly incorporated in process innovations such as machinery, new equipment and facilities which are generated in the supplying industry (Berger, 2001). The adoption process as well as the effect of technical change on the farm level is manifold for which reason an adequate consideration of the adoption process would go beyond this model's intention.[11] However, one effect on the farm level which can be singled out quite clearly is the cost saving effect of technical change associated with machinery and technical equipment. From this it follows that a reasonable way to incorporate the effects of technical improvement on the farm level is to link it with a farm's investment activity. For this we made the following assumptions: First of all, we assumed that the technology underlying each investment opportunity progresses over time, while the investment costs of the object remain constant. Furthermore, we assume that each of the 28 investment options mentioned above incorporates a cost saving potential which depends on the object's size and the number of investments of a certain type to reflect economies of size. An exact empirical foundation of potential cost savings is difficult since this largely depends on the individual characteristics of a farm as well as on learning processes associated with new technologies. Hence, we assumed a conservative cost saving effect by quantifying the cost saving effect per investment between 1 and 1.5% of the variable production costs of the previous period. Another instance of technical change incorporated into the model is the labor saving potential which is usually associated with larger investment objects.

Since technical change is introduced together with investment activities, the expectation of cost savings through technical change does also play a role in the decision whether to continue or to quit farming. If a farm expects to save costs in the next period by investing into better technology, then it might continue farming. Hence at the end of each period, farms calculate their expected profit by taking investments and cost savings into account.

8.3 CALIBRATION AND DATA

Unlike the work presented in Balmann (1997, 1999, 2001) in which a fictitious, hypothetical region was modeled, the present study models the agricultural structure of an existing region. In particular, the following simulations try to capture specific policy impacts on the selected region of Hohenlohe. For this, in a first step, the model is calibrated in order to consider the central characteristics of agricultural production in Hohenlohe. The calibration occurs on two levels: the farm level and the aggregate level.[12]

On the farm level, 12 different basic farm types are defined. These are defined on the basis of data from 12 real farms in Hohenlohe, each of which takes part in the German Farm Accountancy Data Network (FADN). The main selection criterion was that the farms should be typical for the region, i.e. they should be able to cover Hohenlohe's range of farm types with respect to size, main production area, and whether a full-time or part-time farm. Table 8.3 gives an overview. Accordingly, eight full-time and four part-time farms were chosen. Among them are dairy farms, pig farms (sows, hog finishing), turkey farms, crop farms and mixed farms of different farm sizes. The farms operate with selected production techniques that are considered to be typical for the region. The required coefficients regarding investment alternatives, LP/MIP matrices, and the calculations of gross margins and profits are derived from standard farm management data samples published for the German agriculture (KTBL, 1997; Landesanstalt für Landwirtschaft, 2001; Regierungsbezirk Mittelfranken, 2001).

Table 8.3: Characteristics and frequencies of the specific basic farms

Variable	A	B	C	D	E	F	G	H	I	J	K	L
Organization												
specialization	pig/ turkey	pig/ turkey	dairy	dairy	crop	crop	mixed	pig/ turkey	dairy	mixed	Crop	pig/ turkey
full-time	Yes	yes	Yes	yes	yes	yes	Yes	yes	no	no	No	No
Land [ha]												
total	22.5	72.5	67.5	30	37.5	60	50	112.5	12.5	17.5	10	20
rented	15	67.5	55	10	10	45	20	92.5	5	0	0	0
Arable	22.5	72.5	40	12.5	37.5	60	22.5	102.5	5	12.5	10	20
pasture	0	0	27.5	17.5	-	-	27.5	10	7.5	5	0	0
Livestock [heads]												
Cattle	-	-	90	52	-	-	63	25	28	5	-	-
Cows	-	-	39	26	-	-	28	-	12	-	-	-
Sows	40	128	-	-	40	-	64	170	-	-	-	128
fattening pigs	300	600	-	-	-	-	-	0	-	100	-	-
turkeys	-	-	-	-	-	20000	-	-	-	-	-	-
Frequency	480	25	120	244	106	22	231	95	389	154	442	298

As the last row of Table 8.3 shows, each of the specified typical farms is assigned a certain frequency. These frequencies are determined in order to receive a farm structure that reflects the main characteristics of the regional agriculture on the aggregate level. These aggregate characteristics are the number of farms (total and with respect to specialization and size), the amount of hectares of arable land and of pasture land, the land used by farms with a certain organization and specialization, and the number of animals (dairy cows, sows, hogs, turkeys). Table 8.4 shows the obtained adjustment with respect to the selected characteristics.

Table 8.4: Adjustment of the model with respect to main characteristics of Hohenlohe

Variable		Units	Hohenlohe	Model	Error
Farms	total	farms	3013	2606	-14%
	arable	farms	459	570	+ 24%
	dairy	farms	906	753	-17%
	pig, turkey	farms	988	898	-9%
	mixed	farms	516	385	-25%
	full-time	farms	1578	1323	-16%
	part-time	farms	1435	1283	-11%
Land	total	ha	72448	73503	+1%
	arable	ha	55043	54943	±0
	pasture	ha	17405	18560	+7%
Land farmed by	arable farms	ha	9569	9715	+2%
	dairy farms	ha	21683	20283	-6%
	pig, turkey farms	ha	27766	29260	+5%
	mixed farms	ha	14421	14245	-1%
	full-time farms	ha	57464	55565	-3%
	part-time farms	ha	16276	17938	+10%
Animals	cattle	head	60638	51903	+14%
	cows	head	21072	22361	+6%
	sows	head	99787	95718	-4%
	hogs	head	169901	174400	+3%
	turkeys	head	450000	440000	-2%

Since there is a certain trade-off in fitting the different characteristics, the frequencies have been chosen by minimizing the weighted quadratic deviations between the model and the region. A formal presentation of this procedure can be found in Balmann *et al.* (1998a,b). According to Table 8.4, the calibrated model fits the selected characteristics of the real region quite well. Strong differences only exist with respect to the number of farms. This is mainly due to the fact that there is a sample error in the German FADN. In particular, small farms are often not willing to participate and often do not meet the respective criteria for participation. For instance, the smallest farm we found in the FADN for Hohenlohe which fulfilled the selection criteria had an acreage of some ten ha. Thus, it was particularly difficult to represent the many part-time farms which are much smaller.

For the initialization of a simulation, each of the basic farms is instantiated in the model region according to the determined frequency in Table 8.3. However, before a simulation starts, the different basic farms are further individualized with respect to several variables. These variables are the location on the spatial grid and the age of the farms' machinery and buildings. Since there is no empirical data about the variables, these variables are determined randomly. Moreover each farm receives an individual coefficient reflecting a farmer's management ability which affects the farm's variable costs and thus its profitability and competitiveness. This factor ranges between 0.95 and 1.05.

8.4 POLICY SCENARIOS

As mentioned in Section 8.1, this study aims to analyse the possible impact of a policy switching towards a limited livestock density. We have taken the Agenda 2000 as the reference scenario for the alternative scenarios with a limited livestock density imposed. The Agenda 2000 is a political action program of the European Union which was agreed on in 1999 at the Berlin European Council Meeting by the governments of the EU member states (EU Commission, 1999). Starting in 2000, the program is being implemented stepwise until 2006. And – apart from non-agricultural issues, such as to give the European Union a new financial framework for the period 2000–2006 – one of its major concerns is a re-orientation of the Common Agricultural Policy (CAP) by reducing price support and introducing compensation payments related to land and livestock.[13] However, due to the stepwise implementation, at the beginning of 2001 not all measures of the Agenda 2000 were fully introduced.

The farm structure and the considered farms' organization as they presently can be found in Hohenlohe are the result of the pre-Agenda CAP and not of the Agenda 2000 policies. In order to overcome this inconsistency, we first define and simulate a base scenario reflecting the pre-Agenda policy situation. This is done in order to calibrate the simulation model to the actual situation in Hohenlohe. During calibration the assumptions of the pre-Agenda model are subject to adjustments in order to reflect the actual agricultural structure and the main trends of Hohenlohe in an acceptable way. After the model has been calibrated, the reference scenario – i.e. the Agenda 2000 as of January 2001 – is implemented into the calibrated model, as well as two alternative policy scenarios.

In order to consider random effects of the model's initialization, each scenario has been simulated five times with different regional sizes, ranging from the full region with initially 2606 farms to a tenth of the region. For each scenario, *T* denotes the divisor by which the full region is divided, e.g. T = 10 denotes that only a tenth of the full region was simulated. Results of different initializations not only allow to identify random effects of, e.g., a differing spatial distribution of farms, but it also shows the effects of different random numbers. Moreover, it gives an impression on the robustness of the results.

Reference scenario: 'Agenda 2000'

Central changes as compared to a pre-Agenda situation are a reduction of intervention prices for products like cereals, beef, and milk. In return, farmers receive higher direct payments. With respect to crop farming the payments depend on the land used for crop, oilseed, and legume production. In dairy farming the payments from 2005/06 on will depend on the milk quota and in beef production on the number of animals. Intervention prices for cereals are cut by altogether 15%. At the same time direct payments are increased to 324 €/ha for crops and oilseeds and 383 €/ha for protein seeds (reference yield is 51.4 dt/ha). As part of Agenda 2000 the compulsory set-aside is reduced to 10%, but voluntary set-aside is possible up to a level of 33%, and it is compensated for with 324 €/ha. Arrangements for silage maize are maintained, but increased. For beef, intervention prices are cut by 20%. The bull premium is increased to 183 €/head. The annual premium for suckler cows is increased to 215 €/cow. All animals beyond an overall cattle livestock density of 2 LU/ha of forage area are not eligible for premiums. For dairy products intervention prices will also be cut by 15% after 2005. But this is not taken into account in the model, even though 20 periods are simulated.[14] Milk quotas are maintained at current levels.

Alternative scenario I: Limited livestock density ('LU-Agenda')

The first alternative scenario is directly based on the reference scenario of the Agenda 2000. The only modification is the assumption that a farm is only eligible for receiving the full amount of direct payments if the farm's livestock density is below two livestock units (LU ≈ 500 kg alive weight) per ha of farmland. If not, we assume that the payments are cut by 162 € for each LU exceeding the farm's limit of 2 LU/ha times the land which is farmed. Thus, exceeding the limit by 2 LU reduces the subsidies of one ha of grain.

Alternative scenario II: Limited livestock density, unitary premium land ('LU-Premium')

The second alternative scenario ('LU-Premium') also considers the limited livestock density. Moreover, instead of the rather differentiated payments for different crops, it is considered that the farms receive a unitary payment of 250 €/ha of land (arable land and grassland), regardless of how the land is farmed. The payments for animals are those of the reference scenario. Table 8.5 summarizes the change of compensation and premium payments under Agenda 2000 as compared to the pre-Agenda situation.

Table 8.5: Changes of compensation and premium payments under the Agenda 2000 scenarios as compared to the pre-Agenda situation

		pre-Agenda	Agenda 2000 LU-Agenda 2000	LU-Premium
Crops	[€/ha]	279	324	250
Price reduction		-	*-15%*	*-15%*
Sugar beets	[€/ha]	-	-	250
Rape seeds	[€/ha]	484	323	250
Silage maize	[€/ha]	394	458	250
Intensive grassland	[€/ha]	-	-	250
Extensive grassland	[€/ha]	-	-	250
Bulls[1)]	[€/head]	135	183	183
Price reduction		-	*-20%*	*-20%*
Suckler cows	[€/head]	140	215	215
Price reduction		-	*-20%*	*-20%*
Set aside	[€/ha]	363	333	250
Protein seeds	[€/ha]	402	383	250

1) per year, since the premium is granted per head only

8.5 RESULTS

8.5.1 Pre-Agenda and Agenda 2000

As was argued in the previous section, the agricultural structure of Hohenlohe at the beginning of the year 2001 is the result of the pre-Agenda agricultural policies. Therefore, in order to develop an understanding for possible structural and dynamic effects of the Agenda 2000, we first turn to simulations of the pre-Agenda and the Agenda 2000 situation.

Figure 8.3 shows the evolution of the average farm size for the pre-Agenda and the Agenda 2000 scenarios. As can be seen, farm sizes develop rather differently. In the pre-Agenda scenario the average farm size increases by only 1.2% per year. This is below the actual growth rate of the average farm size in West Germany which was 3.2% between 1991 and 2000.[15] Compared to the pre-Agenda scenario, the Agenda 2000 fosters structural change. The average growth rate is 3.2% per year. One explanation for this is a reduction of subsidies as part of the Agenda 2000, i.e. the higher direct payments do not fully compensate for the price reductions.[16] A second explanation is the different set-aside scheme. As a consequence of both, farms with a low productivity are less competitive than in the pre-Agenda scenario and more of these farms leave the sector. As the error bars in Figure 8.3 indicate, results vary for different initializations of the regions. And it shows that the variation of farm sizes and income is higher in the Agenda 2000 scenario where farm size change is more pronounced. Nevertheless, the differences in structural change are significant since there is no overlapping of the error bars.

The stronger structural change in the Agenda 2000 scenario, however, does not mean that the Agenda causes persistently lower incomes. Figure 8.4 shows that, except for the first two periods, average farm incomes in the Agenda 2000 scenario are significantly above incomes in the pre-Agenda scenario.

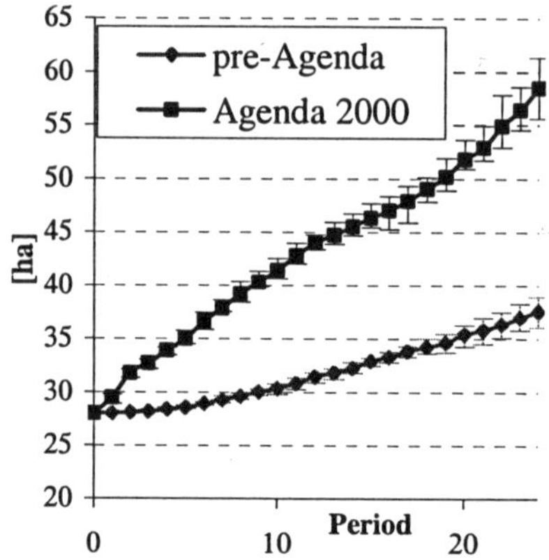

Figure 8.3: Evolution of the average farm size (error bars show different initializations)

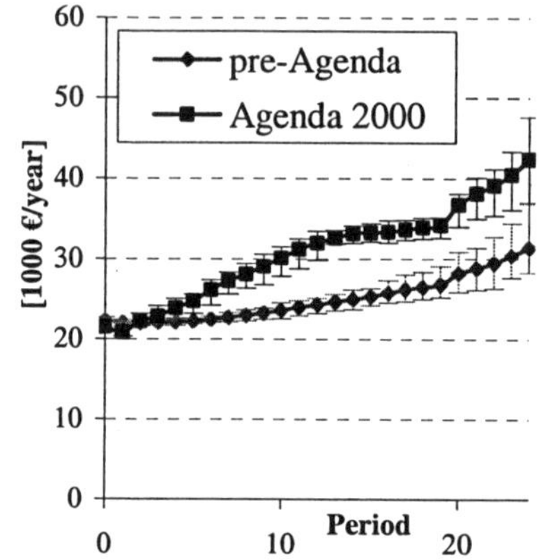

Figure 8.4: Evolution of the average farm income (error bars show different initializations)

There are several reasons for positive income effects of faster structural change. First, farms that leave the sector usually have relatively low incomes, i.e. faster structural change implies a different sample of farms. Second, surviving farms have better opportunities to grow with the consequence of better opportunities to exploit economies of size. And, third, larger farm sizes allow for lower rental prices for land which is particularly relevant since already at the outset of the simulations approximately 46% of the land is rented. On first glance the last two arguments seem contradictory. However, in Balmann (1999) it is shown that unexploited economies of size in small farm agriculture often imply rental prices for land to be higher than average economic land rents (cf. Figure 8.5).[17] Even though economic land rents for the Agenda 2000 remain lower over the full simulation period, rental prices develop clearly below the pre-Agenda scenario where rental prices even increase until period 8. Hence, according to Figure 8.5 significantly lower rental prices for land overcompensate for the lower subsidies after period 5. The Agenda 2000 allows the gap to close between economic land rents and rental prices already in period 10, while it needs 13 periods in the pre-Agenda scenario.

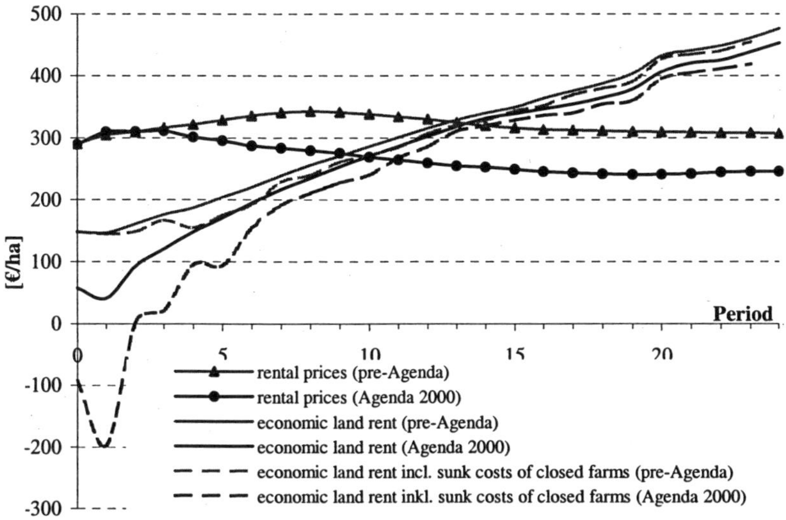

Figure 8.5: Evolution of rental prices for land and land rents (average of all simulation runs)

The existence of the gap between economic land rents and rental prices describes a disequilibrium situation. It can persist for some time due to certain frictions in structural change, such as sunk costs for assets and for human capital. However, closing this gap by faster structural change is not for free. If farms quit despite sunk costs, both assets and human capital are devalued. To take account of this devaluation, Figure 8.5 also shows the average economic rent of land including this devaluation, i.e. we have subtracted the realized sunk costs of farms that quit the sector. As for the Agenda 2000 scenario, during the first 4–7 periods, land rents including realized sunk costs develop far below the economic land rent curve, and are even strongly negative in the first periods after the policy switching. This indicates that on the one hand Agenda 2000 induces a comparatively strong and costly structural adjustment process during which a number of farms quit the sector. On the other hand the remaining farms exploit increasing returns to scale much better and benefit from lower rental prices. Hence, it is easier to overcome the disequilibrium situation in which prices do not cover long-run costs. Summarizing, it can be concluded that in response to Agenda 2000, farms have to adjust considerably to the new policy environment. Adhering to pre-Agenda policies would have changed agricultural structures only little, but at the expense of higher subsidies, lower incomes, and higher rental prices. Hence, compared to the Agenda 2000, the pre-Agenda policy setting describes a less efficient situation in which the driving force towards more efficient structures is almost absent. In this sense, the structural effect of Agenda 2000 is in accordance with its explicitly stated goal, to increase the competitiveness of EU agriculture (EU Commission, 1999).

8.5.1 The Switching Costs of Alternative Policy Scenarios

After having sketched the reference situation, we now focus on the two alternative policy scenarios defined above, both of which require farms to comply with a livestock density of at most two livestock units (LU) per ha in order to be eligible for premium payments. Results of these alternative scenarios are not only analysed on a sector level (which considers only averages), but a more detailed analysis on the farm level is carried out to illustrate differences between individual farms.

Figure 8.6 shows the evolution of average farm sizes for the reference scenario (Agenda 2000) and the two alternative scenarios (LU-Agenda, LU-Premium). All of these scenarios are seen to have specific effects on the structural adjustment process at the aggregate level. For instance, in the case of LU-Agenda, Figure 8.6 shows an adjustment process towards larger average farm sizes. This is due to the fact that farms which initially exceed the limit of 2 LU/ha with their existing production capacities, have to increase their acreage in order to comply with the policy and to keep on using their existing production capacity. The effect of the LU-Premium scenario is somewhat different since it can be seen to inhibit the increase in farm size. This can be explained by the fact that many small dairy farms benefit from equal payments for grassland and arable land, which was not granted as part of Agenda 2000. Thus the competitiveness of these dairy farms on the land market increases as compared to less grassland-dependent farms (intensive livestock, crop farms). Consequently, more small farms survive with the effect that other (probably more efficient) farms have less opportunities to grow (cf. Figure 8.7).

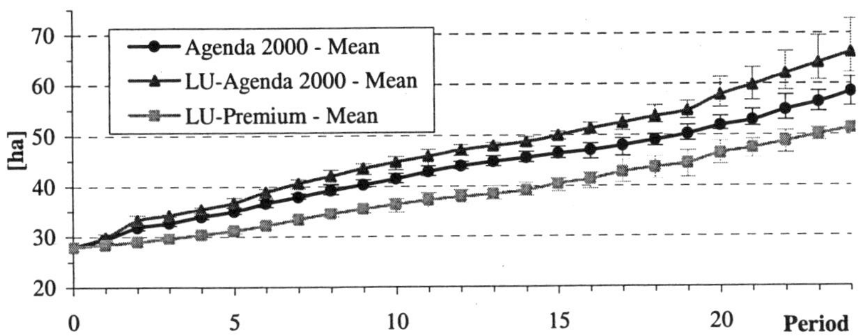

Figure 8.6: Evolution of average farm sizes (error bars show different initializations)

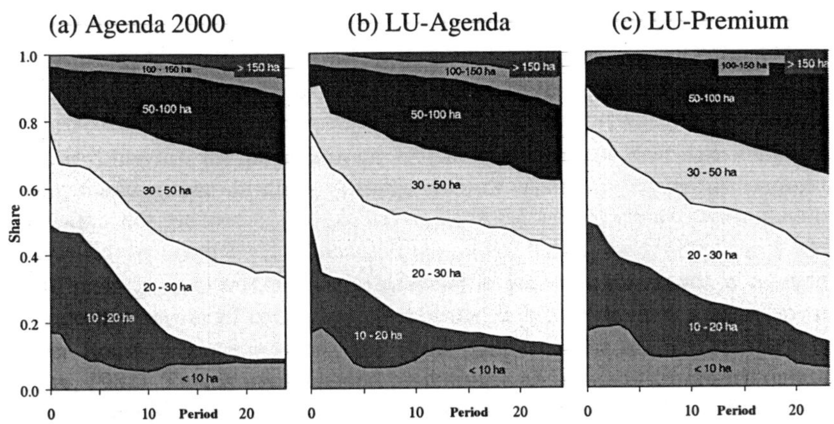

Figure 8.7: Evolution of the shares of farm size classes (T=1)

Figure 8.7 provides a more detailed analysis of the development of different farm size classes over time for the different scenarios. As for the LU-Agenda scenario, Figure 8.7(b) shows that there appears to be a strong tendency towards farm size growth. After 20 periods, more than 10% of the land is managed by farms with an acreage greater than 100 hectares. However, the land share of smaller farms with 20–30 ha and of farms with less than ten ha also increases for certain periods.[18] One possible reason can be found in the specific land allocation mechanism introduced into the model. This mechanism considers that there are potential cost savings associated with renting additional land and investing in new machinery. These savings decline with initial farm size, i.e. initially smaller farms have a higher cost saving potential when renting land compared to initially large farms. The realization of this potential, however, requires significant growth steps if farms are small.

The effectiveness of the different policies with respect to livestock density is presented in Figure 8.8. The scatter plots show the relationship between livestock densities and farm sizes for farms with livestock production. Obviously, the 2 LU/ha limit on livestock density is effective with the great majority of farms. However, two things are worth further discussion: First, as could be expected, the number of farms with livestock production in the LU-Agenda scenario is considerably less compared to the LU-Premium scenario. This is due to the fixed premium payment granted for all kinds of land use which favors dairy farming over arable farming.

A second phenomenon which becomes apparent in Figure 8.8 is that many small farms in the LU scenarios operate at very high livestock densities.[19] This supports to some extent the initial hypothesis that a limit on livestock density may cause inverse responses by some farms which specialize in intensive livestock production. Such an inverse reaction can be understood if the specific policy environment is taken into account. As was mentioned above, the limit on livestock density was enforced by binding it to premium payments. Each additional livestock unit exceeding the limit will result in a premium cut. Furthermore, it is assumed that farms can sell excess manure to the manure market. This specific policy environment creates the basis for the emergence of specialized livestock farms. Since farms showing an inverse reaction operate with a relatively low acreage, the maximum premium payments they can potentially receive is also relatively small compared to their earnings from livestock production. Moreover, the marginal effect on farm performance of a premium cut due to high livestock production is only effective as long as the premium received is greater than zero. Hence, small intensive livestock farms accept losing all their premium payments in order to be able to increase their livestock density.

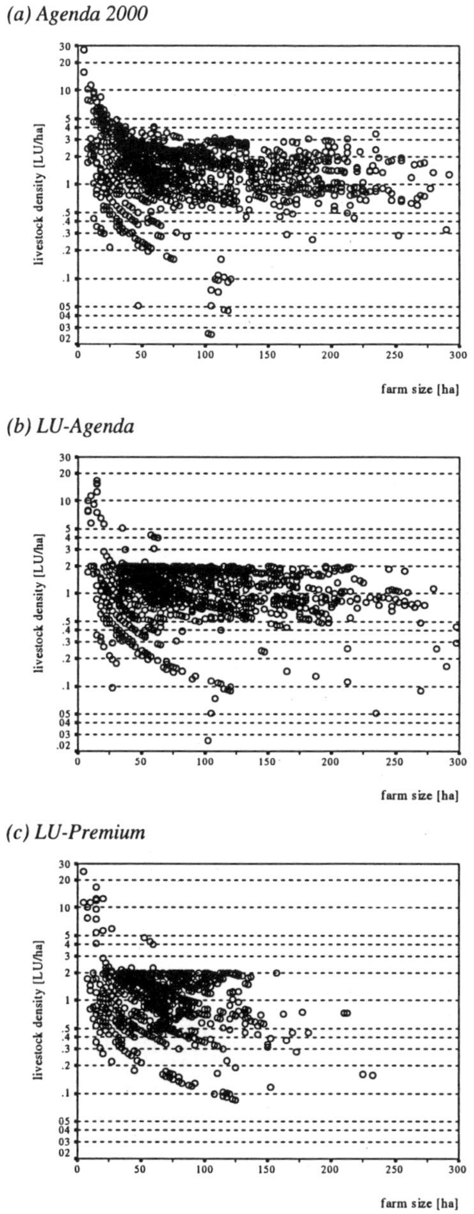

Figure 8.8: Livestock density and farm size of livestock producing farms (T=1)

Surprising is the sharp bend of the curves in Figure 8.9 for the two alternative scenarios after period 19. The reason is the emergence of a small number of specialized livestock farms at the outset of the simulations for both LU scenarios. After about 20 periods the useful lifetime of the livestock facilities is reached. These farms did not reinvest in new facilities since output prices for livestock decreased and opportunity costs of labor increased. Accordingly, this bend has to be explained by a kind of collapsing investment cycle. And, as can be seen in Figure 8.10, it is the share of sow production of the total livestock density per ha which significantly decreases after period 19, whereas the share of sucker cows steadily increases.

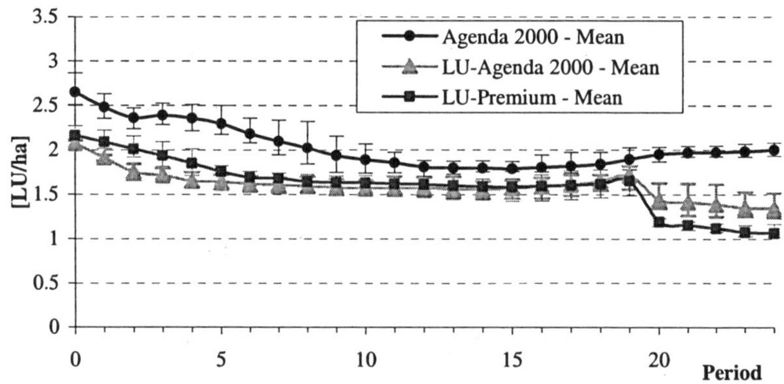

Figure 8.9: Evolution of average livestock densities of livestock producing farms (error bars show different initializations)

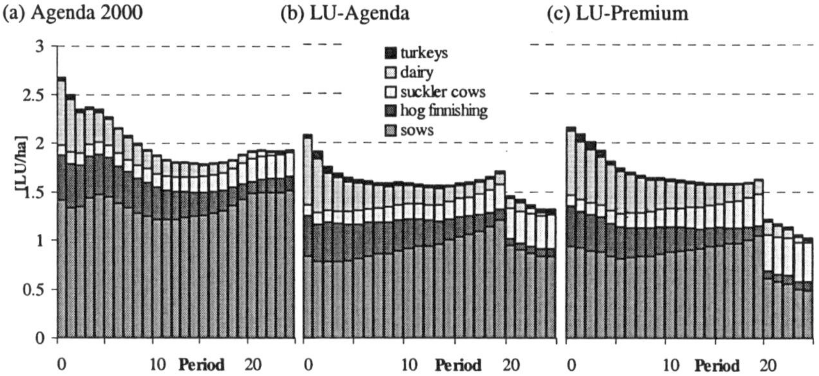

Figure 8.10: Evolution of average livestock densities of livestock producing farms according to livestock types (T=1)

After having shown that the alternative policy scenarios are indeed effective with respect to reducing livestock density, it has to be asked at what costs this happens. A first point of interest is the impact on farm incomes. Figure 8.11 shows that for both alternative scenarios average farm incomes develop significantly below the reference scenario Agenda 2000. The income reduction of LU-Agenda amounts to about 7300 € per year and farm on average, and the income of LU-Premium is about 8400 € per year and farm lower. Moreover, as Table 8.5 shows, the majority of initial farm types incur income losses due to the policy change. But, as the table also indicates, over the time horizon of 25 periods, some farm types can even benefit from the policy change.

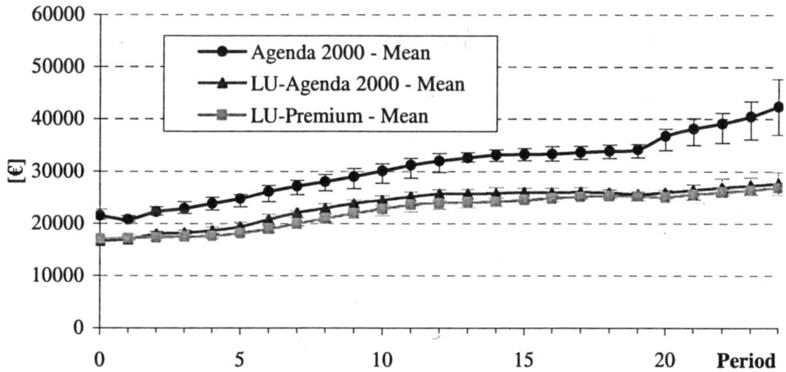

Figure 8.11: Evolution of average farm incomes (error bars show different initializations)

Table 8.6: Average results after 25 periods for the initial 12 farm types[1]

		Dairy farms			Arable farms		
		I	D	C	K	F	E
Agenda 2000							
surviving farms[2)]	[%]	45%	19%	13%	66%	59%	45%
Size	[ha]	66.94	42.13	29.92	45.74	79.73	25.06
Total income	[€]	32633	33107	39299	79977	36578	43805
Farm income[3)]	[€]	21249	22672	22579	56463	20263	25851
livestock density	[LU/ha]	1.06	1.69	2.15	1.84	1.57	1.92
LU-Agenda 2000							
surviving farms[2)]	[%]	49%	23%	6%	69%	36%	26%
Size	[ha]	79.13	45.01	21.22	47.83	66.89	23.23
Total income	[€]	34755	32007	34892	73312	22532	35136
Farm income[3)]	[€]	24673	20561	14377	44750	6905	9362
income change[4)]	[%]	*+ 16%*	*-10%*	*-37%*	*-21%*	*-66%*	*-64%*
livestock density	[LU/ha]	1.07	1.46	2.53	1.04	3.34	1.00
LU-Premium							
surviving farms[2)]	[%]	98%	95%	92%	64%	65%	32%
Size	[ha]	68.43	38.20	14.35	42.71	50.57	18.25
Total income	[€]	44014	36983	34838	70658	20694	35812
Farm income[3)]	[€]	30102	20300	8396	41982	3974	11655
income change[4)]	[%]	*+42%*	*-11%*	*-63%*	*-26%*	*-81%*	*-55%*
livestock density	[LU/ha]	1.03	1.29	1.69	1.05	4.03	1.74
		Pig/Turkey farms			Mixed farms		
		A	B	L	G	H	J
Agenda 2000							
surviving farms[2)]	[%]	83%	84%	57%	81%	46%	34%
Size	[ha]	29.99	64.04	28.11	111.43	71.16	25.03
Total income	[€]	27657	39437	55163	111362	55798	41114
Farm income[3)]	[€]	11834	26079	33692	80745	39145	22073
livestock density	[LU/ha]	1.17	1.68	2.09	1.41	1.43	1.73
LU-Agenda 2000							
surviving farms[2)]	[%]	95%	64%	26%	84%	48%	27%
Size	[ha]	34.31	69.28	41.90	101.72	71.10	19.87
Total income	[€]	28586	38705	48539	100053	55325	33984
Farm income[3)]	[€]	12141	26952	29538	66256	37479	6208
income change[4)]	[%]	*+3%*	*+3%*	*-13%*	*-18%*	*-4%*	*-72%*
livestock density	[LU/ha]	0.92	1.06	1.55	1.06	1.28	0.92
LU-Premium							
surviving farms[2)]	[%]	95%	69%	34%	19%	73%	5%
Size	[ha]	30.60	60.93	29.63	91.69	62.87	16.08
Total income	[€]	27178	36975	48686	97477	63714	35272
Farm income[3)]	[€]	10832	25201	27888	63335	41363	4417
income change[4)]	[%]	*-8%*	*-3%*	*-17%*	*-22%*	*+6%*	*-80%*
livestock density	[LU/ha]	0.94	1.69	1.64	1.16	1.25	0.65

1) farm type changes are not considered here, 2) as compared to the initial number of farms of this type, 3) total income = farm income + off-farm income + interest income, 4) compared to reference situation

The income effects for the two scenarios depend on several factors. One is the initial reduction of subsidies which explains the higher average farm income in the reference scenario at the start of the simulations. As Figure 8.12 shows, subsidy payments in the Agenda 2000 scenario and in the LU-Agenda scenario show a similar development over time, since they increase in period 1 and then (slowly) decline. This development cannot only be observed in the mean, but it can equally be observed for the different model initializations. In both scenarios structural change is more pronounced (cf. Figure 8.7). Particularly farms with subsidy-dominated production activities (dairy, cattle) quit the sector and let their land and quota to the respective markets. With respect to the remaining farms, these opt for production activities that are less dominated by subsidy payments (sows, pig, turkey). This is contrasted by the development of subsidy payments in the LU-Premium scenario, in which, during the first periods, a heavy increase of payments relative to the other scenarios can be reported. Farms adjust quickly in order to meet the payment criteria associated with a fixed land payment.

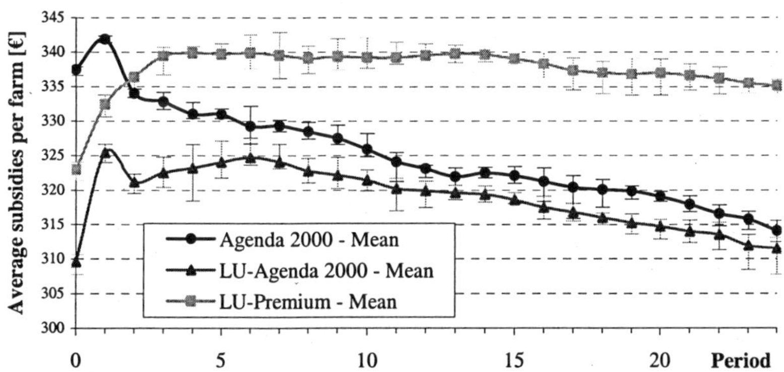

Figure 8.12: Evolution of average transfer payments per ha (error bars denote the variation between differently sized regions)

Another determining factor of the observed income effects is the relation of rental prices and economic land rents which is shown in Figure 8.13. Although the economic land rent is much lower for the LU-scenarios, rental prices even increase for the first periods. The initial increase in rental prices is due to the fact that livestock production capacities are fixed in the short run. Accordingly, the shadow price of the complementary input land is high for those farms which are above the limit of 2 LU/ha. Some of these farms react to the limited livestock density by increasing their acreage with the effect that rental prices increase irrespective of the fact that transfer payments and thus economic land rents are lower. However, not all farms react in this way. As was shown above, some smaller farms specialize in intensive livestock production without land. For the first periods, both reactions lead to an increase in rental prices.

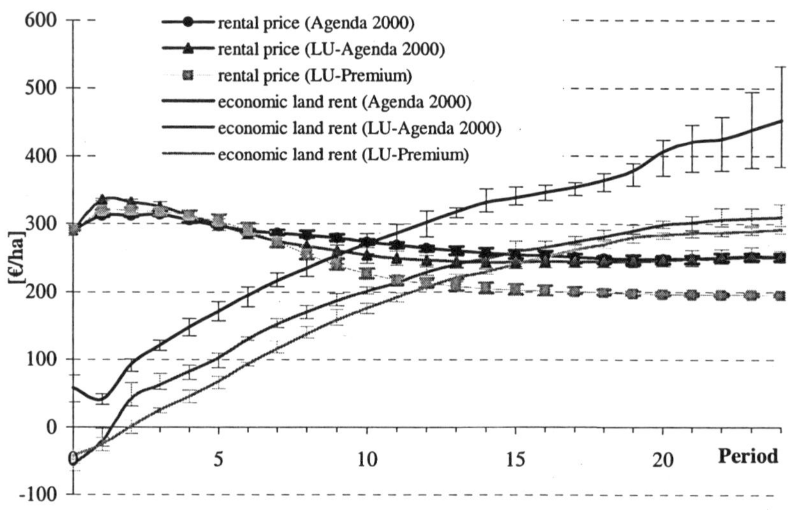

Figure 8.13: Evolution of rental prices for land and land rents

After some periods, rental prices start to decrease significantly. However, this does not compensate the lower economic land rent. The strongest decrease of rental prices can be found in the LU-Premium. This is surprising, because at first sight it would have been reasonable to observe higher rental prices when subsidy payments are higher as happens in the pre-Agenda scenario. The explanation can be found in the different payment schemes. The reduction of direct payments for most arable land uses amounts to 70–100 €/ha. Considering that 76% of the total land is arable land and that 81% of the rented land is arable land, the negative rental price effect of arable land outweighs the increase of rental prices for grassland premiums.

Moreover, there are inefficiencies that arise for LU-Premium scenarios as can be seen in the lower curve of the economic land rents. This is reflected by the development of the economic land rent curves in Figure 8.14. Since, in case of the LU-Agenda scenario more farms with a low productivity leave the sector, the remaining farms perform better on average. The reverse can be said about the LU-Premium scenario, where due to the fixed premium payments a larger share of rather inefficient farms survives. However, as already shown in Figure 8.5, increased structural change is not free of charge: Some farms close down despite having buildings and machinery that could still be used and despite low opportunity costs of their labor. If this realization of sunk costs is taken into account, the net efficiency effect of faster structural change is rather negative.

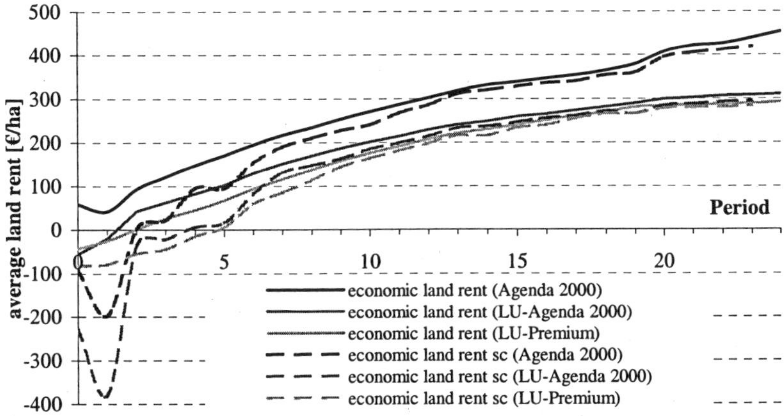

Figure 8.14: Evolution of land rents and land rents adjusted by sunk costs

8.6 SUMMARY AND CONCLUSIONS

The simulation results show that within the framework of the model a far-reaching policy change affects the agricultural system at different levels. In particular, if farms are heterogeneous, they are affected individually and respond differently. This can even mean that more restrictive policies may have positive impacts on some farms while other farms may suffer badly. The alternative policy scenarios presented above fall into this category. A policy that requires farms to meet certain livestock density criteria in order to receive payments can be quite effective because it creates the 'right' incentives – provided that the requested animal densities follow the 'right' goal. However, this is not free of charge. Particularly in livestock production, adjustment costs can be very high and adjustments occur slowly. On the one hand this is due to sunk costs, as illustrated above. On the other hand – and this has not been considered in this study – adjustments in livestock production often require farmers to learn about and quickly implement new and different production technologies. How time consuming this may be can be seen from the transition of East German agriculture. Even ten years after the fall of the Berlin Wall, the successors of former collective farms increase their physical productivity in animal production at enormous rates which cannot reflect normal technological progress but has to be understood as a catch-up process towards what is technologically possible (Balmann *et al.*, 2002).

Thus, from a policy perspective, one has to conclude that policy switchings affecting animal production should either be introduced slowly or should be announced in due time such that farmers can respond without enormous adjustment costs. This is usually not the case, since policies often reflect spontaneous reactions of policy makers to public concerns. From a scientific and from a modeling perspective, one has to conclude that the simulations presented above provide a starting point for discussions as well as for further investigations. Even though the model is already much differentiated with respect to individualization dynamical and spatial issues, many promising extensions are not yet implemented: One may additionally consider more differentiated landscapes, heterogeneous preferences of farmers, etc. Moreover, the model may be applied to different regions and alternative market scenarios.

The presented simulations are based on a model that was developed originally to analyse the particular dynamics of structural change. On the basis of this intention, it also permits the study of long-term policy effects. The obtained results shed some light on policy effects that often are ignored by conventional policy analysis, such as dynamical and distributional impacts on efficiency and incomes. From this point of view, policy modeling on the basis of agent-based systems seems very promising. On the other hand, the question is, how valid and how convincing the model and its results are from politicians' and economists' perspectives. And, indeed, there are some critical points:

- Firstly, one has to concede that validation of such models is difficult. It is neither possible to recalculate all numbers, nor is it possible to compare the results directly with analytical and empirical results. Nevertheless – and this is demonstrated by earlier applications of the approach (Balmann 1999, 2001) – most results are surprisingly robust and did not change significantly during the many model revisions over the past ten years. Moreover, the results fit many empirical observations like slow structural change, persistently unexploited economies of scale, and income disparities very well.
- Secondly, several of the model's assumptions (like the existence of bounded rationality and economies of scale) and results (like path dependence and income disparities) are discussed very controversially

in agricultural economics. Hence one may after the criticism that the model and its applications may only convince those people who have been convinced by bounded rationality and increasing returns before. A critic may argue that the model simulations just reflect what has been put into the model. But, these arguments are too short-sighted from our perspective. If the model's assumptions and results are in line with a particular economic argumentation, this does not mean that it is trivial. Phenomena like bounded rationality, increasing returns, path dependence, and income disparities imply complexity and can often hardly be tackled analytically. Then the only opportunity is either numerical simulation or verbal and qualitative reasoning. The advantage of numerical simulations is that they allow to quantify the effects and to check the consistency of the argumentation which is based on the underlying assumptions. Herewith they enable us to discover inconsistencies in the line of argument and to improve theoretical arguments. They may even permit us to put forward hypotheses which can be tested empirically.

- Thirdly, agent-based models are usually highly dimensional and non-linear. Even for the user it is difficult to grasp the full structure of the results in view of their variability over the different agents and over time. Hence, it is far from trivial to mediate the model's assumptions and its results to third persons. Every presentation requires simplification and it may even occur that the more differentiated and sophisticated the presentation is, the more questions arise on the side of the addressees. In the end, the analysis and presentation of model results may require simplified models of the original model and its results. Sometimes it may even be useful to apply sophisticated statistical methods to study simulation results. For instance, Balmann and Hilbig (1998) applied a factor analysis and a cluster analysis to study conditions under which a predecessor of the presented model shows path dependent behavior.

Summarizing these points, it is clear that there are some problems with the use of complex agent-based models (ABM). But actually these problems should rather be understood as a matter of the research questions to which ABM are applied than one of the method itself. It is reality which is so complex. Models based on agent-based systems allow us to consider and reflect on this complexity. Hence, to renounce this method often means not to use a method that is able to grasp the complexity of the research question. Moreover, ABM have to be understood as a rather young field of research.

When applying ABM we can explore their opportunities and learn about them. This is a valuable byproduct of a method which is a beneficiary of the increasing power of modern computers. The limits of such models are far from reached and they are steadily shifted further.

To conclude, ABM offer the opportunity to look at economic and social processes from new and different perspectives. They allow us to study questions which otherwise could not be tackled at all or which would require rather strong assumptions. Hence, ABM appear to be a promising tool for policy research. Their limitations will depend on the progress in information technology and on the resourcefulness of its users. Thus they are a promising field of future research.

NOTES

1. A livestock unit in Germany corresponds to about 500kg alive weight.
2. However, farms may receive subsidies for investments in pig or turkey production that are undertaken in order to modernize production or for more environmentally friendly production technologies. These subsidies cannot be used to increase production capacities.
3. The detailed presentation of the model would fill a book. Therefore, here the features of the model are only sketched. The interested reader may refer to Balmann (1997) where an earlier version of the model is presented in much more detail. Happe, Kellermann and Balmann (2002) document the model's program code and structure, as well as a description of the main functions and routines.
4. For display reasons only a tenth of the actually simulated region is shown here.
5. Actually, economies of scale can be assumed to exist for much larger units as well. However, this has been ignored in this study because there are hardly any farms larger than 200ha in Hohenlohe and structural change is slow.
6. The foundation of new farms has been ignored because it is not relevant for the region.
7. A very early example of such a normative agent-based model in agricultural economics can be seen in the recursive programming approach developed by Day (1963) that considers a number of interacting farms, each representing a particular farm type. The main differences between this model and modern agent-based models is that the recursive programming approach aggregates all farms of a certain type to a so-called 'group farm'. This causes certain problems. First, it is implicitly assumed, that all these farms behave identically. Second, the so-called 'aggregation error' emerges, i.e. within a group farm, production capacities can be allocated without considering transaction costs and transportation costs that represent usual frictions in the exchange among different farms.
8. In this section certain specific modeling issues are illustrated in more detail because they have not yet been published in earlier work, such as (Balmann, 1997).
9. Berger (2001) has included a direct interaction between agents into his model to reflect negotiations for water rights and land in his study region in Chile.
10 An exception is a situation in which only a few plots are left in a certain period. Then all farms may bid for the same plot. However, in such a case, usually, the highest bids will come from farms close to the plot. These farms have low transportation costs and are likely to take advantage of adjacent plots. Moreover, those farms which are far away usually have already satisfied their demand in previous bidding rounds.
11 For a detailed discussion of the adoption of agricultural innovations see Berger (2001) and Brandes *et al.* (1997).
12 Regarding a detailed documentation of the calibration procedure cf. Kleingarn (2002).

13 The contents of Agenda 2000 can be retrieved under the website http://europe.eu.int/comm/agenda2000/index_en.htm. Several studies analysing its impacts on German agriculture can be found in *Agrarwirtschaft,* Vol. 47 (12).

14 This simplification that takes into account that after 2006/08 the CAP is likely to be changed anyway.

15 The main reason for the slow structural change in the pre-Agenda scenario can be seen in the fact that the smallest part-time farm in the sample of typical farms is much larger than many existing part-time farms in Hohenlohe. Many of these small farms are likely to quit in the next few years. A second reason can be seen in a kind of over-calibration on the farm level such that relatively few farms have to adjust under the pre-Agenda regime.

16 Regarding a quantification of the effects cf. Balmann *et al.* (1998b).

17 The economic land rent is computed as the household income plus rent expenditures minus long-run opportunity costs of the capital and family labor.

18 One should bear in mind that such simple classifications are problematic to interpret. If for instance, some farms with 17.5 ha increase their acreage by 2.5 ha, these farms switch into the next higher size class and increase the share of the higher farm class. If however farms with 12.5 ha increase their acreage by 2.5 or 5 ha, nothing would happen.

19 The few points plotted in Figure 8 should not be misinterpreted. Due to computational reasons only about 10% of the points have been plotted, but the effects remain the same when a larger number is considered.

9. Agent-based simulation of organic farming conversion in Allier département[1]

Guillaume Deffuant, Sylvie Huet, Jean Paul Bousset, Jérôme Henriot, Georges Amon and Gérard Weisbuch

9.1 INTRODUCTION

The evolution toward a better management of the environment often requires that the economic actors change their usual behaviors. This process can be seen as the diffusion of innovative behavior, if we refer here to the definition proposed by Valente (1995): 'Diffusion of innovations is the spread of new ideas, opinions, or products throughout a society, thus diffusion is a communication process in which adopters persuade those who have not yet adopted to adopt.' The diffusion of agri-environmental measures or more sustainable farming practices fits the definition, because the sociological theories of innovation diffusion do not suppose that the change of behavior is technical, or even innovative. The only requirement is that the decision of change is influenced by the others (and especially the adoption of others).

The question of innovation diffusion is the subject of a very large amount of research in sociology, and many sociological models were proposed. The *threshold models* (Granovetter, 1978) suppose that the observation of earlier adopters is a means for reducing the uncertainty about the outcome of the adoption. Therefore risk tends to force individuals to turn to their peers to gain more information and/or reassurance about the outcome of the adoption. Rogers (1983) refers to contagion as the diffusion effect. When the number of adopters increases in the network of peers, the pressure for adoption increases. The capacity to resist this pressure is the threshold defined as the proportion of adopters in the considered group necessary to convince him to adopt. This theory has been applied to very different subjects: farming innovations, family planning practices, medical technology, policy innovation, language (see Rogers (1983) for a comprehensive review and Valente (1995) for more recent research directions and results).

The diffusion of innovation is also an important subject of research in the social simulation research field (Gilbert and Conte, 1995), which aims at reproducing the social dynamics in computer models. The reason is a very interesting convergence between sociologists and computer scientists: the threshold models designed by the sociologists can be directly represented in a computer by a network of simple automata, which is a very popular tool of social simulation research. Each automaton represents an individual in a social system, and the links between the cells represent the links between individuals. The links can represent working relations, friendship or any type of contact. The behavior corresponds to adoption or non-adoption of the innovation. Blume (1993), Ellison (1993), and Blume (1995) consider automata networks implementing the threshold model, in which the threshold is related to an intrinsic payoff to the considered agent. This leads to the definition of the agent's utility as the sum of an intrinsic payoff to the considered agent and the proportion of the agent's neighbors who adopted the behavior. Several variants of this model can be found, some taking into account the advertising exposure instead of the intrinsic payoff (Goldenberg *et al.*, 1999). In general, the updating rule switches the agent to the behavior maximizing its utility. Young and Blume (Blume, 1993; Young, 1998) study a weaker updating rule in which the behavior is chosen according to a probability exponentially decreasing with its utility. This model is called perturbed best reply. This framework led to several interesting theoretical results (Young, 1999), in which the most probable state of the system is shown to be the minimum of an energy function.

However, despite very interesting theoretical and practical results (Weisbuch and Boudjema, 1999), the classical innovation diffusion models appeared quite limited when applied to the diffusion of agri-environmental measures. In particular, a closer study of this process (350 farmer interviews, 100 institutional actor interviews) during a four year European project showed that some important aspects of the decision process are missing in the threshold model: an explicit role of the uncertainty in the decision, the importance of the information diffusion and processing, different stages of the decision process.

We present an agent-based model of innovation diffusion, targeted on the diffusion of agri-environmental measures, which was developed during this European project. The model is the result of an iterative process between modeling researchers and specialists of the agri-environment. It also used several types of data collection from eight study zones in Europe. It represents explicitly the propagation of information and the dynamics of discussions among farmers, as well as its impact on the farmers' decision process. One important particularity of the model is that the opinions and uncertainties are continuous variables (whereas most models of social influences consider discrete opinions). Our modeling approach is hybrid between computer science and cognitive agent approaches (Conte and Castelfranchi, 1995; Ferber, 1995; Muller, 1996; Conte, 1999), and the models of cellular automata networks which are more inspired by physics (Weisbuch, 1991; Blume, 1993; Young, 1998). It builds on earlier versions proposed by Chattoe and Gilbert (1998).

We first present a strictly economic and static perspective on the conversion to organic farming in the French département of Allier. This approach is by nature unable to explain the evolution of adoptions. Then, we present the agent-based model, which combines this economic evaluation with informational and social processes. Then we focus on the simulation results of this model applied to the diffusion of organic farming. We use a decision tree in order to classify a high number of simulation results. The decision tree gives indications about the sensitivity of the model to the variation of the parameters. We analyse in more detail the parameter zones leading to the best results, especially the evolution of the social image of organic farming in the farming population.

9.2 STATIC ECONOMIC MODEL ON A GENERATED FARMING POPULATION

This section focuses only on an evaluation of the financial impact of the conversion for the farms of the considered case study. This model is applied to a population of farms which is generated by combining two sources of data: aggregated data by cantons and a set of prototypes farms. This model is kept as a part of the agent-based model which is presented in Section 9.3.

9.2.1 Organic Farming in Allier

In the framework of the 2078/92 regulation, France proposes incentives for the conversion to organic farming. The objective is to favor less polluting practices in farming and to be able to face the growing demand for organic products through national production.

In order to get the incentives for the conversion to organic farming, the contracting farmer must convert the totality of his farm within a five year period. He commits to comply with the regulations for organic farming, which imply:

- comply with the methods of organic vegetable production defined by the EU regulation,
- comply with the animal production methods defined at the national level,
- submit the farm to the control of an officially recognized organization,
- declare the organic activity to the Direction Départementale de l'Agriculture and to the controlling organization.

Before contracting, an evaluation of the technical and economic consequences of the conversion must be performed by a technician from the Chambre d'Agriculture.

The commitment of the farmer is for five years. The first two years are the conversion process (three for permanent crops). During this initial period, the farmer must modify his production techniques in order to adapt his farm to the constraints of the organic production regulations, and he is not allowed to sell his product on the organic market.

The incentives vary with the production system. In Allier the subsidies evolved significantly between 1993 and 1998. Four periods can be considered:

- Period 1 : no subsidy for the conversion,
- Period 2 : First ceiling of the subsidy (30 000 F per year per equivalent full time work on the farm, limited to two workers)
- Period 3 : second ceiling of the subsidy (60 000 F per year per equivalent full time work on the farm, limited to two workers)
- Period 4: no ceiling for the subsidies.

9.2.2 Generation of the Farm Population

In order to get an evaluation of the number of farms that could convert, we need first to generate a population of farms which is representative of the actual farming population of the département.

9.2.2.1 Method

The method combines two different sources of available data:

- aggregate data by commune or canton (Census or subsidy data),
- individual data, for a limited number of farms, which are representative of the main groups at a regional level (FADN[2] data).

The principle is to generate a population where some of the FADN prototypes are repeated several times, in a way that fits at best the aggregated data. This problem can be put in the form of a quadratic optimization (see

(Deffuant, 2001, for details) which we solved using a specialized software (scilab and Matlab were tested).

9.2.2.2 Application of the method to generate the farming population of Allier

The available data about a FADN prototype comprise a lot of variables: size of the farm, productions, financial figures, etc. Small farms, or for which there is not the equivalent of a full-time worker, are not represented in the FADN data. We used the data for 1996.

The aggregated data of common agricultural policy subsidies by *canton* in Allier in 1996 (number of farms, productions, surfaces for each production, number of heads of different livestock) were provided by the Direction Départementale de l'Agriculture et de la Forêt de l'Allier. These data cover roughly the same population of farms as the FADN.

We selected 201 prototypes among the 350 of the Auvergne region, because of their compatibility with the farms found in Allier. The optimization is done on the following variables: global surface, total number of farms, surfaces and number of farms producing wheat, corn and other cereals, permanent grass, number of cows, ewes, number of farms subscribing to milk quotas.

The optimization method selected 165 prototypes over the 201 initially considered. The highest number of repetitions for a prototype is 270 (4.5% of the total number of farms), but most of the prototypes are repeated less than 60 times.

In order to evaluate the quality of the optimization, we compared the aggregated values by *canton* obtained with our generated population with the data. The results are quite satisfactory: the maximum error is of a few percent.

Moreover, we compared the obtained results to aggregated values which could *not* be used for the optimization (because they were not available at the *canton* level): the farming systems and the classes of sizes of the farms. The results are of course not as good as the variables which were used in the optimization, but the broad tendencies are respected (see Deffuant (2001) for more details).

9.2.3 Financial Impact of the Conversion to Organic Farming

The evaluation of the financial impact of the conversion to organic farming is a very delicate problem because of the very small number of references and the high variability of this impact with the quality of the land, and other parameters on which we do not have information. Our approach comprised the following steps (see Huet and Deffuant, 2001, for more details):

- definition of a simplified model of the financial impact for the period of conversion and after conversion. This model is limited by the available variables in FADN data, because we apply it on our generated population which is made up of such prototypes. It takes as input the characteristics of the farm, the specifications of the measure (level of subsidies), and involves several parameters ruling the decrease of yields, the prices of organic product. Basically, the model computes an evaluation of the costs (decrease of yield, increase of the price of livestock food) and benefits (increase of selling prices, decrease of some costs). An important point is the evaluation of how much food for his livestock the farmer can produce by himself.
- definition of possible ranges of values of variation for the model parameters (decrease of yield, selling prices, amount of food by animal, possibility for growing livestock food). These values come from the literature about organic farming. The combination of all these values of parameters gives a large set of possible models.
- selection of a set of models by testing them on prototypic farms for which we then know that the conversion is financially positive (from the technician).

The selection of models gives a range of possible values of the parameters, and therefore a range of values for the financial impact. This range of possible values defines a mean value and uncertainty for the financial impact.

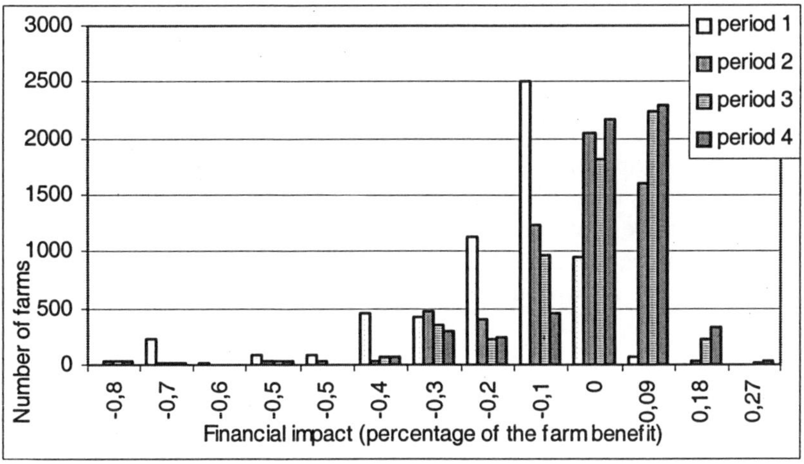

Figure 9.1: Histogram of the most pessimistic financial impacts in the farming population of Allier for the four periods (see paragraph 9.2.1.1 for the definition of the periods). The results are negative for a majority of the population, but they are significantly improved when the subsidies increase.

9.2.4 Application of the Financial Model to the Generated Population

The results of the application of the model on the generated population are shown on Figure 9.1, for the four periods defined in paragraph 9.2.1. We represented the most pessimistic evaluation of the financial impact.

From these results, one would expect the evolution of the subsidies to have a very important impact on the conversions. We see in Figure 9.1 that in period 4, almost 1500 farms have the most pessimistic financial impact which is positive and of more than 10% of the farms benefit. Therefore, one could expect a significant number of these farmers to adopt.

9.2.5 Comparison with the Actual Adoption Data

The conversions to organic farming are given in Table 9.1.

Table 9.1: Progress of the number of organic farmers in Allier between 1993 and 1999 (source: Chambre d'Agriculture)

Year	1993	1995	1996	1997	1998	1999
Organic farmers	32	45	45	68	93	111
in conversion process	5	5	6	16	19	22

We note that the actual number of conversions is much lower than expected from our economic model. It could be that our economic model is wrong because of its simplifications and the lack of standard references from organic farming. However, according to the specialized technician of the Chambre d'Agriculture, the figures given by the financial model are plausible. In this case, the low number of adoptions could be explained by cultural resistance and/or lack of information.

This motivates the design of a model which addresses these social and cultural aspects, as well as the institutional actions in the diffusion of information.

9.3 AGENT-BASED MODEL OF AGRI-ENVIRONMENTAL MEASURE DIFFUSION

9.3.1 Modeling Method and Main Hypotheses

We progressively elaborated the model during a four-year project involving a collaboration between modeling specialists and agri-environmental specialists. The modeling method is based on several loops comprising: data collection on the ground (farmer and institutional actor interviews), model elaboration and criticisms, and then definition of new data needs.

The farmer and institutional interviews exhibited several important aspects of the decision process which are not taken into account in the classical threshold models of innovation diffusion:

- the role of uncertainty is important in the decision and in the discussions;
- the attitude toward information varies according to the a priori interest;
- a higher level of information allows the farmers to better evaluate their personal interest to adopt;
- institutions act in various ways to provide information (meetings, visits, media), which can have an important impact on the decision process;
- the decision process involves more than two stages (yes/no), in particular, there seems to be an intermediate period of reflection and hesitation.

In order to incorporate this complexity, we designed an agent-based model of innovation diffusion in which two types of agents are distinguished: *Institution* and *Farmer*. The model includes two main dynamics: the *discussion propagation* and the *decision process*.

The decision process is based on the following hypotheses:

- The interest state of a farmer is based on the evaluation of an economic and a social criterion (in this, the model is similar to the classical threshold models). The criteria are defined by a mean value and an uncertainty (they can also have no value). The interest state has three possible values: not-interested, uncertain, and interested. Parameter *tau* defines the interest states.
- The farmers send messages containing their criteria and their information when they discuss. The criteria of the received message modify the criteria of the receiver if they are not too opposed, and more certain criteria have more influence. Parameter *mu* rules the intensity of this influence.
- The information of the receiver can be modified if he is interested or uncertain. The probability of updating the information is then parameter *omega*.
- If a farmer has the information about the measure and is interested or uncertain, he performs an evaluation of the economic criterion (based on the economic model of the previous section). If he is still interested after this evaluation he requests a visit from the technician. When this visit takes place, the economic criterion is evaluated again, with more precision (less uncertainty).

- The adoption decision is made when the farmer has been visited by a technician and he stayed interested for a certain time (parameter *ro*).

The discussion propagation is inspired by the two stage diffusion model described in Valente (1995). The main assumption is that the institutions send messages to particular farmers who then talk about it (by sending their own criteria and information level) in their social network. Then the contacted farmers tend to discuss in their turn. We supposed that the number of contacted farmers progressively decreases (parameter *gamma*).

For each institution, we define a scenario of actions which is an input of the model. This scenario specifies how the institution communicates to farmers: the content of the messages (social and personal criteria, as well as information), when and how they are sent (global sending, or regular interactions), whether visits are made or not. The behavior of the institutions is therefore considered as an input of the model. The main components of the model are described in the UML class diagram of Figure 9.2.

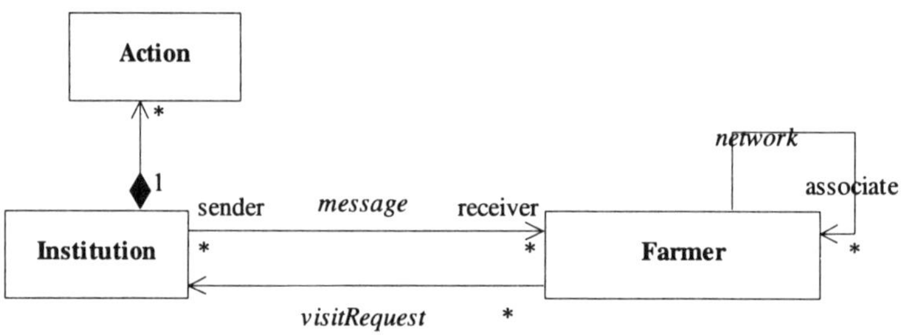

Figure 9.2: Class diagram of the agent-based diffusion model. A scenario of actions (meetings, visits) is defined for each institution. Moreover, the institutions may send messages to the farmers of their network through regular contacts. The receiving farmers are influenced by these messages and tend to discuss the adoption in their networks of associates. After a while some farmers may request the visit of a technician in order to get a better evaluation of his personal criterion.

9.3.2 Farmer Agent

The UML class diagram focusing on the farmer agent is shown in Figure 9.3. An instance of *Farmer* has two main components: an instance of *TalkTendency* (ruling the dynamics of discussion propagation) and an instance of *DecisionProcess* (ruling the decision process).

The operation *talkToNetwork*() uses the frequencies of interactions between farmers to determine whether a discussion takes place or not. These frequencies are supposed to be the frequencies of usual interactions.

The operation *createMessage()* creates a *Message* object which includes the criteria of the farmer decision process and the values of his information (in *dataProcess*). Note that we consider two criteria (social and personal), and that each criterion includes a mean value and an uncertainty.

This message is then received by the *decisionProcess* of the associate farmers. In the current implementation, we consider that the farmers send their values of criteria and information. There is no strategy of communication for hiding anything from the associates. This is of course a strong simplification.

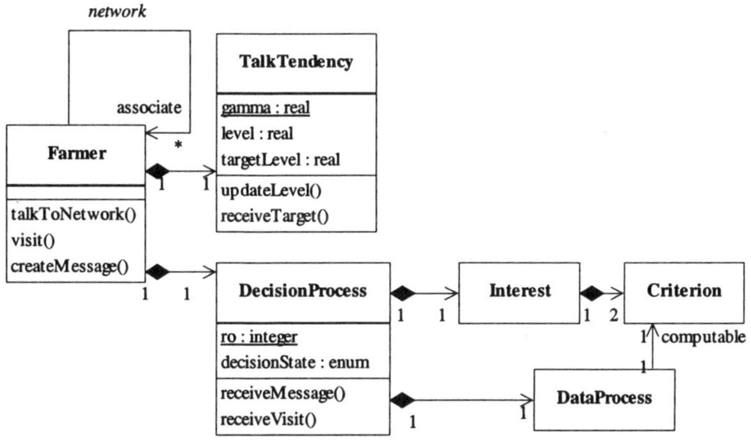

Figure 9.3: Class diagram focusing on Farmer class. Its main components are talkTendency, which rules the diffusion of discussions, and decisionProcess which rules the influence of the messages and the visits on the decision state.

9.3.2.1 The propagation of discussions (TalkTendency class)

The principle is that farmers who are directly contacted by the institutions talk to a large part of their associate farmers, who themselves propagate to a smaller part of their network and so on (see Figure 9.4 for an illustration). The class attribute *gamma* rules the decreasing number of discussions.

The principle for decreasing the number of discussing farmers with time is the following:

- The number of times the first farmer contacted by the institution chooses an associate at random and talks to him is: (1 – *gamma*)* the size of his network.
- The number of times the farmers contacted by the first farmer chooses an associate at random and talks to him is: (1 – 2 *gamma*) * the size of their network.
- The number of times the farmers contacted by the previous farmers choosex an associate at random and talks to him is: (1 – 3 *gamma*) * the size of their network.
- And so on...

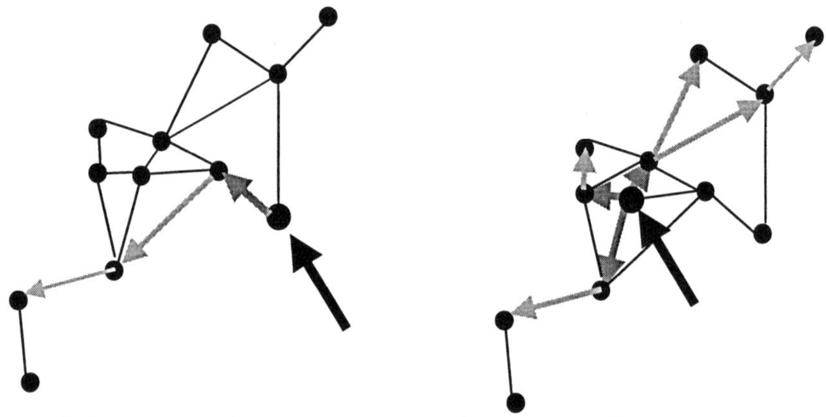

Figure 9.4: Two examples of propagation of discussions with gamma = 0.2 in the same network. The nodes of the graph represent farmers and the links correspond to their relationships. A black arrow points to the farmer who first received a message from an institution. The arrows indicate the interactions which take place. The width of the arrow and its darkness decreases with the propagation in the network. We can see the effect of the first farmer's position.

The number of contacted farmers among these at each time step is defined by the usual frequencies of interactions. The discussions about the measure are considered as particular cases of these usual interactions.

9.3.2.2 The decision process (decisionProcess class)

We chose a simplified decision process which fits particularly the organic farming conversion: the process is linear, and the visit by a technician is a requirement to get the subsidies (for other agri-environmental measures, the decision process can be very different). We suppose here that the farmers request a visit only after a first study of the measure with their own information, and a positive result of this first approach (they are still interested), because the visit of the technician costs time and money.

The *receiveMessage()* operation rules the influence of a message reception on the farmer decision process. The reception of a message may have different influences on the decision process and its components:

- The social criterion is directly influenced by the social criterion of the received message: we propose here a specific dynamics of influence which is adapted to continuous variables with continuous uncertainties (presented in more details in the *Criterion* class). The class parameter ruling the intensity of the influence is *mu* (class attribute of *Criterion*).
- The personal criterion is also directly influenced by the personal criterion of the received message, with the same dynamics, if the decision state is initial. If not, a first evaluation was made by the farmer, and we consider that his criterion is not further influenced directly by the others. The influence is indirect, by the modification of his information.
- The information attribute of *dataProcess* can be directly modified by the information attribute of the received message. In practice, we consider that the information of the received message has an impact only if the receiving farmer is interested or uncertain, and the information is actually modified with a probability *omega*, which is a class attribute of *dataProcess*. A low *omega* can be interpreted as a complex measure which is difficult to communicate.
- If the information is actually modified, the farmer performs an evaluation of his personal criterion by the data processing (a computation which takes as input *farmData* and *measureData* attributes). The first farmer evaluation of the personal criterion triggers a transition of the decision state to *selfEvaluation.*
- The modification of the interest state induced by the modifications of the criteria may lead to a change of decision state (to *visitRequest*, *adopted*).

The *receiveVisit()* operation is directly called by the institution when it visits the farmer. It changes the *decisionState* to *visited* and calls directly the operation *visit*() of the *decisionProcess* (see below).

The *checkAdoption()* operation puts the attribute *decisionState* to the value adopted if the farmer stayed *interested* during *ro* consecutive periods.

Figure 9.5 shows a more complete UML diagram.

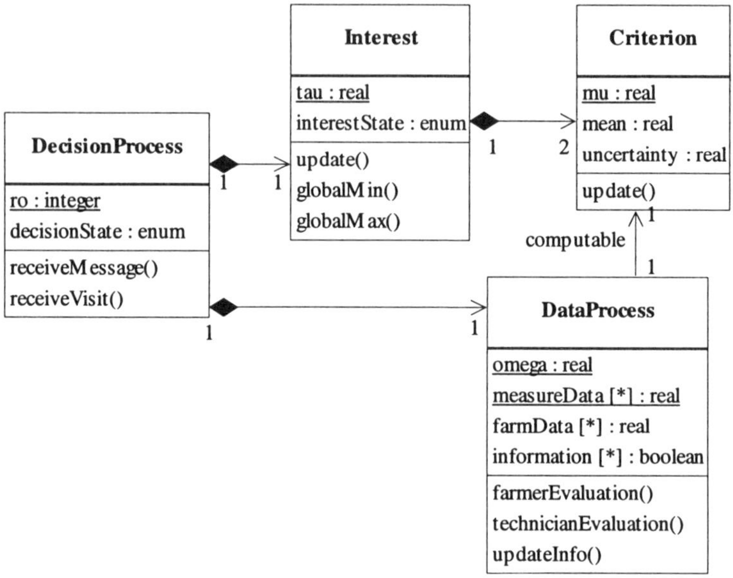

Figure 9.5: Class diagram focusing on the class decisionProcess (component of Farmer). The decision process is based on interestState and on dataProcess. interestSate itself is based on two criteria (social and personal). The data process may modify the value of the personal criterion, which is supposed computable from data and information.

The value of the attribute *decisionState* depends on the interest state and the data process. We consider that the farmer must stay interested during a minimum of *ro* periods after the technician visit in order to make the final conversion decision. During this period the farmer may discuss with colleagues, or institutional representatives, who may make him change his mind (see the state diagram of the decision process in Figure 9.6).

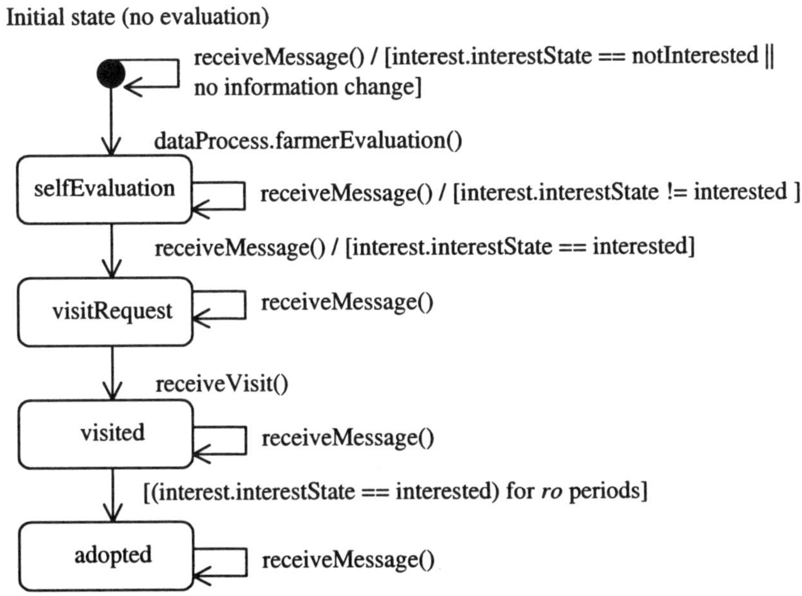

Figure 9.6: State diagram of decisionState. The squares represent the states and the comments near the arrows the conditions for the transitions. In the presented implementation, the decision process is linear: being in all the consecutive states is necessary for the final adoption. The transitions are triggered mainly by the receiveMessage() operation, but also by receiveVisit().

9.3.2.3 Modification of the criteria during discussions (criterion class)

The main operation of this class is updating the criterion when the farmer receives a message. Remember that a criterion comprises a mean value and an uncertainty. The rationale to modify the mean value and the uncertainty lies on the following hypotheses:

- people tend to be more influenced by certain interlocutors,
- opinions which are too distant have less influence, because they seem irrelevant.

In order to implement these hypotheses, we designed a dynamics in which the modification of the mean value and uncertainty takes place only if more than half of the received criterion segment overlaps the receiving segment (see Figure 9.7). This expresses that a minimum agreement is necessary for the influence. This leads to a non-symmetric influence, in which certain opinions are more influential. The class parameter *mu* rules the intensity of the influence. Moreover, when a receiving criterion is nil (no opinion) the received criterion is directly adopted, and if the received criterion is nil, then the receiving criterion is unchanged.

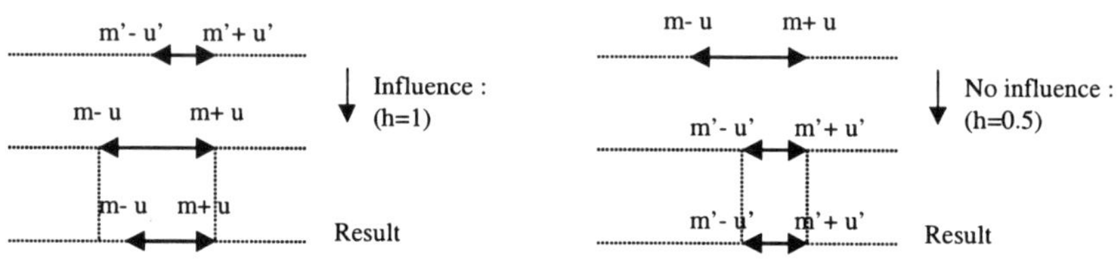

Figure 9.7: Left, the received criterion has a mean m' and an uncertainty u'. The receiving criterion, of mean value m and uncertainty u is influenced by the received criterion because the relative overlap is 1 (the overlap is the whole received segment). Right, the situtation is inverted. The received criterion has a mean m and an uncertainty u. The receiving criterion, of mean m' and uncertainty u' is not influenced because the relative overlap is 0.5.

These dynamics have interesting properties, in particular in the presence of extremists, which we began to study systematically (see Deffuant *et al.*, 2000; Deffuant, 2001). This study exhibits some conditions under which the majority of the population, initially moderate, becomes extremist.

9.3.2.4 Definition of the interest states (interest class)

The role of this class is to compute the global interest for the adoption which is based on an aggregation of the social and personal criteria. The attribute *interestState* of this class can take three values: *notInterested*, *uncertain*, or *interested*. The aggregated minimum and maximum are compared to a threshold *tau*, which is a class variable. The aggregated maximum is the sum of the higher (the mean value plus the uncertainty) values of both criteria. The aggregated minimum is the equivalent for the lower values (mean minus uncertainty).

The rules defining the interest state are (see Figure 9.8):

- if the aggregated minimum is higher than *tau* then the state is *interested*,
- if the aggregated maximum is lower than *tau* then the state is *notInterested*,
- if *tau* is between the aggregated minimum and the aggregated maximum, then the state is *uncertain*.

The value of the interest state is essential in the decision process (see Figure 9.6).

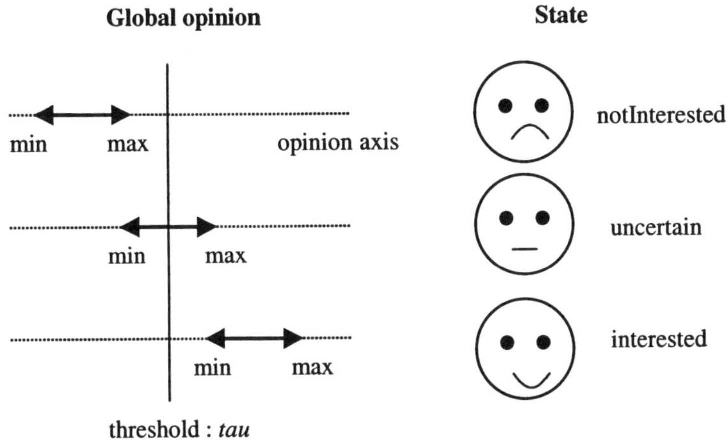

Figure 9.8: The values of the state attribute depend on the position of the global opinion segment compared to a threshold value tau (class variable)

9.3.2.5 Information update and economic evaluation (dataProcess class)

This class rules the modification of information when a message is received, the calculation of the personal evaluation which modifies the personal criterion. The class attribute *omega* is the probability to transmit the information during a discussion. The other class variable, *measureData*, is a table of parameters which are necessary for the economic evaluation (prices, subsidies …).

In our implementation, the information is represented by a table of boolean values. The information is updated (with a probability *omega*) only if the interest state of the farmer is *interested* or *uncertain*. The updated operation is done by a logical 'or' on the tables. If the information of the farmer changed, a new evaluation is automatically performed by the farmer, which can change the decision state (*selfEvaluation()* operation).

The operation *technicianEvaluation()* is called when the institution comes to visit the farmer, which is called the *receiveVisit()* operation in *decisionProcess*. Note that when an institution comes to a farm, we consider that all their information is transmitted to the farmer.

9.3.3 The Institution Agent

A more precise UML class diagram related to the institution is given in Figure 9.9. The institution's main attributes are the frequencies of interactions and visits to the farmers. The institution has a message about the measure, which is delivered during regular pair interactions or mass diffusion (meeting, publication, radio). The message can include mean values and uncertainties about the criteria (social and personal) and a set of information. The scenario of the institution is a set of actions, which the institution executes at the specified date.

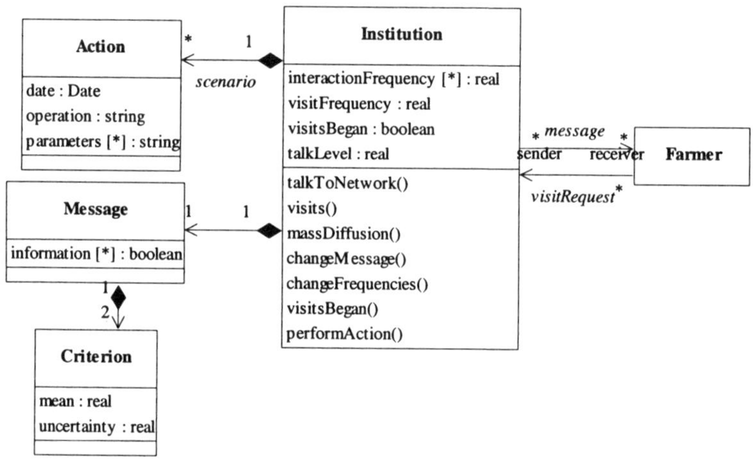

Figure 9.9: Class diagram focusing on the institution agent. A set of actions is associated with the institution. The actions can be of different types and they occur at a given date.

The institution can have a network of farmers with whom it has regular interactions. The operation *talkToNetwork()* sends the institution message to several farmers chosen in the institution network, taking into account the *interactionFrequency* attribute. The attribute *talkLevel* is chosen at 1 in the implementation, which supposes that when a farmer receives a message from the institution, he talks a number of times which is the integer part of :

$$(1 - gamma) * \text{card}(\text{network})$$

The operation *performAction(date)* selects the actions (if any), which date corresponds in the scenario, and these actions are performed by the institution. The possible actions of the scenario are:

- *massDiffusion()* allows the user to specify a set of farmers to which the diffusion is targeted (by default the whole population or the institution network), and the proportion of *notInterested*, *uncertain* and *interested* farmers that will actually receive the message. For instance, one may think that interested farmers are more likely to be reached by the message.
- *visitBegan()* puts the boolean attribute *visitBegan* to *true*, which triggers the operation *visits()* at each period. This operation selects farmers at random within the set of farmers who are in the decision state *visitRequest*, taking into account the frequency of visits, and calls the operation *receiveVisit()* of the farmer decision process.
- *ChangeMessage()* allows the user to specify changes in the message delivered by the institution. This might be the diffusion of new information, or change in the values of the criteria.
- *ChangeFrequencies()* allows the user to modify the frequencies of interactions with farmers of the institution network, or the frequencies of the visits.

This concludes the general description of the model.

9.3.4 Summarizing Table

Before addressing the application of this model to the conversion to organic farming in Allier, we recall the main parameters of the model's dynamics in Table 9.2.

Table 9.2: The parameters of the dynamics

Symbol	Type	Description
gamma	real in [0,1]	Rules the diffusion of the discussions in the farmer network. The target talk level is decreased of *gamma* when it is received. Therefore when gamma is close to 0, farmers discuss a lot, and when it is close to 1 there is almost no discussion.
mu	real in [0,1]	Rules the intensity of the direct influence on the criteria values and uncertainty. This concerns more specifically the social criterion because the influence on the personal criterion is indirect as soon as an evalution took place.
omega	real in [0,1]	Is the probability of transmitting the information during one discussion. A low value means that the consequences of the adoption are complicated to communicate.
tau	Real	Is the threshold defining the interest states.

9.4 APPLICATION OF THE MODEL TO THE DIFFUSION OF ORGANIC FARMING IN ALLIER

The application of the model to the diffusion of organic farming in Allier requires:

- generation of the social networks of the farmers and their institutional networks, which implies a spatial distribution of the farms be defined
- initialization of the criteria of the generated farmers,
- design of a simplified scenario of action of the main institutions which participated in the diffusion.

9.4.1 Spatial Distribution and Social Networks

The definition of the social networks is obviously related to the geographic distance between the farms. Therefore, the definition of a spatial distribution of our generated farm population is necessary for the application of the model.

9.4.1.1 Spatial distribution

We dispatched the farms generated with the procedure presented in Section 9.2 in the commune areas of the cantons, respecting the constraint that for each commune, the sum of the farm surfaces should be as close as possible to the global agricultural surface of the commune. We evaluated this surface from the census data of 1988, and we supposed that it stayed almost constant.

It was also important to locate the initial organic farmers, because we knew that the conversions were more likely in their communes. In 1992, there were around 40 organic farmers in Allier. We got a description of each of these farms, and its location. The description of the farm allowed us to find the most similar FADN prototype.

We added these prototypes to the generated population and localized in their commune. The change to the aggregated statistics is negligible.

These farms constitute a particular network, to which specific messages are addressed by specific institutions (Allier Bio association).

9.4.1.2 Social networks

During the project, we elaborated a schematic model of farmers' social networks, which were inspired by the results interviews. This model is based on the distinction between three types of links:

The neighborhood links link two farmers of the same village, connected to the fact that farming is a public activity. The interactions happen in day-today work, and have a frequency which is typically once to a few times a week.

The professional links connect with higher probability farmers with similar farming systems and in the same commune group, or related to the same larger city. The interactions happen in markets or professional organization meetings, and their frequency is typically around once a month.

Other links connect farmers separated by larger distance. These links can be established for many reasons (neighbors or close colleagues in previous periods, older friendship, family, etc.), and have a random character. Considering that these links are at a larger distance, the frequency of contacts is lower than the previous ones.

We adapted the algorithms proposed by Watts (1999) to generate the social networks. The parameters we need to determine the global networks are:

- The mean neighborhood's connectivity: n_l
- The maximum distance for neighborhood links: d_l
- The mean professional connectivity: n_p
- The equivalence relation for professional links: $\pi(i, j)$
- The mean random connectivity: n_o

The professional links are related to the distance of each farm to the nearest city, or market centre. The rationale is that the professional organizations, meetings and so on happen in these cities and structure the professional links between farmers.

From the interviews made during the project, it appears that an average number of neighbors between 2 and 3 and an average number of professional links between 1 and 2 are plausible values.

A frequency of interactions of around once a week for neighborhood links and once a month for the professional links seems relevant.

We added a small number of random links (0.1 on average), with a lower frequency of interaction.

9.4.2 Initial Values of the Mean and Uncertainty of the Criteria

9.4.2.1 Personal criterion

In the model, we considered that the personal criterion is mainly economic, and that the farmers did not have an economic opinion at the beginning of the process because organic farming was not well known. We only initialized the personal criterion of the initial organic farmers, which we put to 0 (no improvement of the revenue), with a very small uncertainty (0.01). This can be justified because initially organic farming was not considered as a means to improve revenue, but rather as a way of life.

The operations *farmerEvaluation()* and *technicianEvaluation()* of *dataProcess* in the model are based on by the economic calculation presented in Section 9.2. The evaluation by farmer is characterized by a higher uncertainty (which is a parameter of the model).

9.4.2.2 Social criterion

We chose to draw the initial mean value of the criterion from a normal distribution. We can identify the following constraints of the mean and standard deviation of this normal distribution:

- the values of –0.2 or +0.2 correspond to extreme positions, because when they are added with the financial evaluation in the aggregated function, it means that the agent is ready to pay or to loose 20% of his benefit because of his social convictions.
- The mean of the social opinion about organic farming is very probably negative at the beginning of the simulation (1992). The majority of the farmers consider organic farming as marginal practices.

In order to simplify the problem, we decided to initialize all the farmers with the same value of uncertainty, except the initial organic farmers who have a lower uncertainty than the others.

9.4.3 Institutional Process

We modeled the action of two institutions:

- Chambre d'agriculture, which is the local organization in charge of the implementation of the agri-environmental measure,
- the press, which represents globally the society in the model.

9.4.3.1 Chambre d'Agriculture

Chambre d'Agriculture recruited a technician specializing in organic farming in 1992. His role is mainly to provide technical and financial information to the interested farmers. We did not find publications or communications from the Chambre d'Agriculture for or against the principle of organic farming. Therefore, in the model the messages from Chambre d'Agriculture include only information and no values on the criteria. The information messages are boolean values which suggest that the farmer has enough knowledge to make a personal calculation of the impact, and if he is aware of the current subsidies.

The technician was also in charge of an information publication which was sent to the network of organic farmers (Allier Bio association). We considered that there were no regular interactions with the global population because, even though the majority of farmers have regular interactions with the Chambre d'Agriculture, these interactions are not opportunities for presenting organic farming.

We simplified the scenario of the institution Chambre d'Agriculture as follows:

Year	Action n°	Action type	Parameters
1993	1	massDiffusion	Target: AllierBio; message: regional subsidy
1994	2	beginVisits	
1994	3	massDiffusion	Target: population; message: first subsidy; effective diffusion to interested and uncertain
1995	4	massDiffusion	Target: AllierBio; message: second subsidy
1998	5	massDiffusion	Targets: population, AllierBio; message: no ceiling; diffusion according to interest state

9.4.3.2 Press (or society)

We modeled the global trend in the society which gives more importance to the quality of the agricultural products (following the BSE[3] crisis) and to the environment by a global source of messages which reaches at random all the farmers with the same frequency, and delivers a positive social message (for more information about the press articles devoted to organic farming see Deffuant (2001).

This led us to create a second institution, called press, which sends slightly positive social messages, with an increasing frequency. In the model, we considered a message from the press with no economic opinion, and a positive mean social opinion. We modeled the effect of the press as a regular interaction with an increasing frequency. Initially this frequency is 0, and the scenario is the following:

Year	Action n°	Action type	Parameters and considered values
1993	1	changeFrequency	Frequency : 2
1995	2	changeFrequency	Frequency : 6
1997	3	changeFrequency	Frequency : 20
1998	4	changeFrequency	Frequency : 35

We now describe the exploration of the model parameter space.

9.5 EXPLORATION AND CLASSIFICATION OF THE PARAMETER SPACE

9.5.1 Exploration Space

We decided to focus the exploration on:

- The main parameters of the dynamics: *omega*, *gamma*. We fixed the value of *mu*, because previous simulations showed that its role is less important.
- The parameters of the initial distribution of the social criterion.
- The average number of neighborhood and professional links.
- The variations of the institutional scenario.

Table 9.3 shows the parameter values that we chose to fix in the simulations. The reason of this choice is that often their modification is redundant with the modification of one of the explored parameters.

Table 9.3: Fixed parameters in the exploration of the model applied to organic farming in Allier.

Description of the parameter	value
Reflection time for adoption	1 week
Interest threshold (*tau*)	0.05
Average number of random links	0.1
Frequency of neighborhood interactions	0.1 times a day
Frequency of professional interactions	Once a month
Frequency of interaction in random links	Twice a year
Mean value of the press message	0.05
Uncertainty of the press message	0.1
Uncertainty of the initial organic farmers	0.01
Uncertainty percentage for farmer personal calculation	10%
Opinion influence intensity (*mu*)	0.3

Table 9.4 shows the values of the exploration in the parameter space. Most of these values were identified as relevant by thousands of initial exploratory simulations.

Table 9.4: Varying parameters in the exploration of the model applied to organic farming in Allier

Description of the parameter	Values
Mean of the initial mean social impact distribution (*m*)	-0.25, -0.2, -0.15, -0.1, -0.05
Standard deviation of the mean social impact distribution (*sigma*)	0.15, 0.2, 0.25, 0.3
Initial uncertainty on the social opinion of non-organic farmers (*U*)	0.05, 0.1, 0.15, 0.2
Information transmission rate (*omega*)	0.06, 0.2
Discussion diffusion (*gamma*)	0.3, 0.5
Average number of neighborhood links (n_l)	2, 3
Average number of professional links (n_p)	1, 2
Press	Yes , No

Moreover, we wanted to test the influence of the press, and therefore we introduced a scenario where there are no press messages.

9.5.2 First Selection: Parameters of the Initial Social Criterion Distribution

In a study made in 1991, in which the farmers were asked about their priorities for the five next years, 10% of the farmers envisaged converting a part of their production to organic. We decided therefore to select only the initial distributions of the social criterion which led to between 8% and 12% of those interested in the initial population.

We tested the 120 parameter configurations of the initial distribution, and repeated them in 35 trials. The ten configurations which give a number of interested farmers which lies between 8% and 12% for at least 90% of the 35 trials are given in Table 9.5.

Table 9.5: The selected configurations of the initial distribution of the social criterion

Config.	1	2	3	4	5	6	7	8	9	10
M	-0.1	-0.15	-0.05	-0.1	-0.15	-0.25	-0.05	-0.1	-0.2	-0.15
Sigma	0.15	0.2	0.15	0.2	0.25	0.3	0.2	0.25	0.3	0.3
U	0.05	0.05	0.1	0.1	0.1	0.1	0.15	0.15	0.15	0.2

These configurations lead to a variety of initial proportions of *notInterested*, *uncertain*, and *interested* in the population (see Figure 9.10).

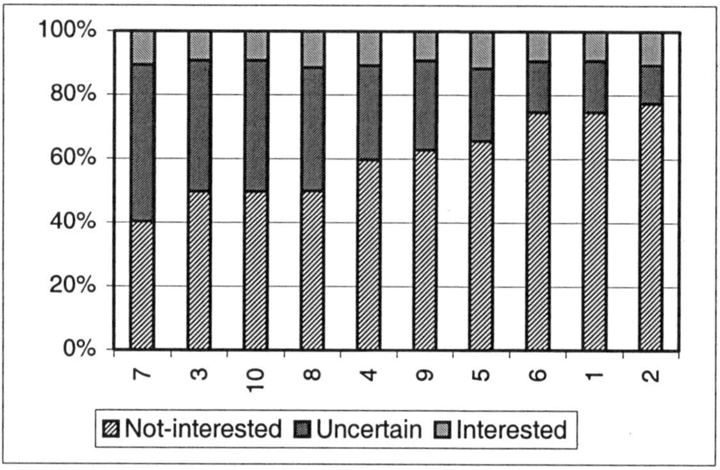

Figure 9.10: The initial proportions of notInterested, uncertain and interested in the population, for the ten considered parameter configurations of the initial distribution of the social criterion.

9.5.3 Classification of the Parameters by a Decision Tree on a Composite Error

For each of the ten selected parameter values for the social criterion initial distribution, we ran all the possibilities defined by the variations on the dynamics, networks and institutional scenarios. Each parameter configuration being repeated 50 times, this exploration led to 23800 simulations.

For each parameter configuration we compute two errors: the adoption error and the error on the proximity of adopters to the initial organic farmers.

The adoption error is the quadratic sum of the relative difference between the number of adopters obtained from the simulation and the real adoption number at three characteristic dates. More precisely, if we call E_{adopt} this error, $s_{ad}(t)$ the number of adopters given by the simulation at the date t, and $r_{ad}(t)$ the real number of adopters at this date, we have:

$$E_{adopt} = \sum\left(\frac{s_{ad}(t) - r_{ad}(t)}{r_{ad}(t)}\right)^2$$

The proximity to the initial adopters was evaluated from the real data of adoption: we counted the probability for an adopter to live in the same commune as an initial organic farmer (we got the commune of each new adopter). This probability is 0.34, whereas on average, a farmer has a probability of 0.14 to be in the same commune as an initial organic farmer. More precisely, if we call E_{prox} the error on this probability, s_{prox} the computed from the simulation and r_{prox} the probability computed from the real data, we have:

$$E_{prox} = \left(\frac{s_{prox} - r_{prox}}{s_{prox}} \right)^2$$

In order to classify the parameter zones according to their performance relative to this error, we decided to use a decision tree (Breiman *et al.*, 1984). This tool has the advantage of providing an easy interpretation of its results, in terms of the sensitivity of the results to the parameters. The algorithm selects the parameters and the values which correspond to the highest decrease of deviance, which can be interpreted as a sensitivity to this parameter.

We defined a learning set with inputs the values of the parameters, and as output:

if $(E_{adopt} < 0.3) \wedge (E_{prox} < 0.3)$ (good simulation): output 1
otherwise (bad simulation): output 0

This learning set constrains the tree to focus on the zones where the simulation has an acceptable performance. The simulations considered as good can have around 50% of difference from the target values. Therefore, we do not aim at selecting simulations which precisely fit our data: with all the approximations made in the model, this would not be coherent. Our aim is rather to identify parameter zones which seem compatible with these data, taking into account the large uncertainty we have on several parts of the model.

The result is presented on Figure 9.11. The values at the leaves of the tree represent the fraction of good simulations.

The most noticeable aspects of this result are, going down in the tree until the best leaf:

- The first most discriminating parameter is the press. Without the influence of the press (the society), there is almost no simulation which gives good results.
- The second most discriminating variable is *omega*, and the best results are obtained for the lower value (0.06). This corresponds to low probability to

transmit the information about the measure (or equivalently, the necessity of many contacts to insure the transmission).

- The third most discriminating parameter is *m* mean of initial distribution of the mean values of the social criterion. The best result is obtained for the lowest value of this mean (i.e. the initial configuration which is on average the most negative against organic farming).
- Finally, the networks which have more links are preferable to get the best result (two neighborhood and two professional links, or three neighborhoods and one professional).

The parameter *gamma* does not appear in the tree, which suggests that its role is not important at this level of analysis.

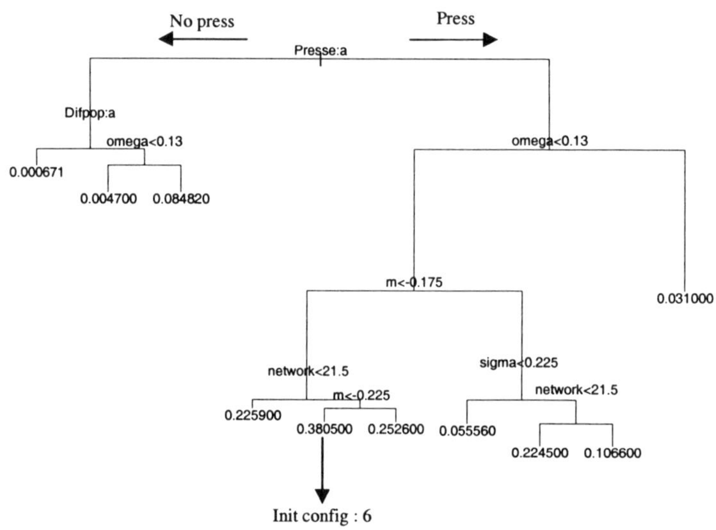

Figure 9.11: Decision tree built on the learning set with inputs the parameters of the simulation, and as output 1 for 'good' simulations (relatively to the error criterion) and 0 for the others. The top left branch corresponds to 'no message from the press'. The leaf maximizing the result (indicated by the vertical arrow) corresponds to the initial configuration of the social criterion n°6 (see Table 9.6).

9.5.4 A Closer Look at Good Simulations

A more refined understanding of the dynamics can be obtained by considering the evolution of the criterion distributions. Figure 9.12 shows an example of the social criterion evolution in the parameter zone identified by the tree.

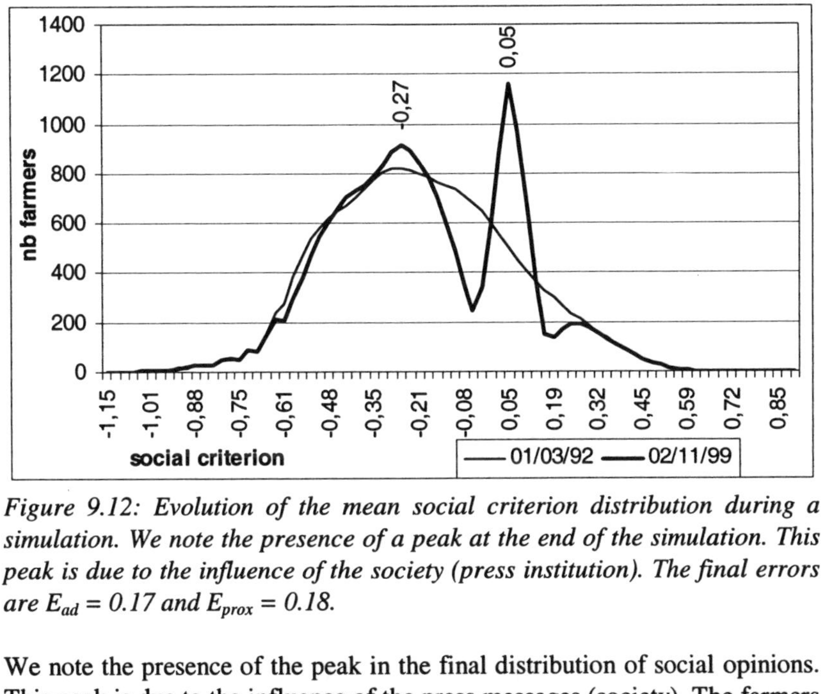

Figure 9.12: Evolution of the mean social criterion distribution during a simulation. We note the presence of a peak at the end of the simulation. This peak is due to the influence of the society (press institution). The final errors are $E_{ad} = 0.17$ and $E_{prox} = 0.18$.

We note the presence of the peak in the final distribution of social opinions. This peak is due to the influence of the press messages (society). The farmers of this group were sensitive to the messages of the society and the specialized press. However, this relative sympathy toward organic farming was not sufficient to make their decision.

It was important to verify if this peak is necessary for satisfying the constraint on the error. To do so, we ran simulations with messages of the press without contents (and the values of the other parameters corresponding to one of the best simulations). We also obtained good results on the error; therefore, the presence of the peak is *not* a necessary condition for getting a good result. In this case, the distribution of the social criterion is much more stable (see Figure 9.13).

The conclusion is that the press is mainly important because it launches discussions among the farmers, through which they are more likely to talk with farmers who have information. Therefore interested farmers who are close to the initial organic farmers are more likely to adopt, because they are more likely to get information, and this insures a good value for E_{prox}. This explanation is confirmed by the study of the number of farmers who got the information at the end of the simulation. This number is around 150 for these simulations, whereas it is about 40 when the press is totally inactive.

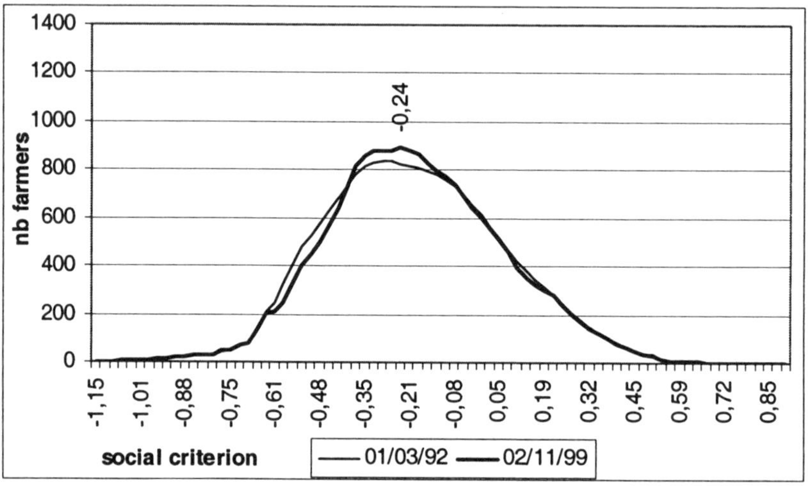

Figure 9.13: Evolution of the social criterion distribution when messages of the press do not contain the social criterion (their effect is simply to increase the discussions about organic farming among farmers). This simulation gives good results (E_{ad} = 0.09 and E_{prox} = 0.08).

9.6 DISCUSSION AND CONCLUSION

The study of the simulations satisfying the error constraints shares the following characteristics:

- The transmission of information about the conversion is difficult. Sustained interactions with an organic farmer are necessary to be able to make a personal evaluation. The number of farmers who really got the information, and are able to evaluate the impact of a conversion on their farm is rather low (about 150).
- The role of the society is crucial, but does not lead directly to people converting. Its main role is to sustain the exchanges among the farmers, and in particular with the initial adopters or members of the Allier-Bio network.

- The results suggest that a group of farmers sympathetic toward organic farming might have been formed, because of their sensitivity to the positive messages of the press. However, this sympathy is not enough to lead them to a conversion. They would be more open to an evolution of their practices, but less radical.
- The farming population in Allier was on average rather skeptical (maybe hostile) toward organic farming at the initialization.

All these conclusions fit quite well our qualitative knowledge of the process. It is particularly noticeable that, using very general constraints (on the number of initially interested farmers, the final number of adopters and their spatial distribution), we select quite relevant characteristics which were not explicitly included in the selection. This result tends to prove that the structure of the model brings some added value, which can be interesting in an ex-post analysis.

However, it is still difficult to evaluate the robustness of these conclusions, and whether they would be sufficient when using the model for a prospective purpose. Some improvement of the personal impact evaluation, including the technical side of the problem would certainly strengthen the global approach. Moreover, a further exploration of the parameter space could bring new perspectives.

ACKNOWLEDGEMENTS

We would like to thank all the members of the IMAGES project: F. Amblard, S. Bernard, L. Dobremez, E. Chattoe, S. Geromin, N. Gilbert, S. Labonne, M. Lazzari, F. Mazzetto, E. Perret, L. Piani, S. Skerratt, for their participation to the elaboration of the model and helpful discussions.

NOTES

1. The research has been carried out with financial support from the *Commission of the European Communities, Agriculture and Fisheries (FAIR) specific RTD programme.* It does not necessarily reflects its views and in no way anticipates the Commission's future policy in this area.
2. Farm Accountancy Data Network
3. Bovine Spongiform Encephalit

10. Scientific measurements and villagers' knowledge: an integrative multi-agent model from the semi-arid areas of Zimbabwe

Tim Lynam

10.1 INTRODUCTION

In this chapter an approach to the development of a multi-agent agro-ecosystems model of a complex agro-ecosystem in the semi-arid areas of Zimbabwe is presented, as is the model itself. The model was developed as part of a research program designed to improve our understanding of the structure and behavior of the agro-ecosystem and to assess the sustainability of human needs satisfaction from the agro-ecosystem. The modeling approach was also part of a developing methodology for the analysis of complex agro-ecosystems where little formally measured data are available. More specifically, the model was designed to examine the interactions among agro-ecosystem components (social, economic and ecological) and in so doing identify those which significantly contributed to a household's ability to satisfy its needs. It was hoped that the model might assist in evaluating policy and household decision options from both a household and a community perspective. The field work was conducted between 1990 and 1992 and the model and analyses were completed in 1993.

In developing this model, it seemed that a model of multiple, interacting agents was necessary – field observations in Zimbabwe had clearly indicated the importance of inter-household interactions as had social scientists working in rural Africa (e.g. Hyden, 1986). The economic models available at that time were of perfectly rational, equilibrium oriented, single entity households whose activities were seemingly uninfluenced by the community of households in which they existed (see e.g., Berdegue *et al.*, 1989; Singh *et al.*, 1986; or Low, 1986). Whilst conceptually an improvement on the economics models, the social or anthropological models (e.g. Hyden, 1986) were for the most part qualitative or conceptual models that did not provide the necessary analytical power required for the analysis of Zimbabwean agro-ecosystems.

It was evident that successful analyses of Zimbabwean agro-ecosystems required an integrated understanding of the social and the ecological dimensions of these systems as well as the many interactions between these two broad system components. Models of that time generally involved single disciplines (usually either economics, agriculture or anthropology) and where there was integration across disciplines (e.g. Fresco, 1986), no manipulatable models were developed – the analysis was either qualitative or descriptive.

For this particular research, a multi-agent systems model was required that integrated the ecological and social aspects of a spatially and temporarily variable and complex agro-ecosystem in which multiple scales of analysis were important. With insufficient resources (including time) for extensive data collection a model was needed that could be developed and validated using the knowledge of local managers as well as knowledge generated through formal scientific data collection and analysis processes. The model presented here is the writer's solution to those problems.

As with many large-scale modeling activities, the process of building the model often alters the perceptions, beliefs and understanding of the modeler. The process of developing large and complex models is one of iterative adaptation; in any given situation were we to start over again, most of us would assuredly do it differently. In presenting this particular model, the researcher seeks to not only describe an approach and a model but also to reflect on the key lessons of that process and what insights the wisdom of ten years hindsight provides.

It is not the role of this chapter to demonstrate the validity of local people's knowledge – this issue is dealt with in greater detail in Lynam (1993) and also in Cumming and Lynam (1997). Similarly, the many field data collection methods employed are not described as these are clearly documented in Cumming and Lynam (1997) and in Lynam (1993). The context of the analysis reflected that of participatory management of natural resources, the underlying principle of which utilized the CAMPFIRE[1] paradigm (Martin, 1986; Peterson, 1991) from Zimbabwe. CAMPFIRE was based on the premise that if local people controlled the benefit streams derived from the resources within their village areas (for the most part this meant wildlife but conceptually was extendable to all natural resources) then they would have the incentives to manage these resources for sustainable use.

In 1988, Guruve was one of the first districts in Zimbabwe to begin exercising authority over its wildlife resources. This was formalized in 1991 when Appropriate Authority status was granted by the central government in terms of the 1975 Parks and Wild Life Act. With Appropriate Authority status, district councils have control over the use of their wildlife resources, with any income generated by wildlife utilization projects being available to the district council

(Cumming, 1991; Peterson, 1991). During 1989, the Kanyurira community of Guruve District, living in Masoka, developed a wildlife land use plan (Kanyurira Wildlife Committee, 1989). Based on this plan, the community chose to fence slightly more than $14km^2$ along the south bank of the Angwa River, and to limit settlement and cultivation to within this area. At the time of this research project (1990–93) virtually all of the households, except those of the VaDema[2] and one of the original Korekore households, lived and carried out most of their cultivation within this fenced area.

The chapter begins with a brief description of the study area to illustrate the variability and complexity with which village managers must cope. The structure of the model and the way it was constructed are thereafter described. The model was designed to enhance understanding, not as a predictive tool. The major use of the model was therefore as a manipulatable representation of a real world system. The chapter ends with presentation of the results of a rigorous sensitivity analysis and concludes with some general lessons on modeling such complex agro-ecosystems.

10.2 THE MASOKA AGRO-ECOSYSTEM

One of the defining characteristics of savanna ecosystems is high spatial and temporal variability. The small-scale agro-ecosystems of the savanna region are similarly variable. The managers of Communal Areas[3] seek to achieve a diverse and complex mix of objectives (Cumming and Lynam, 1997) in socially, economically, politically and economically complex environments subject to continuous change. Kanyurira Ward (Masoka), situated in Guruve District, Mashonaland Central Province, Zimbabwe, between 30° 10' E and 16° 15' S is no exception (Figure 10.1). The political environment has been the subject of several changes in the past two decades with traditional leadership systems being replaced by semi-democratic systems at independence in 1980 (following a 10-year war of independence) only to have these replaced by the traditional systems in 2000. The economic climate is similarly variable and unpredictable. The post-independence Marxist government maintained the existing price controls on most agricultural commodities only to remove these in the late 1980s and early 1990s as part of a structural adjustment program. In 2001 the government reintroduced price controls on many basic commodities.

Climatically the semi-arid Zambezi Valley, in which Masoka is located, has a greater than 30% coefficient of variation on a mean annual rainfall of 725mm and an evapotranspiration rate that exceeds the precipitation rate in most months of the year. The alluvial soils, on which much of the local agriculture is dependent, are spatially variable with some of them becoming highly erodible with cultivation (Cumming and Lynam, 1997). Although it covers a large area

(400km^2) not much of the Ward is suitable for agriculture. The Masoka community set aside most of the Ward for wildlife production (Cumming and Lynam, 1997). The set aside land is leased, through the Rural District Council, to safari companies who pay for the use of the land as well as for animals hunted in the area. The per household incomes from this source at times exceed average annual household income from all other sources combined.

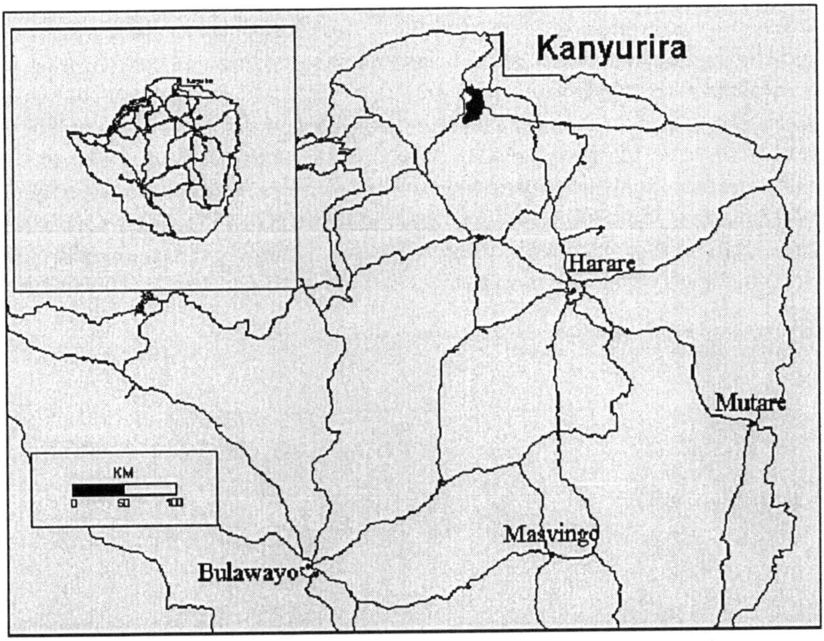

Figure 10.1: Map showing the location of Kanyurira Ward in Zimbabwe

Traditionally, households satisfied their food and cash needs largely from rainfed crop production and illegal hunting.[4] Crop production was based on grain crops, primarily maize and sorghum inter-cropped with cucurbits such as watermelon, pumpkin and cucumber. Less commonly grown were small plots of groundnuts and sweet potatoes. Mean total areas cultivated for small, medium and large households were about 1.3, 1.8 and 2.2ha respectively. Dry season production was limited to vegetable crops grown in small gardens along the banks of the Angwa River that were irrigated from shallow wells dug into the sand of the river bed. Cotton was introduced into Masoka in the mid-1970s and rapidly became a dominant crop, with 73% of the households growing it in 1988 (Cutshall, 1989) but only 57% growing it in 1991. Fewer than 7% of the households reported using inorganic fertilizers in the 1990/1 growing season.

Soil fertility in the older alluvial terraces was restored with a tree fallow dominated by *Acacia tortilis* (subsp. *heterocantha*). In the recent alluvium outside the fenced area, field fertility was restored by the periodic flooding of the Angwa River.

The 14km^2 fenced area (Figure 10.2) is of primary concern in this study. The soils of this area are mostly alluvial deposits of varying ages and were mapped using 1:9750 air photography. Four major land classes, comprising four soil types, were identified with the local (i.e. from Masoka) terminology for soils used throughout this presentation. The mopane woodland area comprises deep, medium textured sandy loams with small, localized patches of fine textured sodic soils. The alluvial or river terraces slope away from the original river bank. Soil texture on these terraces grades from medium textured sands (*jecha*) near the historical river bank to fine textured sandy clays (*bepe*) on the down-slope edge of the terrace. The river terraces (river-*jecha*) have been the focus of traditional agriculture in Masoka. More recently however, settlers have been moving into areas of elevated old alluvium (*jecha*) and even on to some sodic areas (*shapi*).

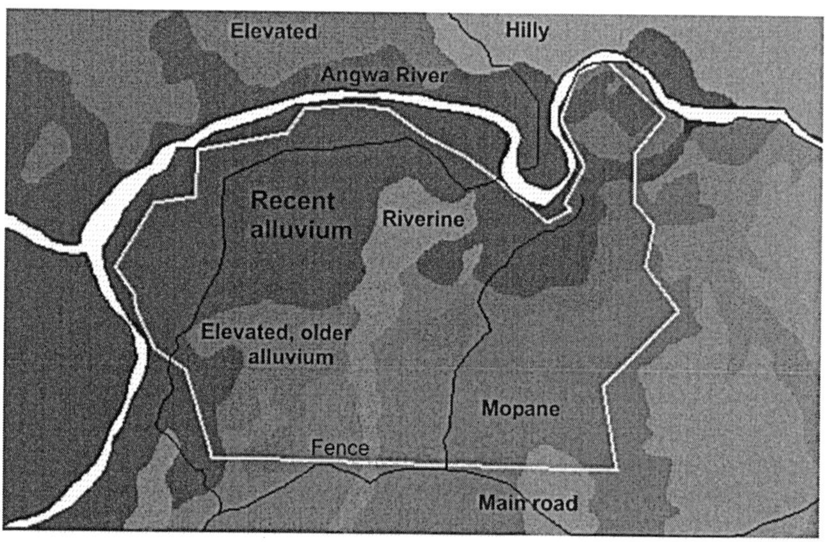

Figure 10.2: Masoka fenced area with roads and Angwa River overlain on different soils types. Basic map data used in simulations.

10.3 MODEL DESCRIPTION

10.3.1 Overall Approach

The temporal and spatial scale at which the system was modeled was expected to have a significant impact on the results that were produced and the interpretations that could be made (Firbank, 1993; Meentemeyer and Box, 1987; Wiegart, 1988; Wiens, 1989; Wiens *et al.*, 1986). The primary scale of interest was the scale at which Masoka residents manage their resources. Essentially, this was the scale of the field, usually between half an acre and one acre.

Spatial heterogeneity was believed to be an important determinant of agro-ecosystem behavior (Baudry, 1993; Sauget and Balent, 1993) and the model was, therefore, designed to use real data of agro-ecosystem spatial structure as far as was possible. Spatial interactions, in the sense of the state of one location in space being dependent on the states of neighboring locations, were not included in this version of the model except in the sense of human (agent) interactions. Households were defined in space, and because households interacted, their interactions were seen as being spatial.

Knowledge of the objectives and experiences of Masoka residents was an essential building block for understanding the agro-ecosystem. It is important to recognize that in conducting analyses of complex systems, about which little formally measured data are available, the use of local expert knowledge may be the only information source available. The approach to developing a complex agro-ecosystem model presented here relied to a large extent on the knowledge of local villagers. Thus although many of the scientific (i.e. formal) measures were made at only one point in time, the local knowledge extended for a much longer period of experience. Wherever it was feasible probability distributions were used *in lieu* of time series data.

10.3.2 Model Overview

The Zambezi Valley Agro-ecosystem Model, (ZVAM) comprised three main sub-models: An initializing model (INITIAL), a rainfall generating model (RAINGEN) and an agro-ecosystem simulation model (MASOKA1). The focus of this chapter is on the MASOKA1 model. Detailed descriptions of the RAINGEN and INITIAL models are available elsewhere (Lynam, 1993). However, a simplified description of the INITIAL model is provided.

Both INITIAL and MASOKA1 used ASCII data layers (maps) developed using the IDRISI geographical information system (GIS). GIS data layers were in raster format with cells (pixels) of 64m (approximately one acre, a common field size in Masoka). INITIAL was used to generate a number of households, each with a randomly generated number of adults, children, upland cells and riverland cells[5] where the random values were generated from field observed distributions.

INITIAL also placed households on the Masoka landscape and located household fields around the household site (non-riverland cells) and along the river. Output files from INITIAL were used as input files to MASOKA1.

RAINGEN was used to generate an annual rainfall amount based on a gamma distribution. The rainfall was written to an ASCII file for use by MASOKA1.

MASOKA1 read in household and biophysical data from the files created by RAINGEN and INITIAL as well as the slope file created in the GIS. Up to 300 households could be simulated at a time over an area of about 1500ha and with an annual time step. The population of households and the number of adults and children in a household were kept constant within any simulation. The number of fields that households owned also remained constant in any simulation. Households could vary the number of their fields they planted in any season and the crops which they planted. Values in each cell (n=3695) of the landscape (soil erodibility, crop yield, yield potential, crop cover) were updated once each growing season (October to April). MASOKA1 used a soil erosion model developed by Elwell (1980) and estimated erosion for each cell.

10.3.3 Initializing Model (INITIAL)

The purpose of INITIAL was to generate households with size and land holding characteristics randomly drawn from observed distributions and then to place these households on a GIS data layer of Masoka that could be read into MASOKA1 as an input file. The total population of households was set, as an experimental parameter, at the start of each simulation. INITIAL also generated random cropping histories (the number of years cultivated) and vegetation cover (%) for each cell in the landscape.

Prior to running INITIAL a household site suitability data layer and a field suitability data layer were created using the GIS. The suitability index of a site for household residence was based on distance from the Angwa river (the primary source of water for most households), the soil type and the slope.

A data layer of a field suitability index was generated based on soil type and slope. Areas with slopes less than 6% were scored with a one and all other areas with a zero. Bepe, jecha and elevated area soils were scored with a one, riverland jecha scored with a two and all other soils scored with a zero. The field suitability index data layer was created by multiplying these two data layers (slope and soil) together. Sites were therefore classed as being zero, one or two. Non-zero classes were suitable for cultivation. By giving riverland and non-riverland cells different identifiers they could be differentially allocated to households.

Once executed, INITIAL searched for the lowest values of the household site suitability scores (i.e. the best sites) and placed households on those cells. Then, in a clockwise direction, INITIAL would search for cells immediately adjacent to the household site that had a cultivation suitability score of one (i.e. were non-riverland) and were not already owned. For households that had riverland fields, INITIAL searched row by row across the data layer until available riverland cells (cultivation suitability score of two) were found to allocate to that household. The data layer or map of fields for every household was saved for use as an input file to MASOKA1.

The number of cells of land to be cultivated and the number of riverland cells to be cultivated by each household were randomly generated from the distributions observed in Masoka. The distributions of areas cultivated by number of adults were taken from questionnaire survey data. The number of fields (hectares) of riverland cultivated per household was derived from data provided by Village Representatives[6] (VRs). These data were a list of all households that owned or used riverland fields, the number of adults and children in the household and the size of the riverland fields. The number of adults and children in each household was randomly generated from distributions derived from the questionnaire survey data. The number of non-riverland cells that a household cultivated as well as the number of riverland cells that a household had under its control could be increased or decreased using a separate multiplier for each land type.

Both cropping history and vegetation cover initialization values were randomly generated from distributions based on field observations and measures taken from air photographs.

10.3.4 Agro-Ecosystem Model (MASOKA1)

MASOKA1 read in five GIS data layer files (slope, soil type, field and household sites, vegetation cover and the number of years of cultivation) and two ASCII data files (household data and rainfall). Conceptually, the model had seven major sub-systems or components: a STORAGE sub-system, an ALLOCATIVE or decision-making sub-system, a LABOUR sub-system, a CROP sub-system, an EROSION sub-system, an ECONOMIC sub-system and an OUTPUT sub-system.

10.3.4.1 The STORAGE sub-system

In the STORAGE sub-system household needs were established[7] and household production (maize, cotton and sorghum) was compared with these needs. Production could be converted from one form to another based on the prices used in the model. Production in excess of needs was placed in storage, from which there was no loss. If household needs could not be met from production an attempt was made to meet them from storage (including using cash stores to purchase needs). If this was not possible the household was considered as having a deficit for that particular need.

10.3.4.2 The ALLOCATIVE sub-system

The ALLOCATIVE sub-system allocated each household to a production and to a tillage strategy. Fixed proportions of households were allocated to each strategy at the beginning of each simulation – the proportions were maintained, within the model, as experimentally manipulatable parameters. Households maintained the same strategy for the duration of the simulation. The production strategy governed what crops a household grew, the proportion of cells that the household planted to each crop and household fallowing practices. Tillage strategies governed whether a household used hand or tractor cultivation (tractor cultivation was only used on cotton and had no impact on yields) and whether a household planted with, or at an angle to, the contour. These factors were important only in their effect on erosion.

Households could be assigned to one of four different production strategies in any simulation. These strategies differed in the parameters that defined: a) the proportion of fields households planted to each crop; b) the number of years a household cultivated a field before leaving it fallow; and c) the proportion of fallowable fields households actually left fallow each year.

Households planted a fixed proportion of their fields to each crop in each year. These proportions were fixed for each production strategy. For sensitivity analyses only two strategies were used: income maximizers (INCMAX) and needs satisfaction variance minimizers (MINVAR). Income maximizing households produced a higher proportion of cotton (50% of cells on all soil types), planted little sorghum (15% of jecha cells) and used shorter fallow periods (minimum cropping period four years, maximum of ten years) than other households. Variance minimizing households planted mostly maize (50% of bepe and river jecha and 35% of jecha cells) with a little cotton (50% of bepe and river jecha cells) and sorghum (35% of jecha cells) and used longer fallow periods (minimum cropping period of four years and a maximum of seven years).

The decision whether or not to leave a particular field fallow was only pertinent on jecha soils; neither bepe nor riverland soils were left fallow. The decision to leave a field fallow was a probability based on the household production strategy and the number of years the field (cell) had been cultivated. In Figure 10.3 the cumulative probability curves of a household leaving a jecha field fallow are shown for each production strategy. An income maximizing household, for example, would not leave a field fallow before the field had been cultivated for at least four years. Thereafter, the probability of leaving a field fallow increased with increasing cultivation until, after ten years of cultivation the probability of leaving the field fallow was one. Other production strategies used different cumulative probability curves reflecting what the researcher believed were reasoned approximations of some likely approaches to fallowing for a given production strategy. Although subjective decisions, these probability curves were based on detailed analyses of fallowing conducted with the Village Representatives.

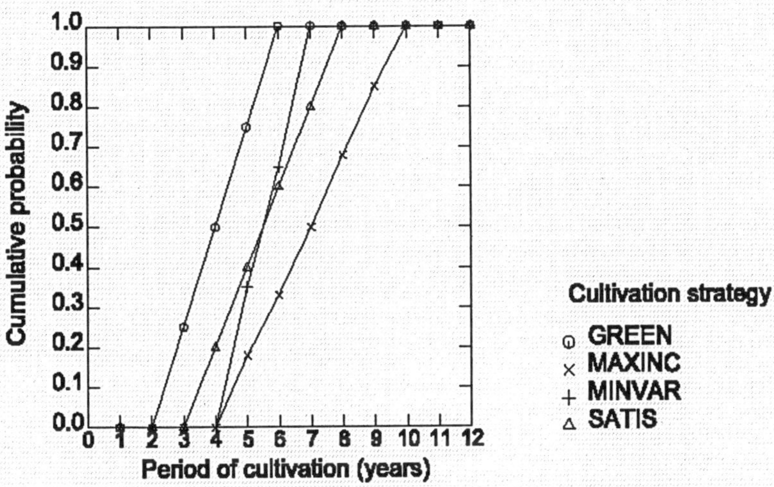

Figure 10.3: The cumulative probability of leaving a jecha field fallow for households using income maximizing (MAXINC), variance minimizing (MINVAR), needs satisfying (SATIS) and environmentally protective strategies (GREEN).

10.3.4.3 The LABOUR sub-system

In the LABOUR sub-system available household labor for each month of the growing season was estimated by multiplying the number of adults by 25 working days in the month. Based on Reynold's data (Reynolds, 1991) 44% of children were considered to be in the age group that contributes to the labor force (10 to 20 years of age). Children were considered to provide labor equivalent to 42% of an adult. This value was a mean of the proportions given by Reynolds (1991) for clearing, planting, re-planting and weeding.

Once a household's fields had been planted to a crop the LABOUR sub-system identified the labor requirements (person days) in each month of the growing season (October to March) based on the Village Representative labor budgets.[8] These requirements were compared to household labor availability. If households were labor deficit and they had cash or grain balances they could search for available labor from other households. Households that had labor greater than their needs and were deficit in cash or maize could supply labor to households that were labor deficit. Labor was hired from other households at the local wage rate (Z$2.54 day^{-1}) in cash or grain equivalent calculated using the Masoka prices for maize. Sorghum was not used to pay for hiring labor.

If households were still labor deficit, either because labor was not available or through not having the resources to hire labor then the yields of each crop were adjusted. Where labor deficits occurred the resulting yield reductions were first allocated to sorghum, then to cotton and finally to maize; these being the local ranking of crop importance. Thus, if the total labor deficit was less than the total labor required for sorghum then only the sorghum yields were corrected. If the total labor deficit was greater than the labor required for sorghum then as much of the deficit as possible was used on sorghum before yield reductions were made on cotton. The reduction in yield due to labor deficits was calculated as the product of the proportion of labor needs that were deficit times the yield reduction for that time period, expressed as a proportion. Yield reduction percentages were taken from the literature for each crop and time.

The local trade in labor was the process through which households interacted.

10.3.4.4 The CROP sub-system

Yields for three crops (maize, cotton and sorghum) were generated in MASOKA1. Households planted their fields (cells) to one of these crops (or left the cell fallow) each year. Yields, for each of these three crops, were a function of soil type, time since fallow (on jecha soils only) and rainfall season. Yield functions were derived from the yield possibility diagrams[9] developed, by the Village Representatives, for each crop, soil and season combination.

To derive a rainfall season from the annual rainfall amount generated by RAINGEN, membership functions for good, average and poor seasons were derived using fuzzy set membership functions. Good, average and poor seasons were considered to be fuzzy sets. The membership functions described the relationship between annual rainfall amount and the degree of membership in each of these sets. Annual rainfall amounts had graded membership, from zero (not a member) to one (full member) in one or more of these sets. The data on Masoka resident classification of each season were used to derive the membership functions for good and poor seasons.

Functions were fitted to each yield cumulative probability distribution generated by the Village Representatives. The general form of these functions was as follows:

$$Y_{ijk} = a \cdot e^{-b \cdot p} + c \cdot e^{-d \cdot p^2} \tag{10.1}$$

where

Y	=	the yield in kg ha^{-1}
i,j,k	=	the *i*th crop grown on the *j*th soil in the *k*th season
a,b,c,d	=	parameters fitted to the function
e	=	the base of natural logarithms
p	=	a pseudo-random number between zero and one.

For crops grown on jecha soils, soil fertility had an important effect on yields. If a soil had been previously cultivated, yields were derived using the following form:

$$Y = Y_{5yr} + (Y_{virg} - Y_{5yr}) / Y_{rsclt} \qquad (10.2)$$

where

Y_{5yr}	=	the yield from a soil continuously cultivated for five year
Y_{virg}	=	the yield from a virgin soil
Y_{rsclt}	=	the number of years the soil had been previously cultivated.

10.3.4.5 The EROSION sub-system

The Soil Loss Estimation Model for Southern Africa (SLEMSA) developed by Elwell (1980) was used in MASOKA1. SLEMSA calculates an index of soil loss at a point but does not estimate the distribution of this loss in space. Measures were made of soil erodibility (Fb) using rainfall simulations for virgin and cultivated soils collected from the field. These measures were then used to parameterize SLEMSA.

Erosion was estimated for each cell in the Masoka landscape and then summed to give a total landscape erosion (t ha^{-1}), a mean and standard deviation (t ha^{-1}) as well as the mean and standard deviation, per hectare of cultivated land for each household. Estimated soil loss did not affect soil properties or yields.

10.3.4.6 The ECONOMIC sub-system

It was assumed (within the model) that households were able to sell all of the cotton they produced, to the Cotton Marketing Board (CMB), as well as all of the maize and sorghum they wished to sell. Maize and sorghum were only sold within Masoka and were traded at the 1990/91 season prices used in Masoka: Z\$0.78 kg^{-1} for maize and Z\$0.55 kg^{-1} for sorghum. Labor was valued at the local (Masoka) daily wage which varied between Z\$2.00 and Z\$3.44 per day. The average value (Z\$2.54 day^{-1}) was used as a constant in the model. Hired labor could be paid in cash or the grain equivalent of the daily wage. All households were assumed to face the same costs, though income maximizing households that hired a tractor for ploughing had to pay the going rate of

Z$148.00 ha^{-1}. Households growing cotton were assumed to use the cotton packs supplied by Agritex which included all fertilizer and chemicals for 0.4096 ha (one acre). In allocating crops to a cell (ALLOCATE sub-system) a test was made to determine if the household had sufficient cash or maize reserves to pay for the basic costs of producing one cell (0.4096ha) of cotton. If the household did not have reserves then no cotton was planted and maize was planted instead. This value was set at Z$155.00.[10] Only the variable costs were used in calculating this amount and it excluded weeding and ploughing costs. A zero/one switch could be set in the model that would disallow/allow households to obtain loans for this amount and up to a maximum number of cells. The number of cells for which loans were obtained were recorded and later used to update the returns to cotton. No interest was charged on these loans.

A partial budget format was used to calculate the economic returns to land, to labor and to initial investment for each crop. Returns to labor were calculated as the returns to household labor and management.

10.3.4.7 The OUTPUT sub-system

The OUTPUT sub-system facilitated sub-sampling of individual households, yields, Fb values or erosion periodically during any simulation experiment. Thirteen output files were produced as required and these included both spatial data (i.e. GIS images) as well as attribute data from the simulations.

10.4 TESTING AND USING THE MODEL

10.4.1 Verification, Validation, Usefulness

Given the objective of an enhanced understanding of the Masoka agro-ecosystem, evaluation of model performance was more concerned with model usefulness than with the accuracy of predictions. The model was considered to be useful if it did any of the following:

- Demonstrate the utility (or lack thereof) of the approach to analyzing complex agro-ecosystems. Did the model provide insights that would not have been likely or possible with any other analytical approach?
- Provide a tool to identify and explore potential management and/or policy actions that could lead to improvements in the satisfaction of household needs and/or community well-being;
- Help identify gaps in the understanding of the Masoka agro-ecosystem or that class of agro-ecosystem of which Masoka was an example;
- Enable the comparison of different agro-ecosystem configurations in a design situation.

It was assumed, *a priori*, that the model was an inaccurate representation of the real world system. As far as was possible major components of the model were tested to establish the degree to which they accurately represented the real world processes of interest. This was difficult however, as there are no measured data on crop production, erosion or on household needs satisfaction.

The model presented here was not expected to withstand rigorous deductive testing. Rather, it was developed as a working hypothesis with which to focus discussion and further research. Given the complexity of the Masoka agro-ecosystem, the model was neither complete nor sufficiently accurate to facilitate making reliable predictions of Masoka's future. The model was expected to provide useful information on the direction and perhaps relative magnitude of changes in the agro-ecosystem. The model was also expected to provide an estimate of the human carrying capacity of the fenced area under defined conditions of technology and locally defined human needs.

Evaluation of the Masoka agro-ecosystem model broadly followed the procedures described by Law and Kelton (1991): Input data were checked for correctness and representativeness. The distributions of output variables generated by the model were tested for goodness-of-fit to the original distributions using Chi square, log-likelihood and Kolmogorov-Smirnov tests. Model outputs were compared to observed system data where these existed or to theoretical distributions where appropriate.

The total area cultivated each year in simulations was found to fluctuate notably in the first few years of a simulation. For all simulations with MASOKA1 the data from an initial warm-up period of 20 years were discarded. The length of this period was calculated following Welch (1983). This procedure requires estimating and plotting moving averages, with different window sizes of the variable of interest. The window size would be steadily increased until a relatively stable plot of the variable of interest against time was achieved. For these calculations the total number of cells cultivated in any year was used.

10.4.2 Sensitivity Analysis

Sensitivity analyses were carried out to identify the sensitivity of model response variables to changes in model parameters and inputs (factors). There were more than 100 factors and decision variables that could have affected model responses. Examining the effects of all of these was not possible. A 2^5 (two to the fifth) fractional factorial design[11] was used with 13 factors (Table 10.1). Where appropriate, model parameter values were selected so that the +1 level was the normal (standard model) mean plus one standard deviation and the -1 level was the mean minus one standard deviation.

Table 10.1: Factors and factor levels used in sensitivity analysis of MASOKA[1]

Factor	+1	-1
Mean annual rainfall (mm)	972	478
Good season classification membership function	$u_{good} = 1$ rainfall >= 1124	$u_{good} = 1$ rainfall >= 1000
Average season classification membership function	Mean ± 1*SD	Mean ± 0.5*SD
Poor season classification membership function	$u_{poor} = 1$ rainfall <= 500	$u_{poor} = 1$ rainfall <= 442
Percentage of children that work	59	29
Loan or dividend	Loan = Z$155 Dividend = Z$0	Loan = Z$0 Dividend = Z$155
Households could employ labor	Yes	No
Maize price	0.80 Z$ kg^{-1}	0.26 Z$ kg^{-1}
Household cash needs	base+0.41*base	base-0.41*base
Proportion of households in production strategies	80% MINVAR 20% INCMAX	20% MINVAR 80% INCMAX
Wage rate	Z$3.07 day^{-1}	Z$2.01 day^{-1}
Land multiplier for riverland and non-riverland cells	1.7	0.63
Minimum and maximum years jecha fields cropped before leaving fallow	Min 0 Max 10	Min 5[1] Max 7

1. The model included an error for these analyses. For the income maximizing strategists the minimum number of years cultivated before a field was left fallow was six and not five.

Ten Monte Carlo runs (replicates), each of 50 years with a 20-year warm-up period, were run for each combination of experimental factors. Data were collected from all Monte Carlo runs. A random sample of 20% of the 100 households was saved for subsequent analyses.

Mean effects due to each factor were computed using Yates algorithm (Yates, 1937, described in Box *et al.*, 1978).

The results of sensitivity analyses were presented as the mean effect of each factor changing from the low (-1) to the high (+1) level. The sign on these means indicated the direction of the effect. Also shown in the results tables were the mean effects expressed as percentages of the overall mean for the response variable of interest. To calculate the value of the response variable at the low level, subtract half of the mean effect for the factor of interest from the overall mean effect for the variable of interest. The value of the response variable at the upper limit would be calculated by adding half of the mean effect to the overall mean effect. To illustrate the use of the model and sensitivity analyses example results are presented from the sensitivity analyses, while fuller results are presented in Lynam (1993).

10.5 RESULTS

10.5.1 Sensitivity Analysis

Model response variables were grouped as needs, yield, economic and resource responses. Needs responses included the mean proportion of households who failed to satisfy their cash, maize or sorghum needs each year.

Yield responses were the mean yields (kg ha^{-1}) of maize, cotton and sorghum, from all fields, corrected for labor deficits, achieved over the 30-year simulation. These means were then averaged over the ten Monte Carlo simulations. Economic response variables included, for maize, cotton and sorghum, mean total returns to a household (total yield * local price - total fixed costs - total variable costs); mean returns per hectare (per hectare yield * local price - per hectare fixed costs - per hectare variable costs); mean returns to own labor and management (total returns / total family labor inputs); and mean returns to initial investment (gross returns / total costs). As with the yield data these were the means of all households averaged over the ten Monte Carlo runs.

The resource response variables were the changes in the total number of years that each jecha cell had been cultivated summed over the whole landscape, the mean erosion (t ha^{-1}) over the landscape, and the change in the total Fb value of the landscape, or the Fb value of each cell summed across the entire landscape at the beginning of the simulation minus the sum of all cell Fb values at the end of the simulation. Lower Fb values indicated a higher erodibility.

10.5.1.1 Social factors – household needs satisfaction

For the most part, the responses of the proportions of households that were cash, maize and sorghum deficit were in the direction and of magnitudes that were expected (Table 10.2). The effects of rainfall and the price of maize are obvious and require no explanation. The effects of factors such as the land multiplier, the proportion of households in each production strategy or the effects of the membership functions used were easily explained or the effects were too small to warrant discussion. Increasing household cash needs had a large impact on the proportion of households that satisfied cash and food needs.

At the +1 level of land availability, households attempted to cultivate, on average, 70% more land than the real world households in Masoka cultivated. Given that in the model households cultivated fixed proportions of the land available to them, irrespective of their labor resources, it was not surprising that more land would increase the levels of deficits as households would have had greater labor shortages and consequently achieved lower yields (Table 10.2).

Counter intuitively the number of cash deficit households decreased when household cash needs were increased (Table 10.2). With increased household cash needs, many of the more marginal cotton producers (those with poor soils or who had a labor deficit) would be excluded from cotton production due to their lack of capital to invest in inputs. These households would have to switch to maize production. The remaining cotton producers would be those households that were more likely to achieve high yields as they would not be labor constrained. They would also be more likely to satisfy household cash needs. The marginal cotton producers (i.e. those who were likely to be labor deficit) who had switched to maize were likely to have been those households that were not satisfying household cash needs, in part because of the high costs of cotton production. These households, relieved of the burden of cotton production, might have been more likely to satisfy household needs through maize production owing to the lower costs of production. It was assumed here that these households were generally smaller (hence the labor constraints) and therefore quite likely to satisfy household cash needs from maize and sorghum production.

The mean percentage of households that were deficit each year decreased when the poor season classification changed from the low to the high level (Table 10.2). This was expected because at the +1 level the season was classified as poor ($u_{poor} > u_{average}$) at higher rainfall levels than at the -1 level. The yield functions for all three crops showed higher yields as u_{poor} approached zero than when $u_{average}$ was small. Thus, at the +1 level of poor season classification higher mean yields were expected and, therefore, a lower percentage of households that were cash, maize or sorghum deficit. The classification of good seasons had a similar trend for households that were cash deficit but the opposite trend for the proportion of households that were maize deficit (Table 10.2).

As expected, when the proportion of households that were classed as variance minimizers was low (level -1) the percentage of households with a cash deficit was higher and the percentage of households that were maize deficit was lower (Table 10.2). Income maximizers planted a greater proportion of their lands to cotton and less to maize than did variance minimizers. At the low factor level (80% income maximizers), therefore, it was expected that fewer households would have a cash deficit but more would have a maize deficit.

The percentage of households that had a maize deficit and the percentage that were sorghum deficit were sensitive to whether the Z$155 subsidy was paid as a dividend or as a loan (Table 10.2). In both cases, the provision of a loan resulted in a greater percentage of households with a deficit than did the payment of a dividend, and for sorghum the provision of a loan resulted in an almost 30% increase in deficit households. Loans were provided to only those households that grew cotton and were provided only as inputs to cotton. Dividends, on the other hand, were paid to all households and were not tied to any productive activity and could, therefore, be used to satisfy household needs. With lower total inputs of cash to the community (not all households receiving loans) and with loans tied to cotton production, it was expected that there would be a smaller proportion of maize and sorghum deficit households with dividend payments than there would be with tied loans.

Table 10.2: Sensitivity of the mean of the % of households that were cash (ACDF), maize (AMDF) and sorghum (ASDF) deficit, each year, to changes in factor levels[3]

Factor	ACDF (% households)			AMDF (% households)			ASDF (% households)		
	Mean effect	%[1]	Sig[2]	Mean effect	%	Sig	Mean effect	%	Sig
Land multiplier	14.2	16	***			NS	6.6	9	**
Rainfall	-13.5	-15	***	-24.0	-54	***	-20.2	-27	***
Cash needs	-13.5	-15	***	-37.4	-84	***	-24.3	-33	***
Maize price	-5.0	-6	**			NS	-11.8	-16	**
Poor season classification	-4.9	-6	**	-1.1	-3	*	-3.0	-4	*
Proportion of MINVAR strategists	3.4	4	*	-2.7	-6	***			NS
Good season classification	-3.1	-3	*	1.0	2	*			NS
Loan or dividend			NS	8.8	20	***	29.0	39	***
Fallow practices			NS	-3.6	-8	***			NS
Average season classification			NS			NS	3.1	4	*
Overall mean	89.8			44.7			74.6		

1. % of overall mean effect for that factor.
2. Significance determined using Student's t test. Effects were different from zero at $p<=0.001$ (***), $p <= 0.01$ (**), $p <= 0.05$ (*) and not different from zero at $p >0.05$ (NS). See text for details of tests.
3. Data show the change in mean response to changing the factor of interest from the -1 to the +1 level.

Household fallowing practices had a significant impact on the percentage of households that had a maize deficit each year (Table 10.2). At the -1 factor level the minimum number of years that a field would be cultivated before being left fallow was five. All yields of maize grown on jecha at this factor level were based on the five-year cultivated jecha function which had a much lower mean yield than the uncultivated jecha function. It was expected, therefore, that at the +1 factor level, where yields could be based on the virgin and the cultivated functions, the mean maize yields would be higher and fewer households would have a maize deficit. This expectation is consistent with the data of Table 10.2.

Counter intuitively, the ability of households to employ other households had no effect on the proportion of households satisfying their cash or food needs. It was expected that local employment would reduce the frequency of needs deficit households through an income multiplier effect but this was not the case.

10.5.1.2 Agricultural responses – crop yields

As expected, mean yields of maize, cotton and sorghum were all markedly sensitive to changes in the mean annual rainfall (Table 10.3). The yields of maize and cotton more than doubled with the change from a low mean annual rainfall to a higher mean whilst the yields of sorghum more than tripled. The move to the +1 level of fallowing practices resulted in the reduction in maize yields but also resulted in increased sorghum yields. The effect on sorghum was expected. At the +1 level a larger number of fields were cultivated for fewer years resulting in higher yields. The reduction in maize yields at the +1 factor level was difficult to explain. Despite the significance of the result, the response (80 kg ha^{-1}) was within the range of error expected from model yield functions.

Mean yields for all three crops were sensitive to the average season and the poor season membership functions used. The directions of the effects were all the same with an increase in yields at the +1 factor level for both seasons.

Table 10.3: Sensitivity of mean annual yields of maize (MZ YLD), cotton (CTYLD) and sorghum (SRYLD) to changes in factor levels[3]

Factor	MZYLD (kg ha^{-1})			CTYLD (kg ha^{-1})			SRYLD (kg ha^{-1})		
	Mean effect	%[1]	Sig[2]	Mean effect	%	Sig	Mean effect	%	Sig
Rainfall	1000	68	***	621	79	***	576	116	***
Fallowing practices	-79	-5	***			NS	65	13	***
Average season classification	11.4	1	*			NS	1.7	<1	**
Cash needs	52	4	***			NS	6	1	***
Poor season classification	25	2	**	66	8	**	15	3	**
Overall mean	1463			784			488		

1. % of the overall mean for that effect.
2. Significance determined using Student's t test. Effects were different from zero at $p <= 0.001$ (***), $p <= 0.01$ (**), $p <= 0.05$ (*) and not different from zero at $p > 0.05$ (NS). See text for details of tests.

10.5.1.3 Economic response variables

Apart from the consistent sensitivity to rainfall and household cash needs there were no patterns of sensitivity to factor levels across all economic response variables.

Not unexpectedly the returns to maize were sensitive to the change in maize price (Table 10.4). Increasing the maize price from Z$0.26 kg^{-1} to Z$0.80 kg^{-1} (i.e. 208%) resulted in increases of 22% in returns to land, labor and initial investment as well as an increase of 17% in total returns to households from maize. Increasing household cash needs by 41% increased returns to all inputs for maize. Rainfall had the expected effect of increasing total returns for maize, for cotton and for sorghum, as well as to all inputs for each of these crops.

The effect of increasing the proportion of variance minimizers from 20% (factor level -1) to 80% produced the results expected; variance minimizers planted a greater proportion of their land to maize so it was expected that they would achieve higher total returns and slightly higher returns to land, labor and initial investment for maize. Only the returns to labor were not significantly improved at the higher level of variance minimizers (Table 10.4). With respect to cotton, the reduction in the proportion of income maximizers from the -1 to the +1 levels (and hence reduced total cotton area) resulted in lower total returns to cotton (Table 10.5).

Maize and cotton returns were notably sensitive to the land multiplier used (Tables 10.4, 10.5). Maize returns (except returns to labor) increased with increasing land availability whilst those for cotton decreased. These trends were not unexpected. With larger land areas households would have been less likely to have had sufficient capital to plant all of their normal proportion of cotton fields than with smaller land areas. This would have been due to households having to plant a proportion of their fields to cotton. More maize and less cotton would, therefore, have been planted, thereby increasing the total returns to maize and reducing those to cotton.

Fallow practices had notable effects on sorghum returns and lesser effects on maize and cotton returns (Tables 10.4, 10.5). For maize the returns to land, to labor and to initial investment decreased when fallow practices changed from the -1 to the +1 level by between 3% (for returns to initial investment for maize) and 6% (for returns to maize per hectare and per unit labor) of the overall mean. Given the reduction in maize yields with the +1 level of fallow practices (Table 10.3) these changes in maize returns were expected. Cotton returns increased when fallow practices changed from the -1 to the +1 level. Total returns increased by 8% and returns to initial investment increased by 5% (Table 10.5). Yields at the +1 level of fallowing practice were expected to be higher than those at the -1 level; at the -1 level the cultivated jecha yield function would always be used (with lower yields) whilst at the +1 level the uncultivated jecha yields function would be used at least some of the time, leading to higher average yields.

Table 10.4: The effects of factor level changes on the means of total returns (MZRET), returns to land (MZHA), labor (MZLAB) and to initial investment (MZINV) for maize [3]

Factor	MZRET (Z$ household^{-1})			MZHA (Z$ ha^{-1})			MZLAB (Z$ day^{-1})			MZINV (Z$ Z$$^{-1}$)		
	Mean effect	%[1]	Sig[2]	Mean effect	%	Sig	Mean effect	%	Sig	Mean effect	%	Sig
Maize price	1059	117	***	788	121	***	9.52	121	***	6.11	122	***
Cash needs	782	86	***	52	8	***	0.6	8	***	1.33	26	***
Rainfall	649	72	***	505	78	***	6.1	77	***	3.18	63	***
Proportion of MINVAR strategists	211	23	***	13	2	*			NS	0.27	5	**
Average season classification	35	4	**			NS			NS			NS
Land multiplier	65	7	**	9	2	*			NS	0.15	3	**
Poor season classification	44	5	**	15	2	*	0.18	2	*	0.17	3	**
Loan / Dividend	35	4	**	-10	-2	*			NS			NS
Fallow practices	-18	2	*	-41	-6	***	-0.5	6	**	-0.17	-3	**
Overall mean	908			650			7.84			5.02		

1. % of overall mean effect for that factor.
2. Significance determined using Student's t test. Effects were different from zero at $p <= 0.001$ (***), $p <= 0.01$ (**), $p <= 0.05$ (*) and not different from zero at $p > 0.05$ (NS). See text for details of tests.
3. Data show the change in mean response to changing the factor of interest from the -1 to the +1 level.

Table 10.5: The effects of factor level changes on the means of total returns (CTRET), returns to land (CTHA), labor (CTLAB) and to initial investment (CTINV) for cotton[3]

	CTRET (Z$ household^{-1})			CTHA (Z$ ha^{-1})			CTLAB (Z$ day^{-1})			CTINV (Z$ Z$$^{-1}$)		
	Mean effect	%[1]	Sig[2]	Mean effect	%	Sig	Mean effect	%	Sig	Mean effect	%	Sig
Rainfall	970	108	***	1274	99	***	9.95	99	***	0.86	91	***
Loan or dividend	-445	-50	***	35	3	*			NS	-0.11	-12	**
Cash needs	423	47	***			NS	0.31	3	*	-0.03	-3	*
Proportion of MINVAR strategists	-190	-21	**			NS	0.34	3	*	-0.07	-7	*
Poor season classification	173	19	**	107	8	**	0.87	9	**	0.14	14	***
Land multiplier	-139	-16	**			NS			NS	-0.05	-5	*
Fallow practices	76	8	*			NS			NS	0.05	5	*
Good season classification	103	11	*			NS			NS			NS
Wage	64	7	*			NS			NS			NS
Maize price			NS	48	4	*			NS	0.06	6	*
Employment			NS			NS	0.75	7	**			NS
Overall mean	896			1282			10.0			0.95		

1. % of overall mean for that factor.
2. Significance determined using Student's t test. Effects were different from zero at $p <= 0.001$ (***), $p <= 0.01$ (**), $p <= 0.05$ (*) and no different from zero at $p > 0.05$ (NS). See text for details of tests.
3. Data show the change in mean response to changing the factor of interest from the -1 to the +1 level.

Total cotton returns increased by 7% as the wage rate increased from Z$2.01 to Z$3.07 (i.e. by 53%). It was suspected that as the wage rate increased more of the marginal cotton producing households (i.e. those likely to experience labor shortages) would have gone out of cotton production. Expecting the mean total returns to cotton production to increase would have been plausible. Mean total returns represented the total returns to cotton averaged among cotton producing households.

Total cotton returns, as well as returns to land and initial investment, were sensitive to the form that subsidies took with the mean of total cotton returns decreasing 50% when subsidies were paid in the form of dividends (Table 10.5). Loans increased the area planted to cotton but also facilitated more labor constrained households, who would otherwise not have been able to produce cotton, to plant cotton. Labor constrained households or those without access to bepe or riverlands faced relatively high risks of low or even negative returns to cotton. The effect of increasing the number of these households that were able to produce cotton would be to reduce the mean of the total returns to cotton. Although the effect was small, the higher returns to land from cotton were however, inconsistent with this expectation.

10.5.1.4 Erosion, erodibility and 'fertility'

A number of the responses of the resource performance measures, mean erosion, change in fertility and change in erosion potential, were completely contradictory to expectations. Through examination of the model it was recognized that many of these responses were an artifact of how the changes had been calculated as well as the large temporal variability in the total number of years the landscape had been cultivated (and hence Fb). These problems made interpretation of the results of factor level changes on resource responses difficult. Basically the landscape that was being simulated appeared to cycle in terms of major soil properties when particular input parameter values were used (particularly the -1 level of fallow practices with a five-year minimum and seven-year maximum before fallowing). The way that changes in fertility and erosion potential were calculated (change = endstate - startstate) was not useful in a cycling world. These sensitivity analyses were therefore, not used.

10.6 DISCUSSION

It is worth remembering that 'in running the models we investigate the consequences of the guesses and assumptions incorporated in them – model output may tell us nothing new about empirical reality'. (Landsberg, 1987). Despite this somber warning we can learn much from developing and using models such as those described in this chapter. Perhaps the first lesson is that the direction as well as the magnitude of effects of policies or management actions can be very different in different components of the system. Increasing the proportion of risk averse strategists, for example, increased the number of cash deficit households but reduced the number of maize deficit households (Table 10.2) and altering the fallowing practices had different effects on maize and sorghum yields (Table 10.3).

The second key lesson is derived from this first one. It is clear that the state of the system – such as the proportion of households that conform to one strategy – has a major effect on the likely impact of a particular policy or management intervention. The example just given is illustrative of this point; increasing the proportion of risk averse strategists increased the returns to maize by 23%. Being aware of the state of the system (e.g. high proportion of risk averse strategists) policy makers could have a far greater impact with focused policies. For management or policy to be maximally effective managers or policy makers should be aware of the current state and dynamics of the system – and in particular those elements of the system that are sensitive to changes in a given factor. Models such as MASOKA1 could be especially useful in this regard. In particular the use of rigorous sensitivity analyses, as presented here, could provide indications of key leverage points for management or policy interventions.

The third lesson is important both for policy and for research. Changes in system variables reverberate through the system in ways that are often very difficult to trace and hence even more difficult to predict. When investigating change in agro-ecosystems such as that of Masoka one would do well to cast the net of performance measures much wider than the few conventional components that discipline-bound academics or results-driven extension agents might have as their focus. Just altering the fallowing practices, for example, could have negative impacts on the proportion of maize deficit households, through decreasing maize yields, positive impacts on sorghum yields and positive impacts on the economics of cotton production.

The model presented here was developed using a combination of local people's knowledge to provide the trend data that would not otherwise have been available, and scientific measures for variables where local people did not know the answers – such as cropping effects on soil erodibility. Although there are certainly many errors in the model it was an important tool for the exploration of the logical consequences of what was known about this agro-ecosystem. With care and use of probability distributions to reflect uncertainty the approach developed for this study has the potential to yield rich insights into the functioning of very complex systems where scientific measures are unavailable. But perhaps just as importantly, by combining local knowledge we are better able to explore the patterns of human behavior, such as setting local needs levels that are very difficult to measure using formal scientific approaches.

Perhaps one of the more important lessons that have been learnt from this and subsequent modeling efforts has been the difficulty of using these very complex models with local managers. Without significant investments in training and capacity building it is unlikely that local managers would be able to use models as complex as MASOKA1 without treating them as 'black boxes'. In the context in which these models are being used – i.e. as generally accepted and manipulatable representations of some aspects of the world, 'black boxing' implies loss of understanding. This is contrary to what is required. To make improved management decisions people need to understand the implications of their actions in an uncertain world. In these instances, simpler, and of course by assumption more easily understood, models are better (see for example Lynam *et al.*, 2002). Complex models provide very rich insights to those who develop them but are unlikely to find widespread application without expensive investments in supporting the user.

Modelers and model users will always be faced with the question 'Is the model correct?' This question of model validation is a thorny one that is unlikely to be completely resolved and unlikely to go away. Through the process of validation we seek to refine the degree of belief or confidence we or the model users have in a) the outputs of a model, and/or b) the structures or processes that produced these outputs. In both cases the degree of belief will depend on the degree to which model outputs, processes or structures are consistent with the outputs, processes or structures of the system they are designed to represent.

There are two key factors which strongly influence the process of validation. The first of these is the purpose of the model and hence of the validation process. If, for example, the model is concerned with predicting the behavior of a real world system then validation should be concerned with the degree to which model outputs, structure or processes are consistent with real world outputs, structures or processes. This is the classic case of comparing simulations and observations. Let us call this 'process validation against the real world' or' real systems validation'.

If however, as in MASOKA1, the model was developed as a representation of a consensus world view (or some other world view) then validation should be more concerned with establishing the degree to which model outputs, structure or processes are consistent with the world views it represents. Let us call this 'validation against the representational world' or 'representational validation'.

The second key factor influencing the validation process is cost. Ultimately the validation procedures adopted should reflect a considered assessment of the benefits of altering the degree of confidence or belief in the model compared to the costs of making that shift. Whilst conceptually a simple benefit/cost analysis, in practical terms the process is likely to be more difficult than that. It follows then that if the risks of making expensive errors through use of an incorrect model are high, then validation should be rigorous and exhaustive. In those instances it will probably also be expensive.

Three trends in the claims of MAS developers as to the purpose of MAS are evident. The first are those MAS that are designed as hypotheses (also called metaphors or thought experiments). Examples are the models of Carpenter *et al.* (1999), Janssen and Carpenter (1999) and Janssen *et al.* (2000). Holland (1995) describes much of his work with the ECHO model in this light. A second purpose is illustrated by those MAS deemed to be common or expert representations of some real world system. Barreteau *et al.*, (2001), Bousquet *et al.* (1999) and Lynam *et al.* (2002) all present examples of MAS developed with this objective. The third, and easily the rarest category are those MAS designed as what Holland (1995) calls 'flight simulators' – tools to guide policy or management through simulating the behavior of actual systems.

The nature and degree of validation of these different classes of models will obviously be different. Those models cast as hypotheses are not formally validated. They are presented as is and the only real check on their validity is the peer review process they undergo prior to publication. Those models cast as representations, either expert or common, should of course undergo representational validation to ensure they are consistent with the representation they are models of. Barreteau *et al.* (2001) and Bousquet *et al.* (1999) discuss their use of local experts in workshop environments to validate model representations. The details of the processes used, however, are not clear. But given the documented fallibility of experts (Shanteau, 1990; Tversky and Khanemann, 1974; 1981) one is left wondering how the process of expert validation might be improved to yield greater confidence in MAS models.

Models cast as 'flight simulators' are rare and we might expect their validation to be very much more rigorous for it is with these models that the risks of costly error are greatest. No examples of this class of model and hence of course the process of validation used were available for inspection. I suspect, however, that many of us would like our models to be used in this sense and indeed several of the authors mentioned above do allude to this. MASOKA1 was definitely designed as an expert representation but with closes flight simulator aspirations. With this context a brief explanation of the validation processes developed and used by this researcher in enhancing the degree of belief or confidence in MASOKA1 are presented.

The case of MASOKA1 represents a situation in which the risks of costly error were small. Both representational validation and a limited set of real systems validation activities were carried out. Given the model purpose of being a useful representation the emphasis of validation was on representational validation.[12] The first stage of the process was to test the degree to which the model produced results that were consistent with people's expressed beliefs. In general these tests took the form of testing for differences in expressed and model produced distributions. This was done for all possible model outputs. The second stage in the validation process was to establish the degree to which the model results were consistent with the 'real world'. Again this was done as far as was possible. Components of the model (e.g. crop yields) were compared to independent, published results for similar areas (none were available for Masoka itself). Although outputs of the whole model (e.g. the frequency of households that were needs deficit) were compared with local perceptions of the degree to which needs were being satisfied this was not done rigorously. No attempt was made to establish the degree to which model processes were consistent with real world processes.

An alternative approach to reducing the high costs of complex model validation that has been used by this researcher has been to shift the cost of validation away from the large complex models to smaller, simpler and cheaper models. A set of large and complex models are maintained as formal statements of the researcher's best bet, current beliefs as to how the system of interest is structured and how it behaves. No attempt is made to formally validate these models. But the understanding encapsulated in these models is used to develop simpler and more tightly focused models of narrower problem domains with which to probe a real or representational system. These probing models are confronted with data or expert opinion and then, if they result in improved understanding, they are re-integrated into the large parent models. These smaller, probe models are of course cheaper and easier to validate. However, the question of how one manages the degree of confidence or belief in the overall model has yet to be resolved.

10.7 CONCLUSIONS

In relation to the objectives originally set for the MASOKA1 model, the model may be considered useful, because:

1) It demonstrated the utility of the modeling and analysis approach and provided new insights. Firstly, the model indicated the general inability of land based production to satisfy household needs – a finding substantiated by the more extensive analyses of Cumming and Lynam (1997). Secondly, it would not have been possible to construct a model of the complexity and breadth of MASOKA1 without the extensive use of local informants. Thus the approach yielded a useful model (i.e. understanding) of a very complex system.
2) The model yielded policy and management relevant information that could lead to improvements in the satisfaction of household needs and or community well-being. For example, the relative impacts of subsidies as loans or dividends or the general failure of land based production systems.
3) The model did help identify crucial gaps in our understanding of the Masoka agro-ecosystem. In particular the relative contribution of natural resources to household well-being was recognized as being important but was unknown. The roles that local institutions played in buffering households from crop failures or controlling their access to production units was also an important

gap in our understanding that was highlighted through the development and use of the model.

4) The model was useful as a mechanism to compare different agro-ecosystem configurations in a design situation. Simulation experiments were conducted in which, for example, the availability of riverine lands was withdrawn or different social configurations were explored. These experiments yielded rich insights into the likely impacts of policies that sought to increase local well-being.

Thus although wrong, the MASOKA1 model was a useful attempt at a multi-agent systems model. But perhaps more importantly, the process that was developed in the study described here has been expanded into a process of using models (not limited to MAS models) not only to represent understanding but to begin to change this understanding where, through error or omission, it is constraining. Through their provision of a common representation, stakeholder-built models provide a remarkable opportunity to rewrite the mental scripts of stakeholder understanding in a co-learning and non-confrontational manner.

It is proposed that in applied, natural resources management settings where local stakeholder acceptance of models is sought, validation of models should first seek to validate the degree to which the model accurately represents the mental models it is a model of (representational validation). Then, if the stakeholders wish to use the model for identifying improved management or policy actions then the model should be validated against real systems (real system validation). Our degree of confidence in these models is likely to be enhanced when the outputs of the model are validated. But an even stronger degree of belief should arise when the underlying structure and process within the model that are responsible for model outputs are consistent with the structures and process producing the same outputs in the real world.

ACKNOWLEDGEMENTS

The author is grateful for the financial and logistical support provided by the World Wide Fund for Nature (WWF) through their Multispecies Animal Production Systems Project (MAPS) in Zimbabwe. Rotary International is also gratefully acknowledged for their support of the author as a Freedom From Hunger Scholar. David Cumming provided wise counsel and an enabling environment within WWF – thank you. Three anonymous reviewers provided useful and insightful comments on an earlier draft of the chapter – their contributions are gratefully acknowledged. The people of Masoka became my friends and confidants – I am very grateful their friendship and for the secrets they chose to trust me with.

NOTES

1. Acronym for Communal Areas Management Program For Indigenous REsources
2. VaDema and Korekore are two ethnic groups within the broader Shona speaking people who form the majority of Zimbabwe's population.
3. The Communal Areas of Zimbabwe are land areas where traditional land rights prevail. They are characterized by mixtures of semi-private land rights and common pool property rights regimes. People have exclusive rights of use over their field or arable areas and common rights of use over the generally larger common resource areas. People in Masoka did not have rights of sale of land.
4. Wildlife in Zimbabwe belongs to the state unless governed by Appropriate Authority status. Hunting is strictly controlled by the Department of National Parks and Wildlife Management (DNPWM). Poachers in the Zambezi Valley face jail sentences if caught.
5. Upland cells were those jecha cells inside the fenced area and on the older alluvial terraces. Riverland cells were those lands outside the fence on the most recent alluvium.
6. Elected villagers who acted as key informants.
7. Household needs were defined with Village Representatives and reflected the average requirement for leading an adequate quality of life in the Ward. These included food of different kinds and cash. Amounts of each required to support an average man, an average woman and an average child were defined and then used to calculate the amounts required for each household.
8. Village Representatives developed detailed labor allocation budgets indicating the total labor day input to each cropping activity.
9. Yield possibility diagrams were cumulative probability functions for yields under defined conditions of soil state and rainfall.
10. This value was derived from the cotton partial budget and consists of seed (Z$30.88) + seed transport (Z$2.47) + fertilizer and chemical costs (Z$345.80) multiplied by the hectare to cell conversion factor of 0.4096.
11. I am grateful to Professor D. Gilliland, of the Department of Statistics and Probability, Michigan State University, for his able assistance in generating, and analyzing the results of this design.
12. There is an important philosophical point here. If we hope to alter or improve people's behaviors through management or policy interventions or through the provision of information, then we do need to know the degree to which people's beliefs are consistent with what actually happens in the real world (real systems validation). Representational validation has limited usefulness in applied settings.

11. Simulating land-cover change in South-Central Indiana: an agent-based model of deforestation and afforestation

Matthew Hoffmann, Hugh Kelley and Tom Evans

11.1 INTRODUCTION[1]

In the early 1800s, the state of Indiana was exclusively forested. By 1900, 90% of the forests were gone. In 2000, the state is back to 20% forest once again and the percentage is continuing to grow. This process of deforestation followed by afforestation is the specific empirical phenomenon we address in the analysis presented in this chapter.[2] As a first step at investigating this pattern we explore a model of individual-farmer portfolio decisions (with both optimizing and boundedly rational approaches) in response to economic, political, and ecological conditions as one aspect of dynamic deforestation and afforestation processes. We would like to understand and explain the dynamics of deforestation and afforestation in Indiana with the expectation that such explanations will lend insight into general dynamics of forestation patterns in other areas.

Our objectives can be broken down into three primary components. First, we would like to explain these forest macro-dynamics within a single agent-based theoretical framework from which both deforestation and afforestation emerge. Second, the macro deforestation/afforestation patterns should emerge from consistent and simple micro or agent level decision 'rules'. Third, any such micro level specification must include agent–agent, agent–economy, agent–political system, and agent–environment interactions/decisions. We believe that the decisions and interactions of agents are a crucial 'driving-force' of deforestation and afforestation and they should therefore be included as a driving-force of our model as well.[3] A more complete problem statement is thus: we would like to understand and explain the dynamics of deforestation/afforestation by specifying a process of agent interactions and decisions that produces both deforestation and afforestation in a similar manner to our observations of the empirical record based on political, economic, and biological components of the landscape.[4]

As a first step toward this end, we present an agent-based model of land-use change in the Midwest (LUCIM).[5] The LUCIM model simulates the decisions of heterogeneous agents on a simple artificial landscape.[6] This stylized model incorporates two types of decision making, three types of land-cover, and a minimally biologically active landscape. With this simple model we have been able to capture some important aspects of the empirical phenomena observed in Indiana in the last 200 years. Our simulations demonstrate agents' price and tax sensitivity, the slope dependence of reforested areas, and the macro-phenomena of quick deforestation followed by slow afforestation – phenomena observed empirically. Our simulations thus lead us to believe we are on the right track, and also point to the many areas where further research is necessary.

In this chapter we first offer a brief discussion of agent-based modeling and land-use/cover change. We then detail the historical land-cover change in Indiana. We follow this with a description of the model: the agents' decision problem, the landscape, and the model specifications. Next, we present and discuss the results of our simulations. We conclude with some thoughts about verifying and evaluating our agent-based model with empirical data, and areas for future work.

11.2 ANALYSING LAND-USE AND LAND-COVER CHANGE WITH AGENT-BASED MODELS

There are many possible ways to approach modeling land-use/cover change including system dynamics models, spatial econometric models, and cellular automata among others.[7] Why then have we chosen to pursue an agent-based modeling approach in lieu of these? Before moving directly to a discussion of the model constructed for this analysis it is important that this question be briefly addressed in light of the many options available and the lack of consensus surrounding such methodological choices.

First, we begin with the assumption that we are dealing with a complex adaptive system.[8] Complex adaptive systems are characterized by interactions, emergent properties, and non-linearity. The actions and interactions of heterogeneous agents shape the land-use and land-cover change observed in the system. The state of the system is thus defined by both exogenous environmental factors and by the actions of the agents that reside within it. Crucially, however, land-use and land-cover patterns cannot be trivially deduced from agent characteristics or even environmental characteristics. Instead, complex adaptive systems exhibit emergent properties, whereby the system, while created by the agents, is more or different than the sum of its parts.[9] In other words, land-use and land-cover patterns, rather than being directly calculable from agent-level characteristics, emerge from the interactions of the agents (and between agents and their environment) in the system. Finally, complex systems are inherently non-linear. The land-use and land-cover properties we would like to explain (i.e. deforestation/afforestation) often change in a non-linear fashion and their relationship with the agent-level properties (the 'causal' variables) is non-linear as well.

Given these characteristics, explaining and modeling land-use/cover change processes requires that we have a method that focuses clearly on the decisions and interactions of heterogeneous agents, allows for spatial explicitness, and can handle feedback between agents and the environment or context in which they are embedded. Of the methods available for studying land-use/cover change processes, agent-based modeling has the most potential to meet these requirements.[10] While the other methods mentioned above have been extremely useful in producing significant insights, we believe agent-based modeling techniques have the potential to incorporate more of the complex processes and characteristics necessary for fully explaining land-use/cover change processes.

System dynamics models are wonderful tools for explaining macro change over time in a system (Gilbert and Troitzsch, 1999). By representing land-cover in terms of stocks of resources and land-cover change in terms of flows governed by differential equations, systems models are useful for modeling discrete change in a specified system. However, they accommodate spatial characteristics with difficulty and there are no explicit agents or agent decision making. As we fundamentally believe that *heterogeneous* agents and their decisions must be modeled in order to fully understand the de/afforestation processes in Indiana, we chose not to utilize a systems modeling approach.

Econometric methods explain land-cover patterns by correlating data on land-usage over time with data for a series of independent variables: crop prices, land prices, population, timber prices, weather data, land characteristics, etc. From this time series model we could get a gross picture of the variables presumably affecting patterns of forestation over time, but we would lose critical insights. First, we would lose spatial information and patterns of land-use, i.e., where the forest might reside.[11] Second, we would be modeling system-level patterns without a clear understanding of the relationship of the system to the agents. We would lose any sense of the individual decision making that is generating the system-level patterns that we observe, i.e., why a particular type of farmer might prefer more forest than another farmer. Thus, we would have no way of predicting or explaining the affect of institutions, laws, or other rules on land-use patterns through their impacts on agents' decisions. While we intend to use econometric techniques (both spatial and non-spatial) in the larger project for verification and validation, these methods are not suited to our primary modeling goals.

Finally, cellular automata (CA) modeling approaches have produced important insights and are closely related to agent-based models (see e.g. Tobler, 1979; Li and Reynolds, 1997; Hogeweg, 1988). In fact, some would classify cellular automata models as one type of agent-based models. CA models are two-dimensional lattices where the state of each cell in the lattice is dependent upon specified transition rules that in most cases depend on the state(s) of neighboring cells. These models, similarly to systems models, are especially useful for modeling macro land-cover change over time and have been used successfully for modeling many aspects of ecological change. In addition, it is possible to consider the transition rules as pseudo-decision making and thus these models can incorporate some of the decision making and heterogeneity needed for our project. However, for this analysis, we wanted to explicitly focus on decision making and thus it made more sense to create a model more focused on agents and less focused on the landscape.[12]

In sum, we have chosen to utilize an agent-based modeling approach to analyse the forestation processes in Indiana for two main reasons. First, this initial analysis is focused clearly on agent decision making. We are explicitly concerned with exploring how agents make land-use decisions and how these decisions come to influence macro patterns of land-use and land-cover over time. Agent-based modeling is the best tool available for this type of agent-centered analysis. Second, this preliminary analysis is designed to build a foundation for more sophisticated, spatially explicit modeling efforts. Agent-based modeling is a flexible enough tool to facilitate spatial models (by incorporating CA modeling) as well as connections with other types of modeling such as geographic information systems (GIS).

Our ultimate goal is to understand how adaptive agents acting on local information and making local decisions 'cause' the emergence of system level patterns of forestation. Thus, we are attempting to 'grow' forestation patterns by specifying rules and behavior of agents. Epstein and Axtell aptly describe our aims in using agent-based modeling:

> We view [agent-based models] as *laboratories*, where we attempt to 'grow' certain social structures in the computer – or *in silico* – the aim being to discover fundamental local or micro mechanisms that are sufficient to *generate* the macroscopic social structures and collective behaviors of interest. (Epstein and Axtell, 1996, p. 4, emphasis in original)

Agent-based modeling, by requiring us to concretely specify decision rules and simulate how actions driven by these rules, when aggregated, create changing patterns of forestation over time, represents the best methodological tool for our project. Ultimately, we need to explore a range of rules (both through empirical investigation of individuals within experimental tests – interviews with human agents – historical economic analysis of the environment – and the comparison of multiple utility functions) until we find a combination that produces an artificial world that parallels the empirical one (how well it parallels the empirical world is discussed below). This chapter presents an initial attempt to do so.

11.3 EMPIRICAL OBSERVATIONS OF LAND-COVER CHANGE IN INDIANA

Indiana, located in the midwest of the United States, has experienced dramatic land-cover change over the past 200 years. These land-cover changes have largely been the result of European–American colonization in the first half of the nineteenth century and the subsequent social and economic changes that accompanied settlement within the context of the biophysical environment. It is estimated from a land survey designed to characterize areas for settlement that, in the early 1800s, more than 87% of the state was covered with forest of some type across a wide range of topographic differences (Lindsey *et al.*, 1965; Lindsey, 1997). Wetland and dry prairie, located primarily in the northwestern quarter of the state, made up the remaining 10% and 3%, respectively (see Figure 11.1).

This combination of agricultural clearing and timber extraction reduced Indiana's forested land to approximately 560,000 ha (~1,390,000 acres), or about 6% of the state by the early 1920s (Nelson, 1998). Since that time, the extent of forest cover has increased to over 1.6 million ha (4 million acres) (Nelson, 1998). Still, nearly all forest that existed prior to European colonization was cleared within the last 200 years. Today, Indiana retains only an estimated 0.06% of its old growth forest from its estimated original forest cover at time of European–American settlement (Davis, 1993; Lindsey, 1997). If selectively logged dry upland forest and glade areas are not included, Indiana may only retain approximately 0.01% of its old growth forest cover (Davis, 1993).

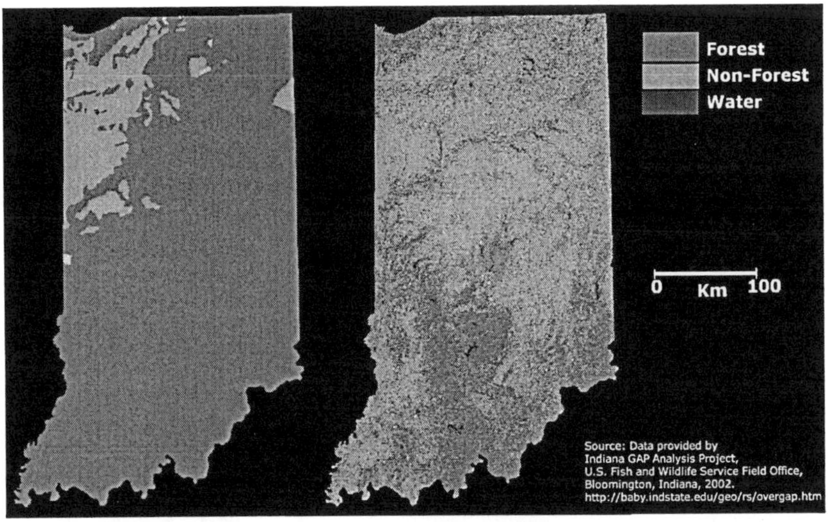

Landcover Circa 1820 (Lindsey 1965)

Landcover 1992 (Indiana GAP Analysis Project, US Fish and Wildlife)

Figure 11.1: Land-cover change in Indiana 1820–1992

The present distribution of forest cover differs dramatically from that of 1820 (Evans *et al.*, 2001a). Figure 11.1 shows a land-cover classification image for the state in 1992 (30m resolution). These vegetation classes were resampled to a spatial resolution of 80m using a nearest neighbor resampling algorithm and integrated with surface slope values derived from a 1:250,000 scale digital elevation model (DEM). Resampling data to a coarser resolution eliminates small patches in the landscape, such as an isolated forest patch (e.g. 20x20m^2) in the middle of a large agricultural field (e.g. 400x400m^2). For the purposes of this analysis the error associated with this resampling is deemed acceptable because: 1) many of these small isolated patches are misclassified artifacts and 2) the aggregate area of these small patches is relatively small and the resampled landscape has roughly the same land-cover class proportions as the original landscape. Currently, forests are preferentially distributed in areas of steeper topography, while agricultural land occupies flatter terrain. Table 11.1 shows the topographic distribution within the state of Indiana. The percentage of land in agricultural uses (which accounts for nearly 60% of Indiana's land-cover today (Table 11.2)) decreases from more than 70% of flat land (slopes < 3%) to about 10% of land at slopes greater than 10%. In contrast, the percentage of land covered by forest increases from 10% on flat to gently rolling topography to more than 65% of land with slopes exceeding 10%.

A series of processes have produced an environment that has experienced afforestation in the nine county study area: 1) establishment of state and federally managed forests in the mid-1900s and 2) agricultural land abandonment and subsequent forest regrowth on slopes marginal for agriculture (especially in the depression era of the 1930s, but also continuing today), and 3) economic changes, transition from economy dominated by timber and agrarian land-uses to a mosaic of land-uses including public and private timber harvesting and residential land ownership. South-central Indiana has also experienced a conversion of forest area to urban area, particularly near moderate-sized cities. However, over the last 80 years forest cover regrowth has outpaced the rate of urban expansion in Indiana.

Table 11.1: Proportion of topographic classes in the state of Indiana

Slope Category (%)	Area (%)
$S < 1$	66.2
$1 \leq S < 5$	28.1
$5 \leq S < 10$	4.4
$10 \leq S < 15$	1.0
$S > 15$	0.3

Table 11.2: Land-cover proportions in nine county area by slope category (all values are %)

Land-cover Class	Nine County Area		Low Slopes	Moderate Slopes	Steep Slopes
	Total	Private	$S < 4°$	$4° \leq S \geq 10°$	$10° < S$
Agriculture/Pasture	25.2	29.0	38.3	18.3	6.5
Advanced Sec. Forest	36.8	30.6	16.3	46.2	69.3
Initial Sec. Forest	22.2	23.8	25.5	23.1	15.3
Residential	9.6	10.4	11.7	9.0	7.0
Commercial	5.4	6.0	8.0	3.4	1.8
Water	0.8	0.2	0.3	0.1	0.1

This empirical information forms the initial 'targets' for our modeling enterprise. At the broadest level (both in terms of geography and time) we see very clearly a long-term pattern of rapid deforestation peaking around 1900 followed by a gradual process of afforestation in south-central Indiana. This is a temporal pattern where the amount of forest cover in the state rapidly decreased and then slowly recovered rather than a spatial pattern. The reforestation in the state has occurred primarily on steep sloped areas, such as that characterized by the hypothetical model landscape. A second target in the empirical data is the slope dependence of the forest regrowth – forests have regrown primarily in areas of moderate to steep topography. The goal of the model detailed in the ensuing sections is to reproduce at some level (at which level is discussed in the final section of the chapter) these patterns evident in the empirical record. We created a model to simulate the emergence of deforestation and afforestation processes as well as slope dependent regrowth in the hope that such a simulation will form the basis for a greater understanding and wider exploration of larger land-use/management questions.

11.4 MODEL SPECIFICATION[13]

11.4.1 The Agents and Their Environment

In order to investigate the empirical phenomena noted above, we constructed the LUCIM model which simulates the decisions and actions of a fixed population of farmers on a simplified artificial landscape. The farmer agents are heterogeneous in their initial wealth and education levels, and their decision making is based on either reinforcement learning or expected utility maximization algorithms.[14] Each agent owns only one parcel of land, and there is not yet any provision for buying or selling or other forms of interdependence.[15] Agent decisions are based solely on current information and previous experience, and only limited provision exists for planning several steps ahead into the future. For example, the devastating effects of cutting trees on soil conservation and soil productivity when cutting close to rivers has to be learned by experience in this model, rather than being anticipated by a more intelligent planning process. As another example, the farmer never anticipates or plans for productivity increases that occur due to fallowing land.

The landscape upon which agents take actions is minimally biologically active. The landscape is arbitrarily divided up into ten parcels, each owned by a single agent. The parcels themselves are further split into ten cells, the minimum unit of alterable land-cover considered.[16] The agents decide among three types of land-cover for each cell: agriculture, fallow, and forest. Biological activity is comprised of endogenous biofeedback and re-growth. The endogenous and exogenous physical characteristics of the landscape influence agents' decisions by altering the marginal productivity of alternative land-uses, leading to land-cover change. Each cell can have a unique slope, but the other physical characteristics of the land are uniform across the landscape. Forest growth rates are also uniform and the forest coverage on the landscape is initially set at 100% to mimic early frontier forest cover in Indiana. Agricultural cells are not differentiated into separate crops, and the costs involved with farming are uniform across the landscape. In addition, the commodity prices to which the agents react are currently generated from a simulated random walk process.

11.4.2 The Agents' Decisions

The agents' decision calculus is an offshoot of the optimal portfolio allocation literature from finance, based on the work of DeLong *et al.* (1990). The general philosophy underlying the use of this methodology is the belief that farmer agents make a portfolio optimization decision about the products or land-uses to pursue based on a capital asset pricing model structure. All agents make a farming decision to maximize or increase their utility which is composed of a portfolio of monetary/wealth items as well as non-monetary items such as the utility derived from subjective preferences for beauty, forests, etc. Through the utility function, agents form preferences about what they want their parcel to look like so as to improve their wealth or utility. The maximizing agents calculate their preferences directly from the utility function, while the reinforcement learning agents use the utility function to gauge how well their previous decisions fared. The agents then take land-use actions on individual cells to achieve the desired portfolio/parcel land-cover distribution.

Both the reinforcement learning agents and the utility maximizers use the same utility function which can be shown to be consistent with a constant absolute risk aversion utility expression, $U = -e^{-2\alpha w}$. This function can be shown to be equivalent to maximizing the expected utility expression,

$$E(u) = E(w) - \alpha\sigma_w^2 \tag{11.1}$$

where w refers to a specific agent's wealth, α represents the agents risk aversion, and σ_w^2 represents the undesirable, normally distributed random changes in wealth. This risk aversion principle is incorporated into the utility calculation for each of the three courses of action (farm, cut, fallow) that we currently include in the model. Whether this risk aversion structure is appropriate is the object of current experimental tests. Earlier literature tends to support this notion (Parks, 1995).

The agents' proposed utility structure has two main aspects: First, the utility of monetary returns from all land-use activities is sensitive to the agent's relative level of wealth. The profit from each period is added to an agent's wealth, so wealth gradually varies over time. As agents become wealthier, they derive less utility from the monetary returns from farming, fallow, and cutting activities, and derive more utility from letting the forest mature. These effects are achieved by scaling components of utility with $(1 - w)$ and w respectively.

Second, risk parameters α 's are unique to a particular agent. These parameters can be specified based on experimental or survey data, or with a priori descriptions of agents that correspond to particular time periods. Risk aversion, a component of capital asset pricing theory, allows optimal levels of land-use to be determined while simultaneously linking this utility architecture to a well-established empirical finding in agricultural and financial decision research; that is, the utility an agent perceives from a monetary reward is decreasing in the variability or variance in the reward level (Parks, 1995). We are not strongly committed to these particular functions at this point. One of the targets of the larger research project of which this chapter is a part is to evaluate these and other competing utility functions with experienced farmers under more rigorous experimental laboratory control.

Each agents' wealth, w, is composed of monetary and scaled non-monetary components for alternative land-uses: farming, fallow, tree cutting, or forest. The monetary components represent simple profit conditions relating current prices and costs to the quantity of farming, fallow, or trees produced. Rational agents actually maximize their momentary expected utility and choose among the three land-use actions to move toward the contemporaneous optimum portfolio of production given current conditions. Reinforcement agents, on the other hand, simply update their actions based on the level of utility obtained only from the actions taken in the previous period, without any forward-looking price, costs, or biofeedback considerations.

An important component of how agents form their expected utilities are the expectations they have for farm product, fallow subsidy, timber prices, and government tax policies. We assume that both types of agents are aware of the fact that these commodity prices evolve in a random way (Shiller, 1981). This assumption allows any given expected price to be represented by

$$p_{t-1}^{e} = p_t + \varepsilon_t \qquad (11.2)$$

where $\varepsilon \sim N(\mu, \sigma)$, $\mu = 0$, and p_t is the last period observed price. An alternative hypothesis, that agents display state dependent bounded rationality, can be introduced by altering the mean of the error term distribution for prices, μ. That is, if agents have a generally bullish sentiment about product prices, they think these prices will increase in the future, one can allow $\mu > 0$.

Bearish sentiment would be represented by $\mu < 0$. Under normal conditions $\mu = 0$ and any errors made when forecasting future prices are random, rather than systematically positive or negative.[17]

The different types of agents use two different decision methods derived from the same utility function. The utility maximizers employ a relatively sophisticated and knowledgeable decision rule. These agents determine the optimal production portfolio (farming, fallow, cutting) that maximizes current expected utility under the constraints of the parcel, perceptions of the current environment, and given expectations of the next period's taxes and prices. These agents are assumed to be aware of the fact that commodity prices follow a random walk process, so their best guess of the price during the next period is simply the current price.

Alternatively the reinforcement learners employ a less sophisticated and perhaps more realistic decision rule. It seems unlikely that all farmers, especially in the early frontier days, were very sophisticated decision makers. The reinforcement learning decision makers simply compare the levels of utility obtained from their actions in the previous period. They determine whether their previous actions generated a positive increase in total utility. If so, they increase the amount of that activity. Alternatively, if the previous actions decreased the utility, then they decrease these activities. To avoid allowing these reinforcement agents from falling into a no action sub-optimal farming portfolio, a 'try farming' shock is introduced into these agents' decision preferences once every ten years. This allows the farmers to test the waters. These agents do not use any forward-looking information. Instead, they use a one-period historical time horizon. These less sophisticated agents may play a key role in explaining why farmers seem to be so slow to react to the effects of their actions on the environment or slow to react to changes in the price structure and competitive nature of the agriculture market.

The specific utility functions used by both types of agents at the parcel level – to calculate optimal levels of land-use activity for the maximizers and to 'check' how past actions panned out for the reinforcement learners – are divided into three components. There is a separate expression for each activity, farming, fallowing, and cutting. Again the agents use these functions to generate parcel level preferences, in order to inform their cell level actions.

First, consider the definition of farming utility:

$$\begin{aligned} U_{farm} = {} & (1-w)[(farm \cdot p^e_{farm})(1-t^e_{farm})/(1+slope) - Cost_{farm})] \\ & -(\alpha \cdot farm^2 \cdot \sigma^2_{farm}) - (\beta_{farm} \cdot Age_{farm} \cdot farm) \end{aligned} \tag{11.3}$$

In this equation, w represents the current normalized wealth, and $1 - w$ represents the distance from the maximum wealth, assumed to be a large monetary value such as $1 000 000. We assume that attractiveness of farming is proportional to the distance from the maximum wealth. As the farmer becomes richer, there is less need or desire to continue the hard work of farming. This principal is also applied later to the utility specifications for cutting trees and fallowing the land. The product $(farm \cdot p^e_{farm})$ represents the expected profit produced by the proportion of a parcel allocated to farming times the expected price obtained by selling those farm products. In terms of real utility or wealth these returns are partially offset by the potentially variable income taxes an agent must pay on farming returns, i.e., $(1\text{-}t^e_{farm})$. Further, these taxes may be variable in as much as they are related to a particular governmental regime. The slope of the cell inversely affects this profit – high slopes are assumed to produce lower farm productivity. The second term in the brackets, $(Cost_{farm})$, is the cost incurred by the farming operation. The third term represents the risk aversion that agents display toward excessive farm product price and farming returns tax rate variance, i.e., $\sigma^2_{farm} = \sigma^2_{pfarm} * \sigma^2_{tfarm}$ respectively. The last term represents the loss in utility from farming due to the decreasing productivity caused by repeated farming of a particular piece of land (a linear decrease is an oversimplification that will be changed in the proposed work, but, currently, a small coefficient β keeps this effect within a reasonable range). This assumption essentially argues that farms will observe diminishing marginal returns to continued farming of the same particular parcel of land, i.e., they should pursue cell rotation. Whether cell rotation endogenously emerges from the system is one particular topic of interest.

Next, consider the expected utility from leaving the land fallow:

$$U_{fallow} = (1-w)[(p^e_{fallow} \cdot fallow \cdot (1-t^e_{fallow})] - (\alpha \cdot fallow^2 \cdot \sigma^2_{fallow}) + (\beta_{fallow} \cdot fallow \cdot age_{fallow}) \tag{11.4}$$

Again, 1 - w refers to a particular agent's normalized wealth level that decreases the desirability of fallowing the land as one becomes wealthier. Next, p_{fallow} represents the Federal set-aside land-subsidy price of abstaining from farming activities, *fallow* indicates the proportion of the agent's parcel devoted to set-aside land, and $1\text{-}t^e_{fallow}$ represents the income tax an agent must pay on these subsidy returns. The variance terms again capture farmers' aversion to excessive variance or risk in both set-aside subsidy levels and income tax rates on subsidy payments, i.e., $\sigma^2_{fallow} = \sigma^2_{pfallow} * \sigma^2_{tfallow}$ respectively. The last term represents the improvement in land productivity, and potential utility from the fallow decision, when it is set-aside continuously for a long period of time due to bio-regeneration (once again, a linear increase is an oversimplification that will be changed in the proposed work, but currently, a small coefficient β keeps this effect within a reasonable range).

Finally, consider the expected utility from cutting trees:

$$U_{cut} = (1-w)[(age_{forest} \cdot p^{e}_{tree} \cdot cut \cdot (1-t^{e}_{tree}) - Cost_{cut})] \\ -(\alpha \cdot cut^{2} \cdot \sigma^{2}_{tree}) - (w \cdot ed \cdot forest \cdot age_{forest}) \tag{11.5}$$

Again 1 - w represents the agent's normalized wealth and the decreased desirability of pursuing a career as a lumberjack and cutting trees as one becomes wealthier. The first term in the brackets is the real return from cutting trees, which equals the price of trees p_{tree} multiplied by the age of the trees, and the proportion of a parcel that is cut, denoted *cut*, and offset by the income taxes one must pay on tree cutting returns. Tree age is included because older trees are assumed to be worth more money. This profit is offset by the cost C_{cut} of cutting these trees. The last term in brackets are the risk aversion factors, $\sigma^2_{tree} = \sigma^2_{ptree} * \sigma^2_{ttree}$. Finally, the last term outside the bracket represents a farmer's esthetic pleasure obtained from forests, which would be subtracted from the utility obtained from cutting trees. Once again, based on preliminary survey data from farmers (Koontz, 2001), this esthetic affect is assumed to depend positively upon on the age of the trees, older trees are better, and on the farmer's wealth and education level.

11.4.3 Simulating Decisions

The model simulations begin by endowing each agent with a parcel that is composed of ten cells. As noted, the current landscape on which these differentiated agents are acting is only minimally biologically active. Each cell can be utilized as one of three types of undifferentiated land-use: agriculture, fallow, forest – though initially all cells are forested. The productivity of the agricultural use is influenced by the parameter 'slope', which is randomly assigned and varies across cells.[18] The productivity of all cells is also influenced by the parameter 'age', which represents the number of consecutive periods that a cell has remained in one land-use. This parameter is also used to simulate a transition between fallow cells and forest cells. When a cell remains fallow for a specified number of periods, it automatically becomes a forest cell with age zero.

The land-use decisions are carried out in the following manner. First each agent determines a parcel level preference. Each agent determines whether he or she would like to do more or less of each action (farm, fallow, cut) over the entire parcel. Rational expected utility maximizers calculate the first order conditions, i.e., marginal utility of the various land-uses from Equations

(11.3), (11.4), and (11.5), for farming, fallow, and tree cutting given their expectations of next period's prices. The optimal proportions of land-use for next period, based on the marginal utilities of each action conditional on expected prices, are then compared with the current proportion of land-use. Agents then hill climb toward the optimum proportions derived from the first order conditions at a rate that is a function of the distance to the optimal portfolio allocation relative to their current portfolio of land-use. For example, if an agent is very far from their optimal portfolio allocation they may change many cells to meets this optimum, while if they are close to the optimum they may only change one cell. The reinforcement learners, on the other hand, do form price expectations but do not perform any utility maximization. They instead use a one-period historical time horizon for utility changes so as to determine whether their previous actions generated positive utility. If so, they increase the amount of that activity, otherwise they decrease it. The rate of adjustment, i.e., how many cells are changed, is related incrementally to the magnitude of the utility change in the previous period.

Thus, at the parcel or macro level each agent begins each year with one of eight possible parcel level preferences: ABC (where: $A,B,C \subset (0,1)$ and 111 = do more of all three activities and 000 = do less of all three activities). The agents use these parcel level preferences to make decisions about individual cells, i.e., at a smaller scale. The individual cell decision takes place in two steps. First each agent searches its parcel for a specified number of cells – the delta[19] parameter – with the desired characteristics to change in accordance with the parcel level preference. If the parcel level preference contains incompatible choices (i.e. the agent wants to farm more and cut more but there are no forested cells available), in order to make a decision, the agent compares the *n*-step ahead expected utility of allowing fallowed land to become forested versus farming it immediately. In the second step, the agent actually alters the land-use of the chosen cells. Once the cell level decisions have been made, the agent calculates the aggregate statistics for his or her parcel (number of cells in each land-use), and uses these values to update wealth and utility, facilitating the calculation of next year's parcel level preference.

11.5 SIMULATION RESULTS

The results from simulation runs with the LUCIM model are encouraging. Even in this initial, simplified model, we have reproduced important observed land-cover change patterns. We are encouraged because our initial model predictions are consistent with some non-linear patterns of land-cover change observed during the frontier and post-frontier periods in Indiana. With fairly simple utility specifications and a very simple landscape we observe a highly non-linear time series of land-cover changes. Agents continuously balance their optimal or preferred portfolios of production. They respond to the vagaries of commodity prices and react to the endogenous landscape productivity changes. Specific features we observe include:

1) Agents deforest the landscape and then begin to let the forest regenerate.
2) Agents are sensitive to prices and the variance of prices.
3) Agents respond to both endogenous and exogenous landscape characteristics. Especially important are agents' reactions to varied farmland productivity across low- and high-sloped farmlands, which roughly correspond to glaciated and non-glaciated areas of Indiana.
4) Agents rotate their crops when exposed to diminishing returns to the amount of time a cell has been used for agriculture.

With the first set of simulation runs we hoped to reproduce a pattern of deforestation followed by afforestation given the decision rules detailed above. Our experiments along these lines consisted of model runs of 200 rounds to match the 200-year scope of the observed data in Indiana (roughly 1800–2000). For these runs we let the agricultural prices follow a random walk and worked with varying combinations of maximizing agents and reinforcement learners. Figure 11.2 shows the results of a Monte Carlo analysis of the output from 100 runs of a configuration with 80% maximizers.[20] It demonstrates that, indeed, the farmer agents do move into the 'pristine' forestland and deforest in order to farm. This is the first performance criterion for any model simulating land-cover change in Indiana. In these runs the agents are responding to commodity prices created through random walk processes. Even with this simple model, agents cut trees to farm, initiating a process of deforestation and afforestation.

The Monte Carlo results gave us confidence that our decision rules are at least reasonable – our agents do deforest to farm and then let the trees regrow eventually. With this foundation, we then moved to a more controlled analysis of the model in order to more fully explore the dynamics it produces. The subsequent experiments employ a sinusoidal price function – displayed in Figure 11.3 – for agricultural commodities and hold timber and fallow prices constant. Having controlled price inputs allowed us to explore other important aspects of the model. We also ran these experiments for more rounds (1000) to explore longer-term learning in our agents. With these controlled conditions we evaluated sensitivity to prices and price variance, slope dependence, and tenure effects.

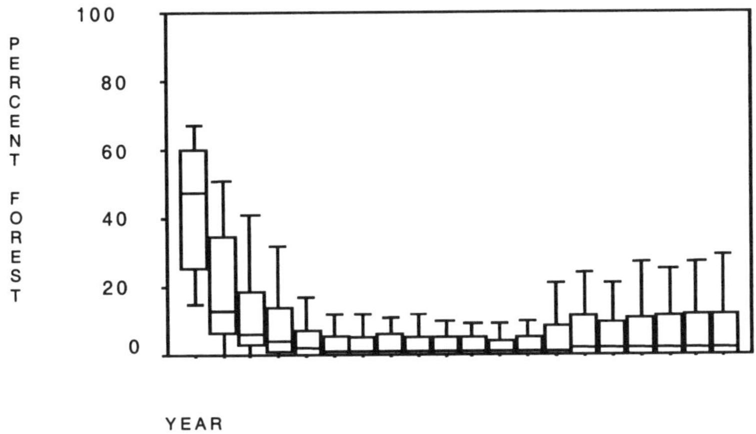

Figure 11.2: Forest on the landscape. Boxplot summarizing distribution of forest cover for 100 model runs of 200 years, in ten year intervals. Vertical axis represents the percent forest cover on the landscape. Agents are 20% reinforcement learners, 80% maximizing; landscape is half flat, half with random slope. (Bars are interquartile range and whiskers are range or range to furthest value less than 1.5 interquartile ranges from the median).

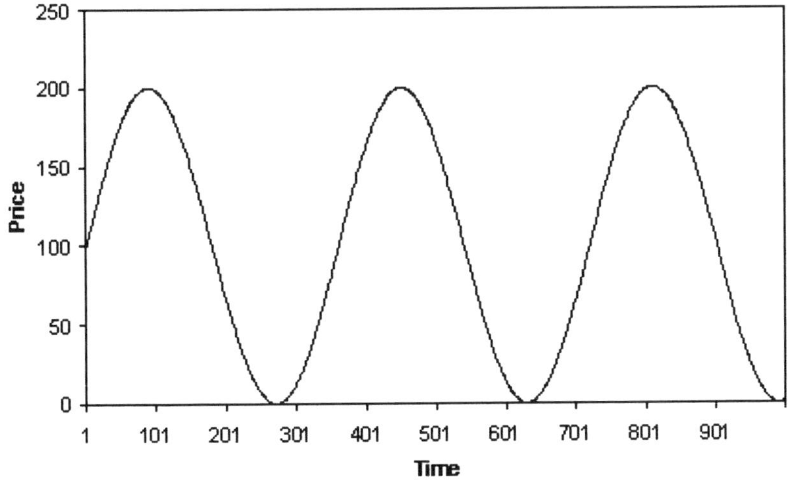

Figure 11.3: Prices for agricultural commodities

Figures 11.4a and 11.4b[21] chart the evolution of farming on the landscape for one typical run of the model. In Figure 11.4a we see that when agents are not concerned with the variance of prices, the number of cells devoted to agriculture across the whole landscape tracks prices extremely well – when agricultural commodity prices are high, agents put a large percentage of their cells in agriculture, and they remove them when prices drop. In contrast, Figure 11.4b reports the results of the agents' actions when an aversion to price variance is incorporated or 'turned on'. This risk aversion produces a noticeable change (this is a signpost result that gives us confidence in the model performance). Given the same price series, agents (who now are concerned with the variance of prices) reduce their exposure to risk by farming less even when prices are high.

Figure 11.4c demonstrates the impact that the agricultural practices of agents (without risk aversion turned on) had on the forest on the landscape. Here we see that while agents respond to varying agricultural commodity prices, the quantity of forest on the landscape varies over time (given constant timber prices). Tree levels begin high, drop off and regrow as farmers respond to agricultural commodity prices.

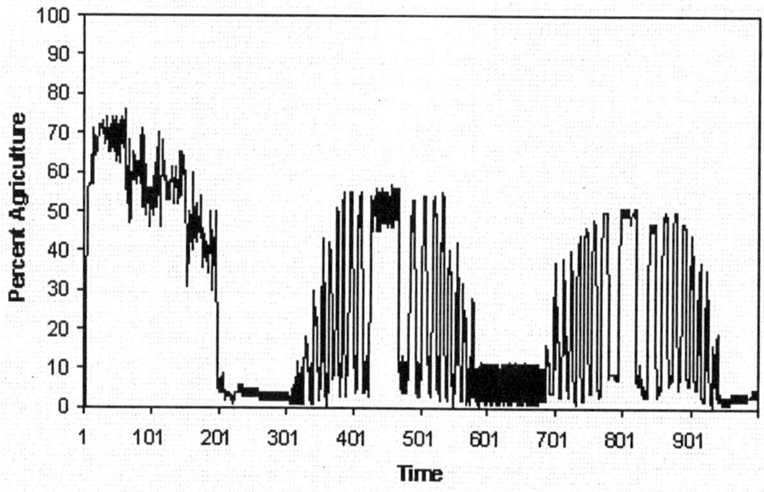

Figure 11.4a: Percent agricultural cells on landscape without risk aversion

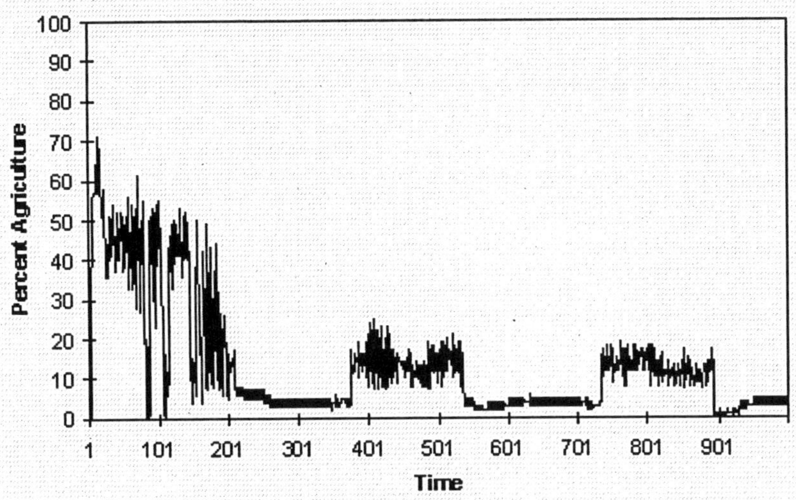

Figure 11.4b: Percent agricultural cells on landscape with risk aversion

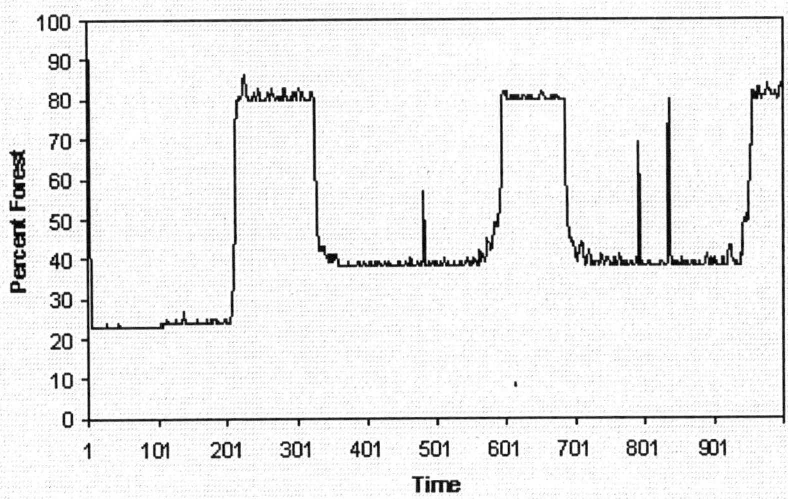

Figure 11.4c: Percent forest cells on landscape

We then turned to explore slope dependence – a crucial aspect of observed land-use, land-cover patterns over time in Indiana. Figures 11.5a–11.5c demonstrate that the model reproduces the observed slope dependence of land-cover in Indiana. In essence, when a high percentage of cells are being farmed, then cells with both high and low slopes are put into agriculture. When a moderate level of cells is being farmed, the agents learn to farm low sloped cells. This is one of the key signposts used to ascertain whether or not the model is producing empirically reasonable outcomes and whether our decision rules are reasonable. Forests in Indiana tend to regenerate first on land with high slopes – i.e., this land becomes marginal for agricultural use before flat land does. When land starts coming out of farms (as it has in Indiana from the 1930s through to the present), sloped lands should be taken out of agricultural use first. Our model was able to 'grow' this sloped–land-cover relationship without the model determining the outcome – i.e., having agents make decisions directly based on slope characteristics. Instead slope only affects farmland productivity, indirectly causing agents to reshuffle their portfolio of farmland toward lower sloped areas. This relationship is robust to changing proportions of reinforcement learning/maximizing agents.

Figures 11.5a and 11.5b report the average slope on agricultural cells across the landscape for the simulation runs displayed in Figures 11.4a and 11.4b. A pattern of slope dependence does not immediately present itself in Figure 11.5a. However close inspection does reveal expected outcomes. In the first 200 rounds, when most cells are in agriculture the average slope hovers near 5% (if all 100 cells were in agriculture, the slope would be 5.5%). When very few cells are being farmed, the average slope varies wildly as agents struggle in farming and keep trying random cells (remember that the reinforcement learners are forces to keep trying farming periodically). From the next episode of major farming (rounds 400–500), we see that the agents have learned to pick better cells as their level of farming has decreased. They are farming less and farming better as for a brief time (rounds 425–475) the average slope stabilizes at a lower value (around 2.5%). A similar result is evident near round 800.

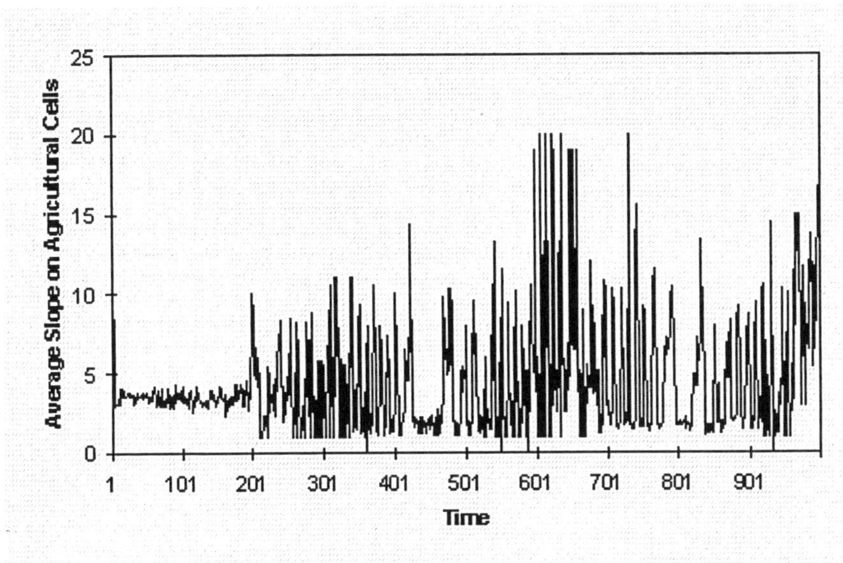

Figure 11.5a: Agricultural cells and slopes without risk aversion

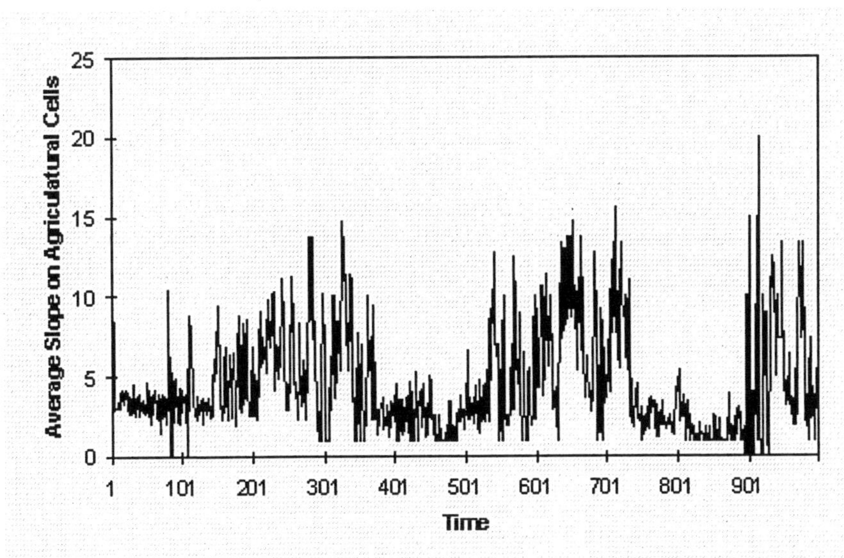

Figure 11.5b: Agricultural cells and slopes with risk aversion

The results are somewhat clearer in Figure 11.5b (refer back to the results from Figure 11.4b). When the agents farm some cells, but not a relatively high percentage of cells (especially in rounds 375–525 and 725–900), the average slope drops significantly and stabilizes at about 2.5% and 1.5% respectively. In sum, Figures 11.5a and 11.5b demonstrate that above a certain price threshold (again the same price series displayed in Figure 11.3 was used for these simulations), it is economically viable to farm cells with any slope (low or high), and thus the average slope of agricultural cells approaches the overall average slope for the entire landscape (if all 100 cells were in agriculture, the average slope would be 5.5). However, when the price drops below the threshold (the level of which is variable and a factor of other model parameters), rapid change occurs in the number of cells farmed, and agriculture only remains on cells with a low slope. A pattern of land-cover dependent on slope emerges from the model without the agents' making conscious decisions about slope – slope only enters the decision process as a 'cost' to doing farming. This is encouraging, as our empirical data points to this exact relationship as well.

In order to more clearly demonstrate the slope dependence that follows from the decision rules of the model, we set up a slightly different experiment. For this new experiment we held all prices constant and altered the slope of each cell on the landscape at specified intervals. The result of this experiment is documented in Figures 11.5c. The left *y*-axis is the percentage of cells in agriculture and the right *y*-axis is the slope of the cells on the landscape. Thus, for the first ten rounds of the simulation, the slope is at 3%, the next ten rounds, the slope is at 4%, and so on up to 21%. What we see with the (very artificial) manipulation is a clear threshold effect and slope dependence. Farming begins to increase, but then drops sharply as the slope rises past 9%.[22] (Some volatility remains at higher slopes because the reinforcement learners are forced to try farming periodically.)

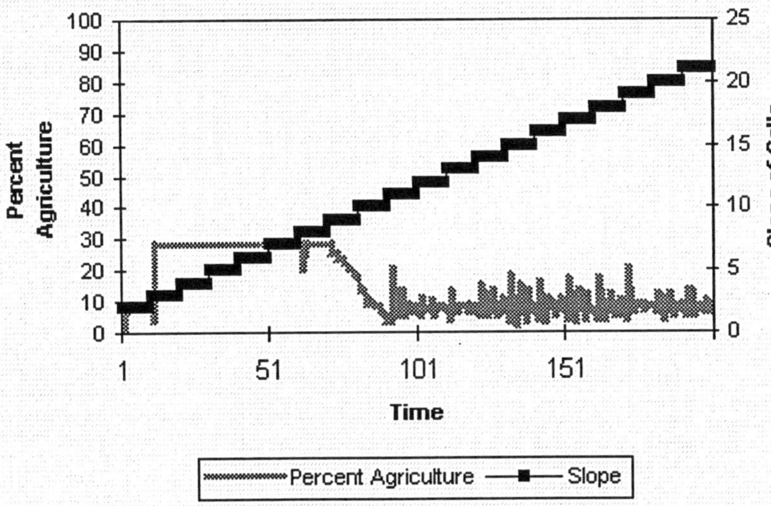

Figure 11.5c: Slope effects with constant prices

Finally we experimented with aversions to keeping a cell in agriculture for extended periods of time, attempting to see if our model would produce crop, or cell rotation. Figures 11.6a and 11.6b illustrate this endogenous emergence of cell rotation as agents respond to the diminishing marginal returns of continually farming the same cells. In Figure 11.6a, there is no 'penalty' for keeping a cell in agriculture. Thus, as the agents farm cells to reap the benefits of high agricultural prices, the average age of agricultural cells increases linearly, while the percentage of cells used for agriculture again tracks the sinusoidal price series. However, this is not realistic because, due to soil degradation, diminishing returns set in the longer a cell is used for agricultural purposes. Figure 11.6b demonstrates the effects of turning on diminishing returns to age. In this figure we observe that cell rotation emerges from the agents' desires to maximize or increase utility. Given the same price series used for Figure 11.6a, while the agents still desire to initially keep a high percentage of their cells in agriculture (especially in the first 150 rounds), they have to rotate the cells that they use periodically to avoid the effects of diminishing returns. The maximum average age of farmed cells drops from over 100 to around 35, and the sawtooth pattern evident in Figure 11.6a occurs more frequently. This indicates that if a farmed cell reaches a large age (say 50), it is swapped out in favor of a new (more productive) cell, reducing the average age of the farmed cells across the landscape.

The experiments conducted with the LUCIM model served two broad purposes. First, we demonstrated the initial reasonableness of our decision-making framework. We were able to reproduce some empirically observed patterns, and demonstrated that our decision rules represent a candidate explanation for some aspects of land-use/cover change in Indiana.[23] Second, the experiments served as a preliminary analysis of the possible. They began the process of exploring what a model based on our particular decision-making framework will produce – an important first step in evaluating the model as a tool for explaining and understanding changing land-use/cover processes in Indiana.

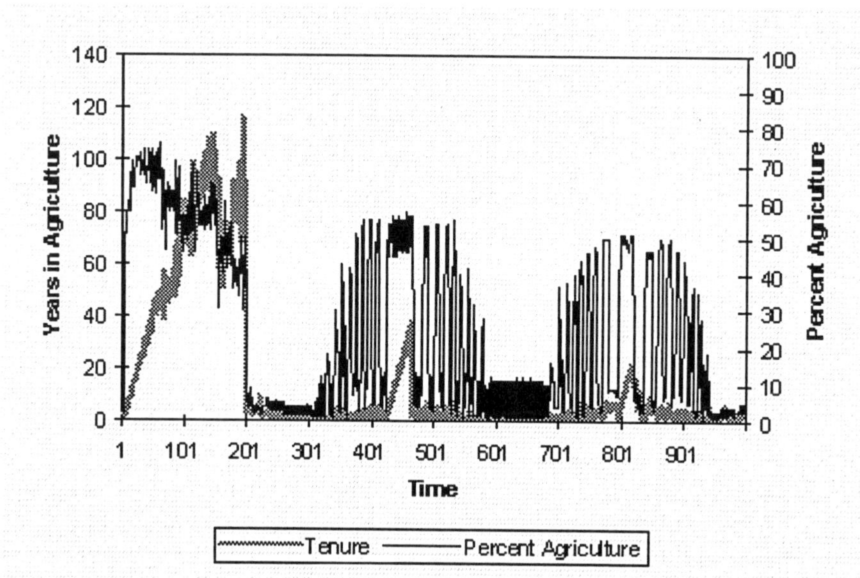

Figure 11.6a: Percent agriculture and agricultural tenure without diminishing returns to age

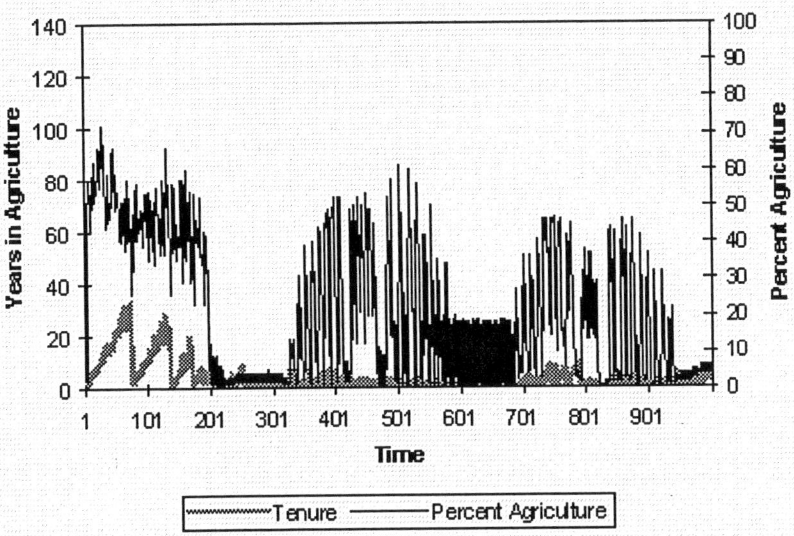

Figure 11.6b: Percent agriculture and agricultural tenure with diminishing returns to age ,

11.6 VERIFICATION AND EVALUATION

The results of our simulations indicate that our agents are sensitive to price levels, price variance, and diminishing marginal returns to continuously farming the same cells. In addition, at a macro level, slope dependence and patterns of deforestation and afforestation emerge from the model. These results are encouraging, but the ramifications of the simulations are not immediately obvious. Indeed, the results raise a larger question: how do we evaluate the results of agent-based models? What are we learning through these simulation exercises? What can we learn through these simulation exercises? These are especially important questions in light of the knowledge that, 'in every case of simulating complex adaptive systems, the emergent properties are strictly dependent on the "rules" preprogrammed by the investigator'. (Fogel *et al.*, 1999, p. 146).

Evaluating agents

One way to evaluate the model is to focus on the agents and their decisions. A helpful set of criteria for this kind of evaluation has been articulated by Rob Axtell and Josh Epstein (1994). They discuss four levels of model performance, used for evaluating the results of agent-based models, as shown in Table 11.3.

Table 11.3: Model performance of agent-based models (Axtell and Epstein, 1994: p. 29)

Level	Performance
0	Model generates agents whose behavior qualitatively matches 'real agents being studied
1	Model generates macro patterns that qualitatively match empirical patterns
2	Model generates macro patterns that quantitatively match empirical patterns
3	Model generates micro patterns that quantitatively match empirical patterns

Moving up the levels of model performance implies increasing understanding of the empirical situation/system being modeled. Thus, once a model reaches level 3 performance, we can be confident that model simulations are generating specific insights about a particular situation. Lower levels of performance imply a more general level of insight.

The LUCIM model results lead us to conclude that our model is currently performing at level 0 and level 1. The agents display price sensitivity and are influenced by aversions to both price variance and diminishing returns to the age of farmland. These results demonstrate that the agents in the model are acting as we designed them to – to mimic the behavior of hypothesized farmers. The fact that the agents are sensitive to prices and aversions gives us confidence that our model is working as designed and that we have created the agents to 'look' like we had hoped. The agents achieved level 0 performance.

The slope dependence, cell rotation, and re/deforestation patterns that emerge in the model demonstrate level 1 performance. The rules embedded in the model produce macro patterns that we expect to see given our knowledge (and data) of empirical forestation patterns in Indiana. Even though LUCIM is simple, contains one class of agent, and displays minimal biological activity, we are able to generate important empirical patterns. Thus, we have learned something general about forestation patterns – the rules we used in the model are sufficient to produce some of the empirical patterns we have observed. The model has achieved some degree of level 1 performance.

11.6.1 Verifying Landscapes

LUCIM can be evaluated through other criteria as well. The purpose of LUCIM is to model the decision-making process of landowners on smallholdings and the land-cover outcomes of these decisions. Model performance can also be evaluated in terms of how well the model produces land-cover compositions and patterns observable from land-cover change data. Land-cover composition is the percentage of the landscape in different land-cover classes (e.g. forest, agriculture). Landcover pattern is the spatial arrangement of these land-cover classes. For example, two parcels can have the identical land-cover composition but very different spatial patterns as is the case with a parcel with a checkerboard arrangement of forest and agriculture compared to a parcel with two homogenous areas of forest and agriculture. In this regard, model performance can be evaluated in these land-cover terms analogous to the agent decision based criteria of performance. Landcover pattern and composition are indicators of the ecological function of the landscape (Forman and Godron, 1986; Turner and Gardner, 1991; Turner *et al.*, 1996). This function can be related to processes/indicators such as species habitat or carbon sequestration.

Table 11.4: Model performance of agent-based land-cover change models

Level	Performance
0	Model produces landscape whose composition qualitatively matches that observed from field data
1	Model produces landscape whose composition quantitatively matches that observed from field data
2	Model produces landscape whose pattern qualitatively matches that observed from field data
3	Model produces landscape whose pattern quantitatively matches that observed from field data
4	Model produces landscape whose spatial function qualitatively matches that observed from field data
5	Model produces landscape whose spatial function quantitatively matches that observed from field data

Models that are not spatially explicit (e.g. Evans *et al.*, 2001b) can only perform at level 0 or level 1. A spatially explicit model that reasonably produces land-cover compositions but does not adequately portray the patchiness or pattern of the landscape would also be evaluated as performing at levels 0 or 1. If the model also does represent the pattern of different land-cover classes then the model may perform at levels 2 or 3. However, the performance of spatially explicit models is highly dependent on the spatial and temporal resolution of the model, and models that operate at very coarse spatial scales cannot reasonably be considered to adequately represent the true spatial pattern of heterogeneous landscapes.

Models that are spatially explicit may be able to construct landscapes that produce land-cover patterns and compositions seen from field data analysis but do not accurately represent specific functions of the landscape. For example, given a model designed to represent the impact of land-cover change on bird habitat, the model may accurately portray changes in the landscape but not the behavior of birds within the landscape. Thus, land-cover change models need to be associated with specific questions of land-cover change implications to perform at levels 4 and 5. There is a lack of consensus over what spatial metrics are representative of landscape function (Frohn, 1998). What complicates the evaluation of models using the criteria described in Table 11.4 is this contention over what spatial metrics should be used to quantitatively compare landscapes. But it is clear that the selection of specific spatial metrics is highly dependent on the objective of the modeling process.

Given the land-cover based system of model evaluation described above, the model presented here operates between levels 0 and 1. Subsequent versions of the LUCIM model will involve more elaborate spatial interactions that will allow us to evaluate the model performance at different levels on this land-cover based scale.

The goal for the model is thus clear – to move up performance levels for both types of criteria (agent and landscape). We are not claiming that this model captures all or even a large part of the dynamic processes of re/deforestation. Instead with simple rules and agents we have been able to generate some noteworthy empirical patterns. The LUCIM model is a first step in understanding re/deforestation in Indiana.

11.7 CONCLUSION

This chapter reports a first step in a necessarily extended process. Patterns of land-use in Indiana and elsewhere result from the interaction of numerous biological, social, physical, economic, and political factors. Sorting through the complexity of these processes is no mean challenge. We have chosen to begin with agents, endowing them with decision rules and putting them on a landscape in an attempt to 'grow' recognizable land-use patterns. We are encouraged with our first steps. The model designed for this chapter has begun to produce patterns that are recognizable and appears to have the potential to go further.

11.7.1 Further Research Avenues

A successful agent-based modeling approach can illustrate how individuals interact within a simple dynamic environment producing emergent patterns that no one individual could produce alone. In specifying such systems detailed information about, for example, the biological, economic, and political landscapes are required. Further, historical, cultural, and family precedents, preferences, attitudes toward risk, education levels, and subjective preferences about environmental protection all combine to yield the complex patterns of forestation. This final section discusses some of the continuing research and additions to the model that we hope to include in future versions.

In terms of the landscape, currently the only biological/physical aspects incorporated into the landscape are slope and forest growth rate. In future versions we will include provisions for more sophisticated forest classification and succession, slope aspect, soil classification, anthropogenic alteration of hydrology, transportation systems, and urban/suburban expansion. In addition, a long-range goal is to integrate this model with GIS models and have the agents interacting on an empirically accurate landscape.

We also plan to modify our agent population. First, we will allow agents to own more than one parcel, to buy and sell cells, to learn from each other through diffusion processes, and to affect each other directly through spatial externalities. These extensions will provide important interconnecting links between agents, and also allow us to predict changes in average parcel size with time. The next step is to include more advanced learning algorithms into dynamic bioeconomic optimal choice models that could allow agents to maximize utility within a present value framework. This allows agents to consider the long run time horizon.

Finally we will add other types of agents: land developers who may buy or sell land based on preferences for urban living, industrial developers who may pursue alternative industrial uses of the land, gentleman/woman farmers who may purchase a farm because of a love of trees or the country but who earn their livelihood elsewhere. Also, in future work we plan to allow agents to migrate in or out of the system. At this point, actions are limited to only three uses of land: farming, cutting forest, and leaving land fallow. In model development, forests may be cut and then left fallow to become forests with later regrowth, or forests may be cut to make way for farmland.

To reiterate, the current model is a fairly sophisticated decision-making simulation that lacks social interaction and biophysical complexity. Yet, it is able to reproduce *some* empirical patterns that are characteristic of land-cover evolution in Indiana. The results point to a robust pattern of slope dependence in the model and threshold effects arising from the interaction of slope and varying commodity prices and government subsidy levels. These results mimic what we know about the frontier and post-frontier time periods in Indiana. We are thus confident that our model captures a component of the dynamics of land-cover evolution in Indiana and that through adding complexity and sophistication to this foundation, we will be able to capture more.

ACKNOWLEDGEMENTS

We would like to thank Lin Ostrom and the other members of the Biocomplexity team at the Center for the study of Institutions, Population, and Environmental Change for input and feedback on previous drafts as well as encouragement throughout the project. This chapter was previously presented as a paper at the Microbehavior Macroresults Conference of the International Fisheries Economics and Trade Association in July 2000.

NOTES

1. The analysis presented in this chapter is drawn from preliminary work of a larger land-use and land-cover change modeling project. This larger project is funded by the NSF Biocomplexity Grant Program.
2. In this discussion, we consider the process of reforestation to involve the active planting of seedlings and saplings to promote forest regrowth. Afforestation is the process of natural forest regrowth as would occur with an agricultural field that was abandoned and left unattended. While south-central Indiana did experience reforestation facilitated by government programs intended to mediate the impact of soil erosion on cleared land (especially ~1930s–40s) we consider forest recovery to be more a process of afforestation for the purposes of this model.
3. This paper reports our efforts at implementing one of these four types of relationships – agent/environment interactions. Continuing work is focused on implementing the other types of relationships.
4. 'Similar' can have many meanings from a qualitative 'look' to a quantitative measure. We will explore notions of similarity and discuss validation in Section 11.6.
5. This model was developed in C++ using Microsoft Visual Studio. The code is available from the authors upon request.
6. As it stands now, the model is not explicitly spatial as the agents do not interact with or have awareness of their neighbors. We are currently working to make the model spatially explicit and to include agent interactions. This preliminary model focuses on agent decision-making processes.
7. For an article that reviews numerous modeling approaches to this subject see Parker *et al* (2001).
8. Good introductions to complex adaptive systems can be found in: Waldrop (1992), Holland (1995), Epstein and Axtell (1996), and Arthur (1994).
9. The idea of more than the sum of its parts is well known, Robert Jervis coined the phrase 'different from the sum of its parts', Jervis (1998).

10. Again the model developed below is mainly centered on decision making and the feedback between the environment and individual, heterogeneous agents. Research and model development is currently underway to make the model spatially explicit and to include agent–agent interactions.
11. This being said, one aspect of the larger project of which this paper is a part is concerned with applying techniques of spatial econometrics to the land-use patterns in Southern Indiana. The parallel modeling projects should dovetail to enhance the results of both.
12. In the larger project, it is more than likely that the final model will utilize CA techniques for the landscape.
13. In this section we outline the model in its 'as is' form. In the conclusion we detail the numerous areas of the model where further research is ongoing or planned.
14. In the simulations we vary the 'mix' of the two types of agent in the population.
15. This is a noted and obvious flaw. At present, each agent is an island, which is unrealistic. In reality, externalities, ecological, social, and economic, are likely have a large influence on decision making. We are working to incorporate these into future versions of the model. However, for this baseline attempt with its very modest goals, we did not feel the simplification negated the utility of the simulations.
16. The choice of using ten agents with ten parcels and ten cells per parcel was arbitrary. Because the agents do not interact, the population size should be fairly irrelevant, but ten agents provided us with enough to experiment with agent heterogeneity. Splitting the landscape into 100 cells was done for ease of calculation. The dynamics should be similar regardless of the number of parcels and cells. Again this was an entirely artificial landscape with a minimum of biophysical characteristics used for the most part to develop the agent decision-making part of the model.
17. Currently the prices are generated through a random walk process, however we have reliable empirical data on actual commodity prices dating back to the mid-1800s and eventually we will be using these data as an input for the model.
18. Empirical studies have shown that the slope of the land is a good predictor of whether or not it will be farmed. However, in this simple model, slope can also be thought of as a surrogate for all the physical determinants of agricultural productivity – i.e. soil type, hydrological factors, as well as slope.
19. Delta is initially one, but after initialization, for maximizers delta is adjusted proportionally to the Euclidean distance of the current portfolio to the optimum portfolio. Alternatively for reinforcement learners, delta is increased or decreased by one when the variance of past utility increases or decreases, respectively.
20. Other combinations showed similar results. The rest of the results displayed are drawn from simulations with this same configuration. We performed experiments with a number of different population mixes between maximizers and reinforcement learners, but have yet to fully analyse the results of these experiments. The general patterns discussed appear to be robust to changes in the agent type mix, but we can only claim that the results shown are illustrative examples of this particular configuration.
21. Unless otherwise noted, the results are drawn from simulations of ten agents (eight maximizers) with the same sinusoidal agricultural commodity price series. The landscape for all simulations consisted of 50% cells with slope 1 and 50% cells with randomly assigned slopes between 1 and 20. This slope mix is designed to very roughly represent Monroe County, Indiana – the site of our larger research project. Northern Monroe County is low-sloped agricultural land, while southern Indiana is variably sloped.
22. The value for the threshold effect is obviously dependent upon the constant agricultural price (i.e. if the price is high, the threshold will be at a higher slope).
23. For more on the notion of candidate explanations see Epstein (1999).

12. Multi-agent systems and role games: collective learning processes for ecosystem management

François Bousquet, Olivier Barreteau, Patrick d'Aquino, Michel Etienne, Stanislas Boissau, Sigried Aubert, Christophe Le Page, Didier Babin and Jean-Christophe Castella

12.1 INTRODUCTION

For several years now, a field of research on simulating societies in interaction with their environment has been taking shape. Methods such as multi-agent systems (MAS) can be used to create virtual societies. The effects of interactions among different behaviors on resource dynamics and associated feedback are simulated. Modelers use such methods to create computer representations of dynamics observed in the field. The multi-agent methodology, which results from research on complexity, corresponds to the evolution of concepts described by Attonaty *et al.* (1990): after using operational research methods to identify the solution to a given problem, researchers turned to ways of representing a stakeholder's strategic and tactical decisions through expert systems or simulation methods. Then, multi-agent systems provided a new tool for the modeling of complexity and particularly the modeling of collective decisions. Our team[1] has been using MAS in several ways.

1. First, we developed abstract models, also called artificial societies that help to understand the generic properties of interacting processes. We mainly developed models on non-merchant exchanges and reputation (Rouchier and Bousquet, 1998; Rouchier *et al.*, 2001b; Thébaud and Locatelli, 2001), models on economic tools for the regulation of economic

exchanges (Antona *et al.*, 1998), and models on spatial dynamics (Bousquet and Gautier, 1998; Bonnefoy *et al.*, 2000).

2. We also developed applied models to understand the dynamics of natural and renewable resources and their management (Barreteau and Bousquet, 2000; Bousquet *et al.*, 2001; Rouchier *et al.*, 2001a).
3. We developed a modeling tool (CORMAS, *common-pool resources and multi-agent systems*) (Bousquet *et al.*, 1998).
4. We also worked on defining the possible uses for these MAS tools within the very wide framework of collective decision support. The objective of this chapter is to propose a method and present our first results.

The use of MAS models is a key issue. As with any kind of representation, these models can be used to increase scientific knowledge about the ecological and social processes at stake. The question of the organization of this learning process remains. There is an old debate between two paradigms (Checkland, 1981; Funtowicz and Ravetz, 1994; Röling, 1996). Schematically, on the one hand, researchers following a positivist paradigm try to discover the objective truth. This knowledge is used to develop and deliver new technologies or new management rules. On the other hand, the constructivist paradigm assumes that reality is socially constructed. As Röling (1996) said, 'Based on their intentions and experience, people construct reality creatively with their language, labor, and technology. Different groups do this in different ways, even if they live in the same environment. The same people change their reality during the course of time in order to adjust to changing circumstances.' Although the scientific knowledge on biophysical and social processes is to be used, this second paradigm is the epistemological context of our research. The collective creation of a common artificial world serves to create a shared representation and to simulate various scenarios with the stakeholders, among which are the scientists. We have been interested in using MAS models in the collective decision-making process for ecosystem management. Unlike the conventional decision-making process derived from Simon's model (Simon, 1947), which defines a decision as a rational calculation on the part of a more or less fully informed decision-maker, our studies see the decision-making process as a series of interactions between stakeholders with various objectives and with varying degrees of importance and influence. Within this framework, any decision, particularly if collective, is context-dependent and should be seen as a stage at a given time *t* in the continuous process of management of a complex issue. The emphasis has shifted from the static to the dynamic and from an individual to a collective level. Several approaches for collective ecosystem management exist.

Adaptive management (Holling, 1978; Walters and Hilborn, 1978) is an approach based on the recognition that ecosystem management requires flexible, diverse, and redundant regulation, monitoring that leads to corrective responses, and experimental probing of the continually changing reality. Although adaptive management is an approach conceived by ecologists, they recognize that adaptive capacity is dependent on knowledge – its generation and free interchange – and the ability to recognize points of intervention and to construct a bank of options for resource management. Thus, interactions with stakeholders for the generation and interchange of knowledge are required. This social process of generation and free interchange of knowledge may lead to new interactions and thus to the issue of devolution of power. Co-management (Berkes, 1997; Borrini-Feyerabend *et al.*, 2000) is defined as a partnership in which local communities, resource users, governmental agencies, non-governmental organizations, and other stakeholders share, as appropriate to each context, the authority and responsibility for the management of a specific territory or a set of resources. Patrimonial mediation is presented as one of the approaches contributing to understanding and practicing co-management (Borrini-Feyerabend *et al.*, 2000). 'Patrimonial' is defined by Ollagnon (1991) as 'all the material and non-material elements that work together to maintain and develop the identity and autonomy of their holder in time and space through adaptation in a changing environment'. A patrimonial representation of a territory, an area, or a set of resources links past, present, and future generations of managers, focuses on the owner's obligations more than on the owner's rights, and promotes a common vision of sustainability that reconciles the needs and opinions of various actors. Mediation is a negotiating method that brings in a third, neutral party in order to obtain agreement among the parties involved in the process; it is an approach in which each party's views on the issue or problem are translated for the others (Babin and Bertrand, 1998).

In our research, we try to develop an approach for the use of simulation models, in particular multi-agent systems, within the context of such patrimonial mediation experiments. This method uses role games to acquire knowledge, build a MAS model, validate the MAS model, and use it in the decision-making process. We named it 'companion modeling' because it is used as a tool in the mediation process – the social dimension of the companion – and it co-evolves with the social process, temporal and adaptive dimension. We started different applications to assess whether models and role games can be used successfully together to support collective decision making, and to explore different uses of these coupled tools. Six tests of the approach have been conducted and they illustrate the article. The goals of the different experiences are:

- To simulate the variability of irrigation schemes in Senegal. A model was created and a role game designed to present the model to the various stakeholders. The model and game were used in villages and at a training centre.
- To help with mediation in biodiversity management processes in Madagascar. A spatialized resource dynamics model was programmed for use as a dynamic support for negotiations between stakeholders.
- To predict the impact of silvo-pastoral development within the framework of forest fire prevention plans in a region in the French Mediterranean. A model of negotiations between foresters and livestock farmers was developed in relation to Mediterranean forest management. It was then simplified into a role game with a view to producing a typology of land management strategies and negotiating methods, and to providing a teaching aid to develop management capacities or explain the constraints relating to each user. This was subsequently used for training purposes.
- To simulate scenarios of adaptation to the spontaneous establishment of conifers in natural ecosystems with high patrimonial values in the South of France. A model of herders' individual practices and actions of conservationists was developed and it led to a role game. During the role game, the model served to simulate the dynamics of the resource.
- To test hypotheses about the differentiation of households at the end of the cooperative period in a region in northern Vietnam. A role game was proposed to collect information and feed a MAS model that is presented to stakeholders for validation and discussion.
- To facilitate discussions between populations and their local representatives in the case of the decentralized allocation of land in the Senegal River delta. During village workshops, role games were used to help stakeholders develop a common model, which was then computerized in the form of a MAS. The model was then used to simulate the scenarios imagined by the stakeholders, which triggered a group discussion of the possible evolution of the interactions between users and resources and of the steps that could be taken to adapt to these changes.

Our chapter begins by covering the principles on which this approach was based. We then go on to present the various tests that demonstrated that it is possible to use models in this way, and indeed that this method is effective in supervising a collective decision-making process. The different uses of role games and models in the mediation process for co-management of resources are then analysed. In some cases, the model simulates the dynamics of a resource and provides a dynamic basis for a negotiating game, while in others the scientists' model has been converted into a role game with a view to validating or more clearly explaining stakeholder strategies. In the remainder of the cases, role games are used for the collective construction of a MAS, which is implemented and then presented for discussion. Lastly, we draw some conclusions from this work and suggest certain prospects, particularly the issues of institutional arrangement and assessment of the process.

12.2 MODELS, GAMES, AND STAKEHOLDERS: A REVIEW

The development and use of MAS models and role games for collective decision making in natural resource management (NRM) is very new. In this review we start presenting contributions on interactive modeling which do not use these tools but give interesting conclusions or adopt different approaches. Firstly we present selected contributions in participatory modeling (with other tools than MAS and role games) to support problem solving. Then we present contributions on the use of models and experiments to explore theories on human behavior. We end with the few contributions on the joint use of models and role games for education and management of resources.

12.2.1 Participatory Modeling to Support Problem Solving

The relationships between simulation models and collective decision making in natural resource management occupy a large part of the literature on adaptive management (Holling, 1978; Walters and Hilborn, 1978). Scientists from different disciplines and some decision makers develop a model and explore various scenarios. However, the model is often a biophysical one and very few papers mention the participation of stakeholders from different organizational levels in one of the modeling steps (from conceptualization to scenario simulation). We present here the ecosystem simulation models developed by Costanza and Ruth (1998), two references on integrated assessment models, and some Australian experiments (see Abel *et al.*; Chapter 13).

- Costanza and Ruth (1998) advocate the use of a three-step modeling process, with all steps involving the stakeholders. They use graphical STELLA programming, which allows the simple design of non-linear dynamic systems. The first step is to develop high-generality, low-resolution scoping and a consensus-building model involving broad representation of stakeholder groups affected by the problem. The second-stage research models are more detailed and realistic attempts to replicate the dynamics of the system. It is still critical to maintain stakeholder involvement and interactions at this stage of model progress. The third stage of management models focuses on producing scenarios and management options based on the earlier scoping and research models. They tested this methodology for Louisiana coastal wetlands, South African Fynbos ecosystems, and the Patuxent River watershed. The main lesson drawn from these experiments was that participation of various stakeholders has to be maintained through the various steps of the modeling activities. The second lesson was that these are long processes lasting for several years.

- Abel (1999, Chapter 13) presents the main principles of his approach on rangelands as complex adaptive systems. To understand these systems, one needs to elicit the mental models of the stakeholders. A mutual exploration of these mental models in a collective learning process is necessary for better management. In the project, stakeholders (aboriginal peoples, pastoralists, conservationists, and the minerals and tourism industries) developed scenarios of sustainable patterns of land use and policy makers designed the institutional changes needed to support changes in those scenarios. Researchers organized the stakeholder and policy processes and estimated the sustainability of land-use scenarios using conventional models. In this approach, stakeholders guide the questions and at the same time gain an understanding of processes, such as climatic changes or regional economic growth, with which they may not be familiar. Their scenarios were developed using a process and software developed by Cockes and Ive (1996): a spatial information-based mediation support process for resolving resource-use disputes, SIRO-MED. It is supported by the software WinLUPIS (Land Use Planning and Information System). It uses rules proposed by stakeholders to allocate land uses to mapping units. Stakeholders enter negotiations in which their land uses compete for land with the land uses of other stakeholders. The mutual adjustment of land allocations generally leads to an efficient and politically acceptable mix of land uses. WinLUPIS offers a range of tools to aid this process. WinLUPIS is not a simulation model, but it is an interesting experiment reflecting a shift in the use of decision support systems. The authors state that there is public demand for the procedural legitimacy the process offers. The policy process was also participatory: practitioners worked with researchers to build influence diagrams as a basis for understanding and then proposing changes in laws, policies, and organizations.
- Van Asselt *et al.* (2001) have identified two examples of participatory modeling in the field of integrated assessment. The RAINS (Regional Air Pollution Information and Simulation) model started as an IIASA project

(IIASA, 1999) and was subsequently used as the worldwide standard model, with scientific experts and policy makers participating in the model design. QUEST (Quite Useful Ecosystem Scenario Tool) is a computer game (Rothman *et al.*, 2001). The approach focuses on learning through scenario construction rather than through scenario results. Both of these models were not participatory models when their design started.

Little has been done in participatory modeling but lessons can be learned from participatory geographic information systems (GIS) research in NRM. This is a more advanced field of research. Today, manuals are available for extension agents in rural areas (CBFMO, 1998). Participatory GIS have demonstrated the ability of many illiterate people to use high-tech tools. The following authors have survey papers:

- Fox (1998) presents the principle of community mapping. Several responses followed his article. The use of spatial technology for community mapping and counter mapping is very useful but has some weaknesses. Constructive criticisms have to be borne in mind when using simulation models. Among others, the design of maps destroys the fluid and flexible nature of boundaries, increases the potential for surveillance and violation of privacy and ownership of information.
- Abbott *et al.* (1998) review participatory GIS. The same kind of constructive criticisms are made. GIS, like many technological innovations, both marginalizes and empowers people and communities simultaneously. Issues of confidentiality and privacy might be exacerbated: there are risks that local knowledge will become recorded and centralized.

One interesting experiment, among others, is the work of Gonzalez (2000) on watershed management among the Ifugao (Philippines). The author presents GIS as tools for interactive learning, to facilitate what Giddens (1987) calls 'double hermeneutics' in science, that is, learning from one another. Furthermore, these experiments with participatory GIS are interesting because one can find experiences for scaling up the results obtained at the community level.

12.2.2 Models and Experience to Explore Theory

Modeling interactions between players is the objective of game theory. Game, here, has not the same meaning as role game. Game theory is an approach used mainly in economics: an interaction situation between two or several players is represented as a strategic game. Payoffs are associated with the different potential choices of the players. Thus it is possible to compute the expected results of the interaction according to the strategies of the players. Many models have been developed, particularly in the field of resource sharing. Reference papers in the field of common pool resources (CPR) management such as (Hardin, 1968; Ostrom, 1990; Stevenson, 1991; Ostrom *et al.*, 1994) are based on a game theory approach. We do not cover that literature because the stakeholders' strategies are modeled but they are not involved in the process. Some experiences in the field of experimental economics are more relevant for our review. Economists involved in institutional and policy research link game theory, institutional analysis, and laboratory experiments (Ostrom *et al.*, 1994). In this field, players are usually students but recent experiments have involved concerned actors such as farmers in Colombia (Cardenas and Ostrom, 2001). Players are usually interested in game results with a salary depending on their results in the game. The objective is to observe whether human beings behave as various economic theories predict. There are several experiments (Mason and Philipps, 1997; Sell and Son, 1997; Gintis, 2000b) and discussions on experimental economics and NRM. One important conclusion of these experiences is that game theory is relevant to model situations in which no communication between the players occurs. From their experiments with stakeholders, Cardenas and Ostrom (2001) conclude that a better understanding of the composition of groups is needed if we are to understand and forecast the efficiency of cooperation. Other works based on the comparison of experiments with managers or students have also pointed out that the expertise of the players in the field of the game improves their strategic behavior (Cooper *et al.*, 1999).

Agent-based modeling has been linked with experimental economics (see Jager and Janssen, Chapter 6)). Duffy (2001) uses an artificial agent-based computational approach to understand and design laboratory environments. Similarly, Deadman *et al.* (2000) describe the development of a series of intelligent agent simulations based on data from previously documented common pool resource experiments. These simulations are employed to examine the effects of different institutional configurations and individual behavioral characteristics. These results are compared with those experienced with human players. For instance, in non-communication simulations as well as in experiments, high performance leads to overinvestment in the CPR and the resultant drop in performance causes a reduction in group-wide investment in the CPR. As in the lab experiments, communication in these simulations does frequently result in the discovery of the optimal level of investment.

These experiments show a linkage between the human players involved in experiments of social interactions and agent-based models. The MAS models may be useful devices for both understanding and designing human-subject experiments. But the objective of the experiments is to test very theoretical rules given by economic theory and not to simulate a real-life situation. Consequently, the players do not participate in the design of either the model or the experiments. The objective of experimental economics, coupled with MAS or not, is to assess theoretical models when decision centers are more than one and psychological patterns are activated. We use role games both to elicit knowledge and to simulate it in a less controlled way. These two steps, collectively achieved, aim at setting up an agreed-upon representation of reality rather than assessing a theoretical pre-given one.

12.2.3 Models and Role Games for Education and Management

Our approach stems to a quite large extent from work with role games and policy exercises and negotiation (Mermet, 1993). This work has shown the suitability of such kinds of tools to support negotiated processes such as local development choices (Piveteau, 1995) as well as for educational purposes (Burton, 1994). This work has also pointed out some difficulties encountered in their use: slow and heavy design, difficult comparison of results, and difficult control of parameters as well.

Therefore, these authors plea for coupling with some modeling tools aimed at alleviating these difficulties. The several potential parallels between role games and MAS listed in Table 12.1 led us to try such joint use with MAS and role games.

Table 12.1: Correspondence table between role games and MAS

Role-playing games	Multi-agent systems
players	agents
roles	rules
game set	interface
game session	simulation
turn	time step

There are a few experiences with the coupled use of models and role games for ecosystem management. Meadows and Meadows (1993) developed a famous role game called *Fish Banks*, which is used for educational purposes. Human players play the role of fish companies that share a common resource (two fishbanks). A simulation model simulates the dynamics of the resource that the human players harvest. The objective of the *Fish Banks* game is to illustrate and teach the tragedy of the commons principle: free access to resources leads to biological depletion and consequently to economic overexploitation. This game is played by students or resource managers. Another game associated with models is Burton's game (Burton, 1994). It aims at simulating the management of an irrigated system. These role games and simulation models have been used for educational purposes. These two models use the systems dynamics approach. Knowledge is represented as stocks and flows. Chapman (1987) compares gaming simulations (role games for us) and systems analysis. He states that systems analysis can cover only a very small part of the complexity that games can handle: 'Analytical statements are limited to a few components and the unwarranted assumption of a trend towards some equilibrium point ... Systems are reflexive, acting on component decision makers, who through their own actions and perceptions act on the system. These perceptions are by definition partial and imperfect. The only simulations which can include self-transformation and reflexive thought are ... gaming simulations.' We think that, in 1987, Chapman was unaware of new modeling methodologies such as MAS, which can capture the cited dimensions. The assumption is that MAS and role games share a common representation of complexity and can thus enrich each other.

Recently, following the experiments presented in this chapter, experiments started in Switzerland within the context of the Firma (Freshwater Integrated Resource Management with Agents) project (Hare *et al.*, 2002). The objective is to mediate coordination among stakeholders sharing water resources.

12.3 COMPANION MODELLING: THE USE OF MAS AND ROLE GAMES

In 1996, we proposed to use MAS and role games (e.g. Bousquet *et al.*, 1999) for better understanding and modeling of the decision-making process and for better management of natural and renewable resources. We presented the principles and the steps of a methodology which has been used and developed by the authors of this chapter.

One of the classic uses of simulation is for prediction, but this is not the option we have chosen. The very long term cannot be predicted in the economic and social field, though it is partially decidable. This is the hypothesis underlying the patrimonial approach (Montgolfier and Natali, 1987; Ollagnon, 1989). As Weber and Bailly (1993) said, 'Because the very long term is beyond the scope of prediction, if we wish to take it into account in the analysis of environmental problems, we must give ourselves very long-term reference points or objectives to guide the possible or impossible pathways of development. The long-term approach must inevitably be based on a scenario.' Because the rules result from interactions among stakeholders, they are legitimized in the eyes of all stakeholders and they incorporate particular perceptions. It is on the basis of a shared conception of how the present situation should evolve that stakeholders are able to 'decide' on very long-term objectives. On that basis, the scenarios that enable the objectives to be reached can be discussed. The entire mediation approach presupposes the establishment of an initial situation, in which the stakeholders are clearly informed of the issues that divide them and of their common dependence upon a solution to the problem at the origin of the mediation process. The challenge of the initialization phase is to enable stakeholders to express their perceptions of the present situation and of its evolution. When a 'map of perceptions', all equally legitimate and equally subjective, has been established and discussed, the stakeholders are asked to discuss the acceptability of prolonging existing tendencies.

How can simulations be involved in this process, that is, how can they help stakeholders to govern? The various stages of the patrimonial approach were given by Weber (1998) and presented by Borrini-Feyerabend *et al.* (2000). Below we present the different stages when a simulation model is used:

- Construction of an artificial world. The first stage, involving one or more researchers, is to gather information on the system under study. We suggest that fieldwork and modeling be performed in unison. The task is to identify the different stakeholders and perceptions and to use multi-agent systems for modeling. In a highly complex world, MAS provide a means to identify the most acceptable form of simplification by focusing questions on problems of representation the agents have on their biophysical and social environment, communication and exchanges among the agents, and control of the interactions. Simulation raises questions in the field that provide new data for the model.
- The second stage is a feedback stage that could also be referred to as validation of the knowledge represented in the model. The aim is to present the model proposed for the decision-making process to the stakeholders. It involves a thorough analysis of the representations and interaction processes among agents. Indeed, it is difficult to explain what has been 'put into the machine'. On the other hand, it is possible to put a stakeholder in the situation of the agent who is in the machine, with the hypotheses of representation, communication, and control that constitute the model. To perform this operation, we propose to use the role game methodology tested by the authors mentioned above. The artificial world is evaluated by plunging the stakeholders into it, that is, by creating a

world similar to the model. These stakeholders may be actively involved in the management system as users of the resource (farmers), regulators of this management system (managers or administrators), or observers of the system (researchers). Does the artificial world inhabited by these stakeholders resemble the real world? The aim is to validate a simulator in the same way as, for example, a flight simulator. A good flight simulator incorporates the same components of the decision-making process as in reality, rather than simply reproducing an actual flight. This stage may be included in the initialization phase of the patrimonial approach as it provides a means to establish a map of the various types of stakeholder and the different perceptions and interactions and to convert them into shared knowledge.

- The third phase is the simulation phase. Simulation shows how the dynamics of the system arise out of interactions among stakeholders with different weights and representations. We can divide this phase into two subphases. Initially, the simulation can be performed in the form of role play, which enables stakeholders to validate the fact that it is indeed in the interactions among different representations that the motor driving the dynamics of the system is to be found. This first subphase also brings to light the different scenarios that are worth testing. Then, once this phase has been completed, the multi-agent model can be used to make simulations based on different scenarios. Simulations, both 'in ludo' and 'in silico', are also involved in another phase of the patrimonial approach in which, after long-term objectives have been defined, the various scenarios liable to lead to these objectives are tested and their results discussed.

12.4. SHADOC[2]

The first operation was conducted under Olivier Barreteau's PhD thesis on the viability of irrigation schemes in Senegal. The issue was the low outputs of the irrigated schemes in North Senegal. The initial objective was to explore whether the viability of these systems was in relation to the coordination modes among farmers within them. For that we had to design a virtual irrigated system in which to conduct experiments based on scenarios dealing with collective rules, individual behavioral patterns, and environmental parameters. This work could not be done in real systems for ethical and practical reasons. The focus on coordination methods and the viability of irrigated systems led us to use a multi-agent system in building the model. The desire to concentrate on the processes involved in the coordination methods rather than on the specific case of each type of irrigation scheme led us to build a virtual irrigated system typical of the central Senegal River valley, but which did not represent any particular scheme. It was not until later that we converted the model into a role game and presented it to stakeholders.

12.4.1 The Model

The model developed is an experimental aid for use in studies of real systems, which overcomes the constraints of size, duration, parameter checking, and potential negative effects on those parameters. In particular, it can be used to simulate the evolution of the system represented according to different hypotheses concerning initial collective and individual rules (Barreteau and Bousquet, 2000). The MAS produced, SHADOC (Figure 12.1), is based on a representation of an archetypal irrigated system in the central Senegal River valley. Computer agents represent mainly elements of the scheme (plots, watercourses, main canal, and pumping station) in relation to a society (farmers, credit access, and water allocation groups). Social agents have rules and partial representation in the system, notably representation of friendship links. The processes represented deal with the circulation of water and credit and with interactions about their allocation and access to them. After sensitivity tests, the model obtained was deemed appropriate for research purposes and capable of discriminating among scenarios according to their viability, which in this case was defined as the life span of the irrigated system without external intervention.

However, it is not necessarily easy for stakeholders outside the model creation process to grasp its logic. Even if they are familiar with information technology or with the area represented, the model is still somewhat of a 'black box'. We therefore had to find a way of opening the box in order to show farmers the work that had been done. Three purposes lay beyond that need of opening the black box: telling interviewed people what had been done with the collected information, partially validating the model, and testing its potential as a negotiation support tool (Barreteau *et al.*, 2001).

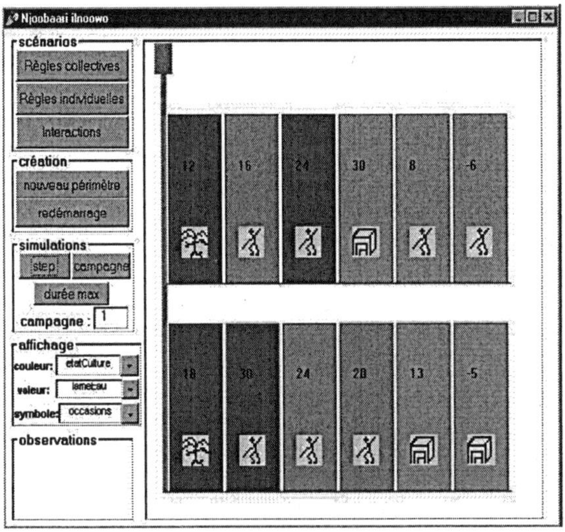

Figure 12.1: Interface of the model. The main window holds a view of the archetypal irrigated system with 12 plots, each belonging to one farmer, distributed among two watercourses. Colors (here levels of grey) represent cultivating stage, numbers indicate water level, and symbols are linked to random events occurring at each time step, of obligations to stay outside the plot. On the left, different buttons are available to entail user-specifying scenarios 'live'.

12.4.2 The Role Game

The game is played with 10 to 15 players and comprises three categories of trilingual cards (French, Pular, Wolof) describing possible behaviors for each player: the aim of planting for the first category (Figure 12.2), social status for the second, and the farmer's propensity to repay loans for the third.

Each player draws a card at random from each category, which describes his or her behavior during the game. The players are in a place representing the village. In another place representing the irrigation scheme, a structure corresponding to the number of players is drawn on a board, with the name of the player to whom the plot is attributed, while the amount of water and the choices made at the time of planting are taken into account when calculating yields (Figure 12.3). The game is played over a half-day in three stages: presentation of the game and the roles, the game itself, and discussion of the game. Each season is subdivided, in both the game and the model, into three phases that occur in strict succession: obtaining credit, irrigation operations, at the end of which potential yields are calculated, and evaluation, after which players may change cards if they wish. This role game was initially tested in the villages on which the thesis was based and, a year later, was presented to farmer associations and technical staff at a workshop organized in Senegal. The workshop lasted three days. On day one, the players discovered the role game, enabling them to complete one or two seasons. Day two was given over to a long game over five or six seasons to observe the long-term effects of the decisions made. The last day was spent on computer simulations.

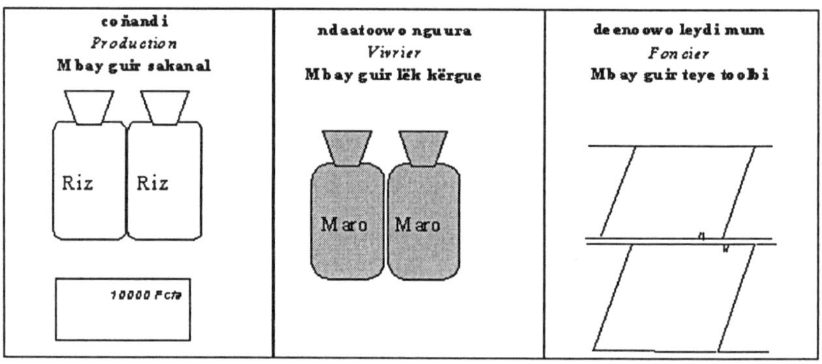

Figure 12.2: The role game: the set of cards from which each player draws one at random. The left one gives the role of a farmer interested in production for self-sufficiency and income, the center one is for a self-sufficiency strategy, and the right one for a land tenure strategy.

Figure 12.3: Blackboard featuring the irrigated scheme with plots, names of players, and water level during a game session in the Senegal River valley, here at the very beginning of a simulated cropping season. The participants are in discussion (exchanging services and information) while the supervisors update the resource dynamics (amount of water in plots).

12.4.3 Some Conclusions

The representations used in these two tools are both relatively abstract in terms of their relation to reality. However, this abstraction did not pose any particular problems when presenting the results of our work, either to field stakeholders or to researchers.[3] The game was well received in the survey villages, clearly showing the relevance of the two tools in representing irrigated systems. During game sessions, the players associated the roles with real farmers and, in some villages, our transportable game board had to be erased before we left to prevent other villages from seeing the 'mistakes' that had been made. This therefore enables us to go some way toward validating the MAS and the representations they include through play.

The tools were also warmly welcomed in the game sessions conducted in the field: several players asked if they could keep the game for use in their own village. What now needs to be done is to distribute it and monitor how it is used, which sends us back to the question 'How can stakeholders be brought to the negotiating table?'. Even if the process has not yet fully reached this stage, the discussions prompted by the game show that the launch of a discussion process can lead to real negotiations, including a comparison of different points of view.

12.5 STRATAGENES

In Madagascar, plant genetic resource management is sometimes transferred to a local level under contracts between the state and local communities. This transfer calls for coordination among the various stakeholders concerned before negotiation of a contract. Local management of plant genetic resources involves biologically, economically, and legally complex processes.

The development of a game called 'Stratagenes' made it possible to integrate knowledge of use in managing certain species and to make it available, albeit in different forms, to the stakeholders concerned about their management (local representatives, farmers, foresters, NGOs, collectors-exporters, scientists, etc.). The completion of a fictitious negotiation, supervised by a mediator, enables the players to see the short- and long-term issues at stake in their individual and collective decisions. The players are challenged to negotiate, within a limited time, a transfer of responsibility enabling viable management of plant genetic resources and a fair share-out of the resulting benefits.

12.5.1 The Role Game

The role game was created interactively and iteratively. Several game sessions were conducted with experts in each role (sociologists, biologists, foresters, economists, etc.) to identify the elements that would be necessary and useful to model. The game was then tested with non-specialists to check whether it was fun to play. After each session, it was adjusted and improved.

12.5.2 The Role of Multi-Agent Modeling

Role games use multi-agent modeling to create the biological, economic, spatial, and temporal dynamics of the decisions made by the players. At each stage, the decisions made by the players have an effect on the resources modeled and have economic repercussions for each stakeholder. Modeling can be used to test management schemes on a decade scale to assess the long-term effect of contract conditions.

12.5.3 Some Conclusions

This tool can now be used to identify useful knowledge and ensure that it is integrated into a formal model. The plan is to use it to help train those involved in the transfer of management and local negotiators before negotiations proper begin. To this end, it is currently being adapted semiotically and linguistically in conjunction with our Madagascan counterparts, and there are plans for tests with local stakeholders.

The development of such a tool is an attempt to integrate knowledge, innovations, and practices into a decision-making process in which every stakeholder is involved and benefits from the representations of his or her partners or adversaries. The first game session with those in charge of plant genetic resource management nationally in Madagascar demonstrated the merits of such role games in fitting administrative decisions into a decision-making process involving other stakeholders. It also brought researchers from different disciplines face to face with use made of their expertise and it was their responsibility to identify and prevent persistent misapprehensions.

12.6 SYLVOPAST

The Sylvopast model was developed to represent the spatial dynamics involved in silvopastoral development and to describe possible strategies for striking a compromise between livestock farmers and foresters. The issue is the silvopastoral development within the framework or prevention plans in French Mediterranean region. Four fundamental questions were tackled:

- How can the evolution of a silvopastoral system governed by contrasting pastoral and silvicultural management rules be simulated?
- How can a representation be produced that is shared by all the stakeholders in the dynamics of the forest area being developed?
- How can the acceptable forms of cooperation be evaluated in the medium and long term?
- What type of spatial structure would best satisfy the objectives of each stakeholder?

12.6.1 The Model

The basic principle behind the design of MAS is that any forest area can be defined by combinations of plant structures (tree/shrub/herb), whose attractiveness differs depending on the type of user, and which are subject to a specific dynamic linked to management methods. At the same time, the management methods used in the different spatial units are the result of a compromise between the different agents concerned, based on a hierarchy and weighting of the objectives attributed to those units.

In the Sylvopast model, eight vegetation structures were chosen, based on the possible combinations of trees, shrubs, and herbs, whereas two types of agents whose objectives sometimes diverge but who are subject to the same main risk (fire) intervene in the area: foresters and their laborers, and herdsmen and their herds.

The virtual forest is a closed space of 2,500 cells subject to a westerly wind, and with some probability of fire and climatic hazards. Each cell is randomly defined by its plant formations and the age of the dominant trees it contains. It also bears traces of the human interventions it has experienced (date of last grazing, clearing, sowing, etc.), which affect the temporal evolution of its vegetation structure. From time to time, it is affected by fires that destroy some of its vegetation and trigger a new vegetation dynamics.

Herdsmen and their herds see the forest as a resource area defined by the fodder potential determined by the plant formations, the crop management sequence, the climatic conditions, and the number of pasture zones capable of feeding the herd for at least ten days. Herdsmen undertake to remain in the area for at least 250 days a year, and are paid if the effect of their herds helps to reduce the risk of fire.

Foresters manage a protection and recreation area defined by the patrimonial value linked to the dominant species and the type of stand, and by the fire risk determined by the plant formations. They need to break down the forest into enclosures of a sufficient size to be able to apply silvicultural rules and ensure effective protection against fires. Their aim is therefore to minimize the fire risk at the lowest possible cost, while ensuring the persistence of the forest and fostering the diversity of structures and species, within the framework of a 20-year management plan.

12.6.2 The Role Game

The Sylvopast role game was developed once the development of the MAS finished in order to support silvopastoral training programs. However, it rapidly became a means of enriching and improving the representations within the MAS of the negotiations and interactions between livestock farmers and foresters. Numerous adaptations have been made, some of which are specific to the creation of a teaching aid with a fun aspect: it proved necessary to make the playing time as short as possible, look for a substitute for the computer tool used, and simplify the rules, without fundamentally altering the MAS. Other adaptations were necessitated by the need to simplify the representations: it was necessary to work with smaller areas, shorter time lapses, and indicators that were easy for the players to measure. Finally, it soon became necessary to replace the graphic interfaces in CORMAS with appropriate 'scorecards' so as to keep track of the progress of the game easily.

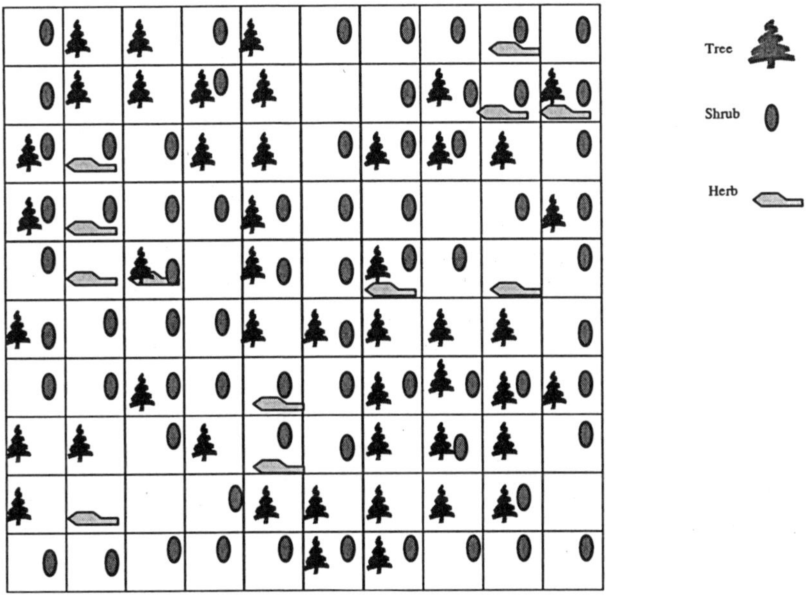

Figure 12.4: The interface of the Sylvopast model and role game adapted for publication. In the real game, specific colors correspond to combinations of land cover. The initial state presented in this figure is calibrated: proportions of each vegetation structure are respected and forest cells are grouped.

The change in the rules regarding vegetation dynamics primarily concerned long-term processes such as the encroachment in abandoned areas by trees or the substitution of forest species, which were not included in the role game. However, the negotiating methods were left entirely open, with the only constraint being the need to develop a way of recording the sequence of actions on the part of each agent. The method consisting of asking each player to justify each of his or her actions in writing was abandoned since it excessively reduced spontaneity. The new method chosen involves a confrontation between two pairs of players who are required to explain their choices and discuss them with their partner, under the watchful eye of an observer. The observer also has a computerized version of the role game, which facilitates monitoring, carries out automatic calculations, and keeps a full record of all the plays made and their sequence.

12.6.3 Some Conclusions

The role game was developed with a view to three objectives adapted to three different targets:

- In playing with managers, the aim was to create a teaching aid to enable forestry practitioners to appreciate the constraints on herdsmen, and vice versa. The game thus enabled players to put themselves in the other players' shoes and better appreciate their needs and difficulties.
- In playing with technicians, the aim was to 'switch' the development operation based on a broad sample to an area they were used to managing or evaluating. In this case, the game served to identify the main spatial organization strategies developed and to establish a typology of the negotiating tactics used.
- In playing with amateurs with no knowledge of either animal production or forestry, the aim was to test the clarity of the rules of the game, their ease of application, and their ability to express the problem of silvopastoral development clearly. The game in this case served to make players aware of the complexity of forest development, and was a means of teaching everyone to test, and even improve, their own natural resource management capacities.

Many positive points showed that some of the objectives of the role game had been achieved. The fact that the players understood the rules quickly and easily adopted the markers provided demonstrated that they obtained a satisfactory representation of their reality. The fact that they commonly used, for discussing and negotiating, the codes proposed as a way to set up a shared representation of the land pointed out the effectiveness of translating multiple resources into a common language. The fact that the stimulation of the negotiation process by gaming led to a set of satisfactory final solutions proved that role games help to achieve reasonable collective solutions to collectively shared problems.

However, several questions remain. Is it better to use tokens that are seen as fun (playing cards, coins, pawns, dice, etc.), which stimulate the desire to win rather than the desire to act appropriately? Even if it is less fun to play, isn't the computerized version more explicit in terms of the shared constraints and processes? For the categories of players, should each game be played by players with the same level of knowledge, or should the categories be mixed to increase the range of experience and rationales? In searching for a socially satisfactory solution, isn't it necessary to replay a game whose final result was unsatisfactory for one of the two players? Alternatively, should a single person play both roles at once to find out what should be the best compromise for him or her?

12.7. MEJAN

The Mejan model was developed to simulate contrasting scenarios of behavior in front of the encroachment of conifers in natural ecosystems with high patrimonial stakes. Its main two aims were to compare management scenarios and enable the different categories of protagonists to exchange their views on an ecological process over which they have little control. If successful, the approach can enable the development of joint management strategies for natural areas. An initial MAS serves to represent the system based on scientific knowledge of the ecological processes in play and the practices with which the agents are familiar. A simplified MAS is then drawn up to make this formal expression accessible to agents and to share the representation of the system with them. The simplified MAS is then used in support of a role game that serves to make all agents aware of the ecological process and react to its overspreading. The role game permits putting the agents in a virtual context that is neutral enough to minimize local social tensions but sufficiently similar to their reality to push them to anticipate the probable changes in their practices as a result of the spatial dynamics likely to occur in the near future.

12.7.1 The Model

The basic principle behind the design of the MAS is that the different categories of agents found in the isolated area constituted by the South of France Causse Méjan (farmers, foresters, and conservationists) are all involved in managing a natural area affected by the strong encroachment of Scotch pines and Austrian pines, but on very different temporal and spatial scales. All three categories are concerned about this biological process, which affects their development objectives (meat or milk production, timber production, conservation, and preservation of open spaces) in different ways.

The natural resources available are defined by vegetation structure, topographical position, land tenure status, and patrimonial value (fauna, flora, and landscape). This land, which was represented by rasterizing a real vectorial map, evolves as a result of the natural spread of conifers and the range management and timber harvesting regimes (Figure 12.5).

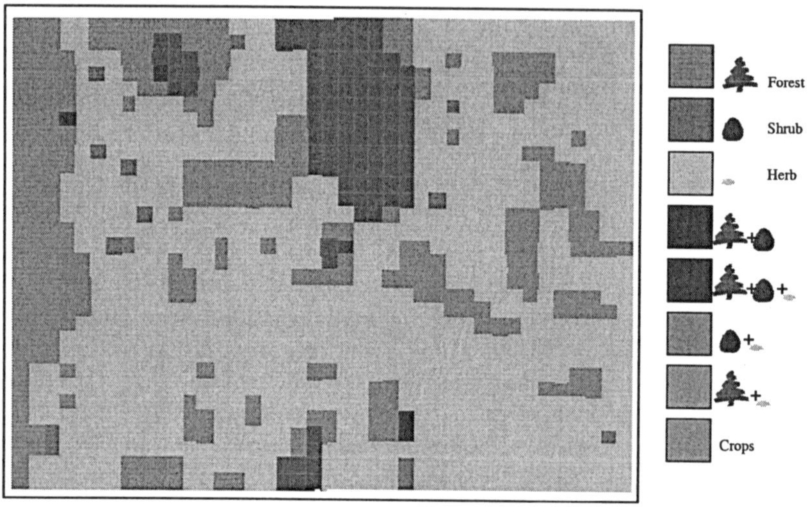

Figure 12.5: Map of land cover

Agro-pastoral strategies depend on the ratio of cultivated land to rangelands, grazing pressure, the position of the livestock farm, and the farmer's degree of ecological awareness. Forest management strategies depend on land tenure status and on whether the area is covered by specific regulations. Faced with the effect of the combination of these multiple strategies on conifer dynamics, the conservationists asked for a MAS to test different intervention scenarios in support of, in addition to, and even in opposition to the operations conducted by farmers and foresters.

12.7.2 The Role Game

The Mejan role game sets out to make farmers, foresters, and conservationists (the Cevennes National Park staff) react to spatial dynamics likely to occur in the near future. It is based on a representation of a natural area sufficiently close to reality for players to be able to relate to it, but sufficiently distant from it for the game to take place in a neutral context, with no preconceptions. The role game relies on a combination of a simplified MAS and a GIS, enabling the rapid production of updated thematic cartographic representations on different scales.

The temporal and spatial scales are different from those of the MAS model, with the rules of conifer dynamics and plant and animal sensitivity to changes in their habitat remaining the same. The participants playing the role of farmers have a limited view of the area, reduced to the area covered by their farm. Within that area, they distribute their animals according to their livestock production system and the cover of conifers on their farm. They also cultivate all the flat lands. The participants playing the role of conservationists (Figure 12.6a) have an overall view of the natural area, in the form of a patchwork of holdings used for farming, forestry, or hunting. From several patrimonial viewpoints (fauna, flora, landscape), they can define the plots with priority patrimonial stakes and develop a control strategy of pine encroachment. The participants playing the role of foresters define the priority objectives for the forest holdings and propose a management and exploitation plan for those holdings. At the end of each round, negotiations occur concerning the technique to control pine encroachment, the location of the plot to be cleared, and the labor required. A cartographic representation of the effect of the practices adopted by the players on natural resource dynamics and land occupation in each agricultural holding is produced for use in the following round (Figure 12.6b).

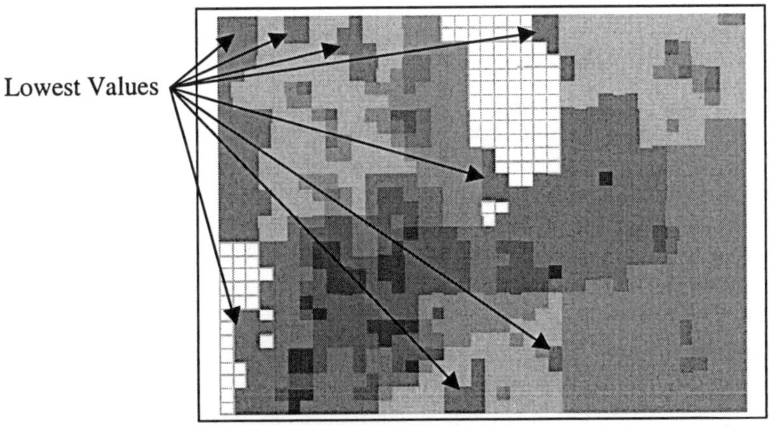

Figure 12.6a: The map of the conservationist (full area): lowest values are indicated; the darker the patch, the higher the value

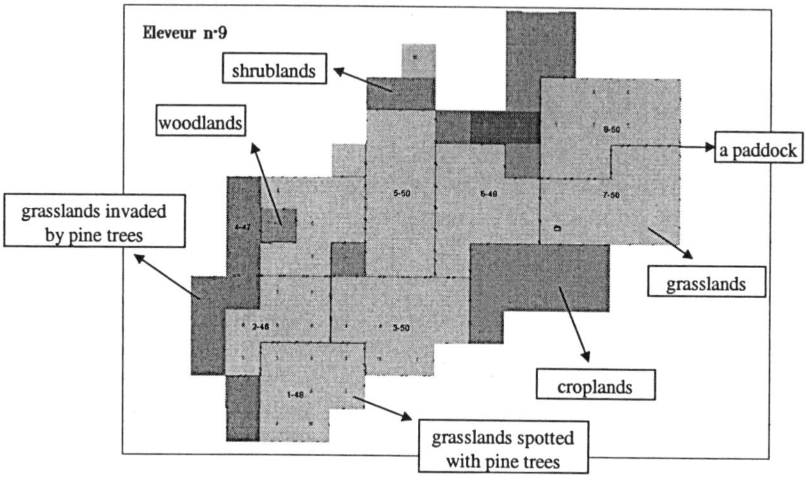

Figure 12.6b: The map of a farmer

The minimum number of players is four farmers, two conservationists, and two foresters. It can be played only by local managers playing their own role, which means that the players playing farmers have to include representatives of each of the main livestock production systems included in the model. Likewise, it is important that the fauna viewpoint and the flora viewpoint be clearly dissociated by being played by two different conservationists. However, the role game makes it easy to formalize the functioning of the interactions among agents and makes it possible to produce a representation of the conifer establishment process rapidly, to be shared with them. Above all, it gives free rein to the players' inventiveness in developing a strategy for action or negotiations, both for agents at the origin of the process and for those subject to it.

12.7.3 Some Conclusions

The MAS entirely fulfils its role as an aid in testing different scenarios of intervention on natural dynamics and in representing different types of collaboration among conservationists, farmers, and foresters. It is innovative in that it shows how agents adapt their behavior according to the changes in the areas concerning them, as a result of both natural dynamics and day-to-day management of those dynamics.

The role game makes an essential contribution to the MAS in that it helps to determine the range of possible strategies through which farmers and conservationists can adapt to a process with which they have not yet been confronted. It also helps to explain and share issues that arise on different temporal and spatial scales.

Although the spatial environmental characteristics are much more closely linked to the local level than in Sylvopast, the overall structure of the model and the way in which it is simplified into a role game can be transposed to many other resource-sharing simulations. Moreover, the generic value of the problem posed (the threat to patrimonial stakes by pine encroachment) permits users to easily adapt the approach to similar ecological situations. Finally, the potential for using the approach under the developing trend of contractualizing the interactions between agricultural practices and environmental protection is particularly promising.

12.8 SAMBA

The 'Regional' component[4] of the SAM (Mountain Agrarian System) program is working in the mountainous area of Bac Kan Province in northern Vietnam. It aims at understanding agricultural dynamics and their consequences on land-use systems in order to promote more sustainable natural resource management from the village to the regional level. In this framework, several methodologies have been combined using and/or developing different tools from traditional fieldwork using on-farm surveys to remote-sensing data interpretation through computer modeling and role games. These various tools, methodologies, and the way they are used pursue different objectives, especially in the relationship between researchers and their objects of investigation, as well as between researchers and stakeholders (Castella *et al.*, 2001a, Castella *et al.*, 2001b).

A first multi-agent model called SAMBA has been developed for testing hypotheses about the differentiation of households at the end of the cooperative period in Xuat Hoa commune, Bac Kan Province (Castella *et al.*, 2001b). The way this model has been developed could be seen as 'armchair modeling'[5] as it was built upon data collected in the field but was constructed and run without interaction with stakeholders. This kind of model serves as a tool for representing, testing, sharing, and discussing hypotheses among researchers from different disciplines and thus having different representations of the world.

The problems experienced in validating this kind of model (Axelrod, 1997) led us to adopt a new methodology in the process of building a model and sharing it with stakeholders. Following a previously described experiment in Senegal (cf. Section 12.9), we adopted role games similar to gaming-simulations (Greenblat and Duke, 1981) as a way to collect information and 'feed' a second multi-agent model that is subsequently presented to stakeholders for validation and discussion.

12.8.1 Role Game and Model

This methodology, which is still under development, follows four steps that take place over a one-week period:

Step 1: collection of preliminary information and selection of participants (one day)
From one to three villages in a commune of Bac Kan Province are selected based on their original characteristics and the topic pursued.[6] Village headmen are interviewed to collect some basic information and identify participants representing the diversity of the village.
Step 2: role-playing (one day)
The 'SAMBA' role game is conducted around a game board composed of 1,600 wooden cubes, each one having six colors representing different land uses. The aim was to use an environment as close as possible to the one in the SAMBA multi-agent computer model developed under the CORMAS simulation platform (Figure 12.7). Around ten players and a few observers (e.g., village headmen, forest wardens) usually participate in the role game. The players draw household cards describing the composition of their virtual household and thus their food needs and available labor force. They also get an initial endowment of lowland rice fields and buffaloes. Then, the players have to propose and manage activities in order to feed their family and develop their own strategy. Actions of the players are situated on the board game and registered by facilitators. During the one-day sessions we have already organized, players could simulate around six years (i.e., six rounds of play). The meetings ended with a debriefing session during which the sequence of the game was analysed by both participants and facilitators.
Step 3: individual interviews and model construction (three days)
During the third step, two activities are carried out in parallel. On the one hand, a multi-agent model aimed at representing the sequence of the role game is developed. On the other, individual interviews are conducted with participants in the role game session (step 2). These individual interviews build upon the actions of the players during the role game. Through a comparison with the real situation, the surveyors investigate the individual behaviors of the players and the various strategies developed during the role game.

Step 4: participatory computer simulation (one day)
A new meeting with participants in the role game is organized and the model constructed during the third step is presented. The similarities of the environments in the role game and in the model help participants get comfortable with the computer. A first simulation similar to the role game session is easily described by the participants. Inconsistencies with and/or alternatives to this first model can thus be discussed for possible simulations.

12.8.2 Some Conclusions

The methodology described here serves several interrelated objectives. First, the role game is used to gather information on selected topics. It does this very efficiently. More traditional interviews usually face problems of confidence between the researchers and the stakeholders, with the interviewees sometimes being reluctant to give information requested in a formal way. On the contrary, the gaming aspect of the role game helps create a more natural and friendly environment, which in turn facilitates communication. Information can then be collected at two levels: (1) when the players are playing on the board, questions can be asked about their actions, and (2) through direct observation of conversations and interactions among participants in a more ethnographic way. Information gathered during the role game is completed by subsequent individual interviews and discussions around the computer model. Such information is used to explore the evolution of land-use systems.

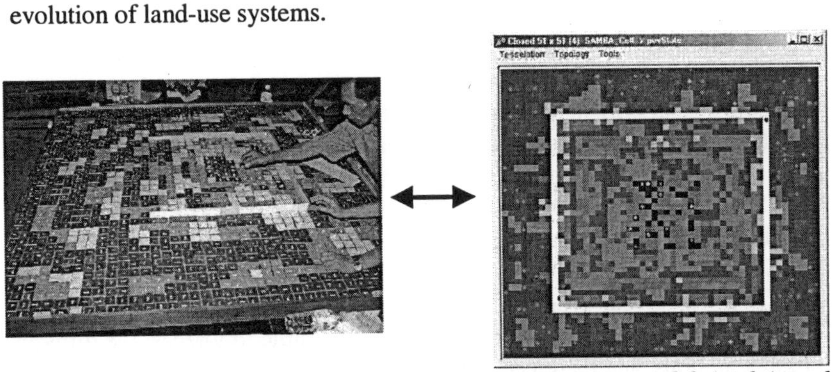

Figure 12.7: Similarity between the SAMBA computer model (right) and SAMBA role-play (left) environments. The white boundaries of the square are the limits of an area protected from unattended livestock. The limits were set during the game and reproduced in the model.

The second objective is to enable the rules (the behavior of the players) to change and new rules to emerge. When conducting the role game, an initial situation is assigned to the players but no rules, especially social rules, are predefined. These rules have to be set by the players themselves in the course of play, whenever they feel the need for them. The players may, for example, ask for the election of a village headman or organize collective pasture areas. Thus, if such rules are enacted, one can say that they are emerging from the convergence of individual actions in a particular environment. Emergence is also a basic principle of multi-agent modeling, where macro-phenomena are seen as emerging from micro-actions following a bottom-up approach. The combination of role games and multi-agent modeling thus appears to be a powerful tool for studying the emergence of collective rules that can be identified in the role game and more precisely explored by the computer model (Boissau, in prep).

Finally, through the information gathered and the discussions with participants, problems faced by local people in their production systems can be identified. Role games and multi-agent systems can then be used as communication platforms for the introduction of technical innovations. The need for accompanying organizational changes is explored through the construction of scenarios to ensure successful diffusion of the proposed innovations.

12.9 SELFCORMAS

The third experiment has been under way since 1998 at the POAS (Plan d'Occupation et d'Affectation des Sols) experimental unit in northern Senegal. In support of a local decentralization policy, the aim is to test simulations that will help local rural authorities and the people under their jurisdiction to organize sustainable land management, taking into account the different forms of land use (agriculture, animal production, the environment, etc.). The management scale considered is around 2,500 km² and 40,000 people. The simulation developed needs to help stakeholders at every stage of the decision-making process: when zoning the area, identifying rules of access to a given type of area, and evaluating possible social and environmental effects. It should also make it possible to forecast the different possible options, and therefore needs to be flexible.

Two precise objectives were set right at the start of the experiment. The first was to test the direct design of the MAS model by the stakeholders from the very first stage (unlike the SHADOC process), with as little prior design work by the modeler as possible, hence the 'self' added to the CORMAS name. This means that upstream modeling work had to concentrate on producing an environment enabling the stakeholders to express themselves in designing their model. The second objective was to test the use of a GIS managed by the region's public development company,[7] to sustain the decision-making process and modeling work regularly if necessary.

Rather than developing a specific model, the modelers developed tools within and around CORMAS (cf. role game, GIS, etc.) to formalize as accurately as possible the knowledge and views expressed by the stakeholders (including the GIS) during the continuous collective decision-making process.

12.9.1 The Role Game and Model

The 'self-design' experiment was organized in the form of discussion workshops. Three three-day workshops were organized by the rural council for the village of Ross-Béthio, a local partner that volunteered for the experiment. The first workshop, in French, concerned a group of source-persons from the rural community, chosen directly by it (a teacher, a young person who had moved back to the village, etc.). The other two workshops were a test, in the local Wolof language, conducted directly with illiterate members of the local population. The subject, which was chosen by the local stakeholders, was coordination between the crop and animal farming in the area. Each workshop set out to have the stakeholders construct and then test a model enabling them to make progress in solving this local management problem. The modelers, who attended the workshops as mere spectators to the discussions, were then to model the progress of the discussions in terms of the collective construction of the model, change and regulatory scenarios, hypotheses as to changes in external factors, etc.

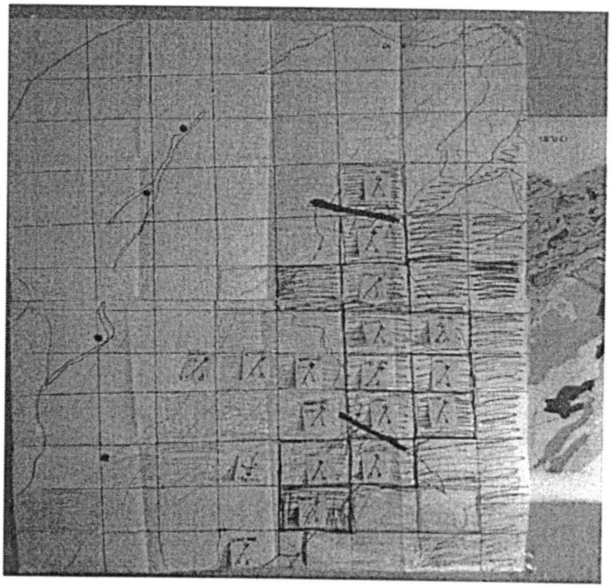

Figure 12.8: The initial map of Ngnith

Around 25 people attended each of the workshops, which were held over three days as follows:

Day one:

The first stage was the identification by the participants of the different stakeholders' requirements in terms of crop and animal farming (soil quality, water salinity, distance from a water supply, distance between plots, etc.), in other words, the definition and above all the collective approval or *legitimization* of the minimum requirements of each stakeholder. This was also the first stage in the design of the MAS model, the form of which was thus dictated to the modelers by the stakeholders.

The second part of the day was given over to the introduction of the cartographic support: the representation of the 'resources' in the zone concerned, with resources defined from criteria chosen by the stakeholders in the previous stage. Maps were produced based on the farmers' perceptions in the first stage, using data from the SAED GIS. Data were amended and approved jointly.

Day two:

A role game was proposed to represent the dynamics of the model designed the previous day. The cartographic support was based on the maps produced the day before, but broken down into regular rectangular polygons (Figure 12.8). In the game, month after month, all players decide on their activity and position within an area they know well, according to a definition of their requirements established jointly on day one. To show their position at every time lapse (a month), all players use a post-it note that they move around the map. After several rounds, a discussion is held to analyse significant phenomena and events that have occurred during the game (conflicts, resource degradation or shortages, dissatisfaction of a particular stakeholder, etc.). The discussion then turns to the joint definition of possible futures (scenarios) in the form of either trends (for instance, population growth) or events (for instance, the digging of new canals).

Day three:

Between days two and three, the multi-agent model corresponding to the role game is programmed for presentation on day three. The first simulation concerns the role game played the day before (Figure 12.9). After that, the scenarios identified the previous day and the new ones that have emerged since are simulated and discussed.

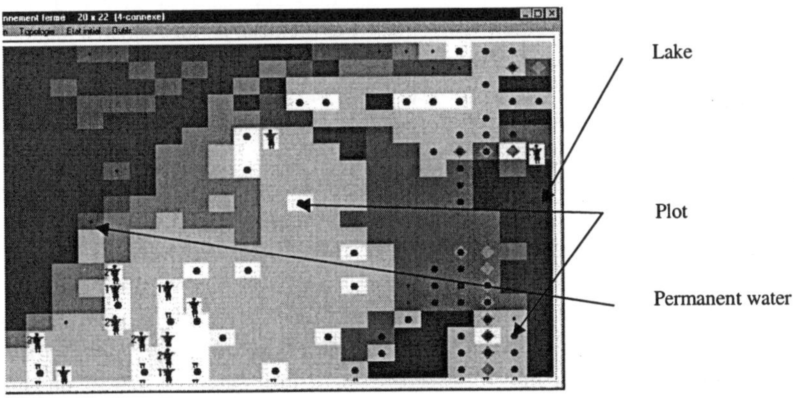

Figure 12.9: The map at the end of the simulation. Zones in which the resource has been consumed are shown in white. This screen shows that the livestock farmers who do not have access to water move west.

12.9.2 Some Conclusions

A good deal of information can be drawn from this experiment, which is continuing. Developing a role game in conjunction with stakeholders seems to be an interesting way of enabling novice stakeholders to play an active part in designing a multi-agent model. The role game serves in this case as a sort of dialogue interface between computer modeling, the 'machine', and stakeholders.[8] It enables them to interpret the behaviors and interactions in the simulation correctly. The stakeholders who developed and played the game were fully capable of interpreting the results of the model and linking them with reality. As they were themselves the initial designers of the simulations carried out, they were also aware of the distance between the model and reality, and of the way in which simulation results should be used. Having largely designed the model, they were well aware that it was not a black box capable of seeing into the future.

Simulation made it possible to go much farther than the role game. For one thing, it would have been physically impossible without computer simulation to 'play' the different scenarios selected by the stakeholders and to observe their multiple effects over sufficiently long time lapses. Furthermore, a sufficiently flexible modeling platform offers many more possibilities of modifying the rules on request than cumbersome game sessions. Simulation thus multiplies the effectiveness of the role game and can take the decision-making process much farther by taking into account the long-term future or through the feasibility of the decisions made. Concerning this last point, the permanent link with a GIS for data restitution and their precise mapping enabled the local stakeholders to shift from a mere exercise to a heated discussion of the future of each area (and of each stakeholder) in a zone they knew well. The satisfaction expressed by the participants largely stems from this return to a precise local reality, using a GIS fed with quite detailed relevant information.

This operation provided us with confirmation of the feasibility of using computers in such socio-cultural situations. By progressing gradually over three days, the stakeholders are capable of following the different scenarios on screen, discussing them, and proposing new ones. There is no need for sophisticated agents for the main land and resource management problems to appear easily in the simulation. As the simplifications were instigated by the stakeholders themselves, they reflect the most important aspects of actual behavior. As regards cartographic representation, we were surprised to see how easily the stakeholders adopted the concept and used it as a negotiating tool, even in the case of a simplified map in grid form (CORMAS), which was just as well received and used by the participants. As a result, if we measure the educational progress made, it is clear that a CORMAS-type simulation platform can be used with participants with a very limited level of schooling. As a result, in a nevertheless conflict-laden local context, our simulation enabled new progress to be made in local negotiations. For instance, at one workshop, the establishment of rangelands was debated, whereas it had previously been a taboo subject, and a decision was even made to ask the local technical services to talk about the issue. In another area, in which the problem was access to water, simulation resulted in an agreement on new collective regulations to protect animal production.

The emphasis placed on the flexibility of the platform also accounted for the progress made in the local decision-making process. For instance, it enabled a response to a wide range of situations. At one workshop, the problem posed and discussed was that of agricultural diversification and its consequences for animal production and environmental areas. At another, it was rangeland development, and, at another, water access regulations.

The model was not designed to provide solutions to problems, but to encourage discussion of the different alternatives, to improve the effectiveness of a collective decision-making process and even to change the behavior of local stakeholders with respect to their technical partners. In our approach, recourse to technical expertise is the stage that follows, and not that which precedes, the collective choice of scenarios that can 'reasonably' be envisaged by the community. Thus, for the project itself, one of the main achievements of this experiment is to have reversed the relations between local populations and technical staff. After the interactive simulations, the representatives of the local populations themselves identified the priority types of support they required within their decision-making process (information on rangeland development, appraisals of certain crops on sandy soils, etc.).

12.10 MAS MODELS AND ROLE GAMES: SOME PRELIMINARY LESSONS

Several conclusions can be drawn from this set of experiments. We start with some practical conclusions and propose a discussion on the different processes these tools can be involved in. The practical conclusions are the following:

- Confirmation of the results obtained by participatory GIS: people are capable of following simulations on a screen and interpreting and discussing them.
- The role-game simulation model seems to be appropriate to favor the involvement of the stakeholders in the exploration of scenarios (see SHADOC, SAMBA and SelfCORMAS). Through the game, participants gain an idea of what the model is. They are then capable of following simulations. Better still, they are able to understand the relation between the model and reality. Having played a role game, they measure the simplification of the process. The associated MAS model is then also considered a simplification. As a result, participants can better interpret the results of the simulations. As the aim is not to predict but to encourage

and enrich discussion, simulations and role games serve to identify and formalize problems for discussion.

- Usually, each actor plays his or her role in the game. In certain cases, it can be useful to switch the roles. Each actor playing the role of the other can better understand the constraints the other faces and the resulting behavior. This sometimes prepares the discussion very well.
- The environment of the MAS and the role game is often different from reality. It is structurally similar to reality (proportions, neighborhood), but we do not use the map of the case studied. The stakeholders have no difficulties with that; they are freer to discuss and this preserves some confidentiality on the knowledge.
- KISS ('keep it simple, stupid') is one of the principles of complex simulations: the interactions between simple processes are a result of complex dynamics. As a rule, the argument is that this principle is valid on paper, but is not operational. In this particular phase, the collective identification of problems, it appears that simplicity is required. For instance, when developing SHADOC, huge efforts were required to make the model simple enough to play. The same was true for the Sylvopast model. On the other hand, for SelfCORMAS, a collective discussion process was triggered based on a simple model (designed by the participants themselves), which increased in complexity as the discussions became deeper.
- The game has to be played with players of each level represented in the model. This aims at validating the behaviors and interaction processes. In the case of separated validations (with stakeholders belonging to different levels surveyed separately), it is better to start the validation process with the players at the lower level. Once validated, this part of the model can serve to implement the behaviors and interactions of the upper levels.

Although all authors of this chapter are using the role game and MAS for integrated NRM, some differences can be pointed out. We present some conclusions here and more can be found in Aquino *et al.* (2002) and Barreteau (2002). We think that the decision-making process for NRM can be viewed as a collective learning process. This process involving representations and emotions has been conceptualized by Varela *et al.* (1993) and Maturana (1996) and advocated by Röling (1996). The stakeholders, including the researchers, exchange their knowledge and generate new knowledge that is used to monitor or transform the environment. This is a constructivist approach in which an agreed representation is the objective. However, we experienced three possible modes of interaction:

1. Interactive modeling for the creation of a better model of the reality. The modeler, alternating modeling activities and field work, conceptualizes a model of the studied ecosystem. The knowledge is elicited with the stakeholders and the model is validated through role games, among other methods. The model is then used within the scientific community or for educational purposes. This is the case, for instance, for SHADOC or Sylvopast or Stratagenes. For example, the Sylvopast game on Forester-Herder negotiations has been played about 40 times. The scientist now has a classification of individual strategies correlated to the forms of negotiation.
2. Interactive modeling to give back the representation of the scientist. Scientists work on a model, often based on biophysical processes, and then return and share their results and conclusions with the other stakeholders. The scientists are pro-active in the decision-making process. This is the case, for instance, of the game Mejan, in which the objective is to help coordinate against pine invasion.
3. The model for mediation. Scientists act as mediators in a collective decision-making process. The model is one of their tools and they will develop their model accompanying the social process; more than an accurate model of reality, the purpose is to reach an agreement and models are used as a mediator toward this agreement. This is the case of the SelfCORMAS experiment.

The status of the modeler and the use of the model are different for each of these cases. They depend on the context. Sometimes, there is no need for intervention in a decision-making process, for instance, when institutions correctly regulate the use of resources or if there is no demand from the stakeholders at all. The model can be developed interactively for a better understanding of the dynamics of the system. In other cases, there will be a demand for the intervention of the scientist, either as an expert or as a mediator.

There is no strong relation between the status of the scientist and the linkage between role games and MAS. At first, we thought that the scientist who wanted to build a model for understanding should start with a model to be transformed into a role game. But we realized how much the role game was efficient in obtaining knowledge on social dynamics and even on individual decision-making (see experiment). At first, we thought that the mediation process would be facilitated if no previous model existed before the role game. But it happened that a role game based on a model could be used as a mediator tool to facilitate discussions and decisions in emerging organizations. This was the case for the SHADOC game.

As noticed by many other authors, it appears that the efficiency of the tools lies much more in the social process at stake rather than in the combination of tools. If institutions for mediation, also called platforms, already exist or are emerging, these tools are positively perceived and can be appropriated. Furthermore, placing these tools in such institutions lowers the risk of sidelining the responsibilities attributed to decision makers.

12.11 CONCLUSIONS

To conclude, the supportive approach on which all these experiments and CORMAS are based can be seen as a bottom-up rather than a top-down modeling method. These experiments enabled us to explore an interesting way of using MAS for better understanding of the decision-making process and, in some cases, in support of negotiations among stakeholders. A new horizon has now opened up and other applications are under way or are planned. Now that this feasibility has been demonstrated and the elements of a method are in place, it is important to study the contributions and effects of such tools over longer periods. How do the models evolve with a decision-making process? Do they disappear? Do they change status? It is now necessary to establish ways of monitoring such research-action dynamics, either in relation to their final purpose or, more generally, in relation to their effect on the local community (impact on stakeholders in the decision, ethics of intervention, etc.). New research is starting on this issue (Dare, 2002).

Because they allow users to consider agents of a very different nature with their own perceptions, modes of communication, and control, MAS have a considerable potential in integrated natural resource management research for the modeling and simulation of complex processes among stakeholders, as well as between social and ecological dynamics.

ACKNOWLEDGEMENTS

The authors wish to thank Louis Lebel, Nick Abel, John Wolfenden and Nils Röling for their comments and Bill Hardy for editorial help.

NOTES

1. The various experiments described in this article, and some others still under way, originated from a process set in motion at CIRAD (Centre de cooperation internationale en recherche agronomique pour le développement) in 1994 by its 'Green' (Renewable resource management, Environment) research group. http://cormas.cirad.fr
2. Simulateur Hydro-Agricole Décrivant les modes d'Organisation et de Coordination
3. However, as far as the researchers were concerned, the more they specialized in a given issue concerning the viability of irrigated systems, the less comfortable they were with the representation in the model form, which they believed did not go far enough.
4. SAM-Regional is a joint research programme of the Vietnam Agricultural Science Institute (VASI, Vietnam), the Institut de Recherche pour le Développement (IRD, France), and the International Rice Research Institute (IRRI, Philippines).
5. In reference to the 'armchair economics' denounced by Herbert Simon (1986).
6. So far, topics have been the allocation of the labour force between paddy fields and upland crops, livestock management systems, the saturation of uplands, and the emergence of forest land as a scarce resource.
7. SAED, Société d'Aménagement et d'Exploitation des terres du Delta du Sénégal et des vallées du fleuve Sénégal et de la Falémé (a public regional development company).
8. The low educational level of the participants in the workshop (a very high rate of illiteracy) is no obstacle to the use of such computer simulations.

13. Institutional change for sustainable land use: a participatory approach from Australia

Nick Abel, Art Langston, John Ive, Bill Tatnell, Mark Howden and Jacqui Stol

13.1 INTRODUCTION

This chapter describes a participative approach to the promotion of sustainable land use at the regional scale. Compared with other decision support methods for land use change (Mintzberg *et al.*, 1979), our approach is unusual in these respects (with acknowledgement to an anonymous referee):

- our focus was upon institutional change rather than land use planning as the driver of land use change;
- we used stakeholders' values to generate options for land use change, while constraining those options within the bio-physical limits of the region;
- we used policy makers to generate options for institutional change;
- the approach is based upon an analysis of the role of political power in maintaining or changing resource allocation patterns.

We define an institution as a social rule and the organization that implements it. In this project this definition includes laws, policies and the organizations that administer them. The approach was applied in the rangelands (known as the Western Division) of the State of New South Wales (NSW), in the federation of Australia (Figure 13.1). The project 'Sustainable Use of Rangelands in the 21st Century' ran for four years. Its interim website is in the list of references.[1]

The region occupies 325,000 km^2. Mean annual rainfall varies from 150 to 450 mm/y from West to East, with high variability in time and space. There are 251 land systems, each representing a different pattern of topography, soils and vegetation, each with internal variation in soils, vegetation, and resilience to grazing and drought (Walker, 1991). The region contains around 52,000 people, 6 to 7 million sheep, and as many kangaroos. Animal numbers fluctuate widely. Minerals, crops, wool and sheep sales are the main primary contributors to gross regional product (NIEIR, 1999).

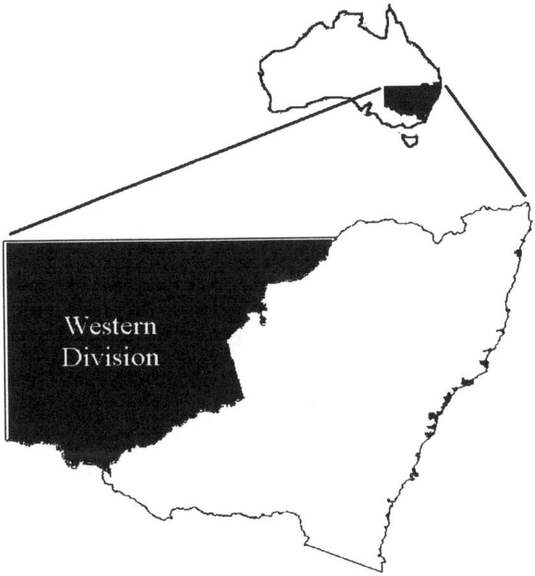

Figure 13.1: The region

Problems of the region that threaten sustainability include slow decline in an ageing population, land use conflicts, declining terms of trade for agricultural products, loss of indigenous species, shrub encroachment in some areas, and over-dependence on a single agricultural commodity – wool. These problems have historical origins, so we begin with an outline of the region's history. Next we present the objectives of the project, before outlining some theory on political economy, land use and social change, and resilience. This is followed by an explanation of the rationale of the participatory approach, a description of the stakeholder process, an outline of the sustainability scenarios, and a discussion of the policy process (the process for designing institutional changes). We conclude by discussing implementation and communication of the proposals.

13.2 LAND USE HISTORY

The region was occupied for around 60,000 years by *Aboriginal* hunter-gatherers. Their culture was based on intimate links between particular tracts of land, and the clans and tribes that occupied them. These links were mainly broken when their land was taken by invading pastoralists and their livestock, beginning in the mid 19th Century. New diseases reduced Aboriginal populations drastically. Loss of land undermined their cultural base. The Aboriginal population of the region is now 11.5% of the total, and they live mainly in rural towns where, because of historical dispersals, clans and tribes are mixed. Aboriginal peoples had no citizenship rights until 1967. In 1983 Land Rights legislation was passed and a fund and organization established to enable Aboriginal peoples to purchase unoccupied publicly owned land, or to purchase pastoral leases. During the 1990s Native Title legislation was passed that allowed Aboriginal peoples to lay claim to native title on pastoral land. If successful they would have the right to use their rights in parallel with the pastoralists' activities. Claims may be tested in court. There is therefore conflict between pastoralists and Aboriginal peoples. There is also conflict among Aboriginal peoples because of overlapping Native Title claims, and because of Land Rights and Native Title claims to the same land. Much of the intra-Aboriginal conflict arises from the dispersals that occurred after the pastoral invasion.

Pastoralists hold around 93% of the region under perpetual lease tenure, and produce mainly wool. Their history is marked by adaptation to harsh climate and wool price fluctuations, steadily rising costs, the widespread replacement of perennial grasses by annuals, and the growth of shrubs that compete with grass and make herding costly. The shrubs have increased because of the cessation of burning, which was standard practice among Aboriginal peoples. It is now difficult to re-institute burning because of legal liability, fire regulations, and because grazing pressure rarely allows sufficient fuel to accumulate. Grazing pressure is spread widely because artificial water points have been constructed throughout the region.

Pastoral activities are monitored by, among others, the Western Lands Commissioner, the Department of Land and Water Conservation and the National Parks and Wildlife Service. The wool price collapse after 1992 has left many pastoralists in financial difficulties (Figure 13.2), but they are restricted in their freedom to clear native vegetation and plant crops as a means of increasing incomes. They are not allowed to carry out non-agricultural activities except farm-tourism, and they do not own the timber or minerals on their leases. They feel threatened by over-regulation of vegetation clearing and cropping, by Native Title claims, by the large number of kangaroos that compete with sheep for pasture, by feral rabbits and pigs, by ovine Johne's disease, and by the attempts of conservationists to increase the area of National Parks and Nature Reserves. The increase in prices for goat meat is beginning to lure pastoralists into this market. Feral and farmed goats thrive on the shrubs that continue to spread.

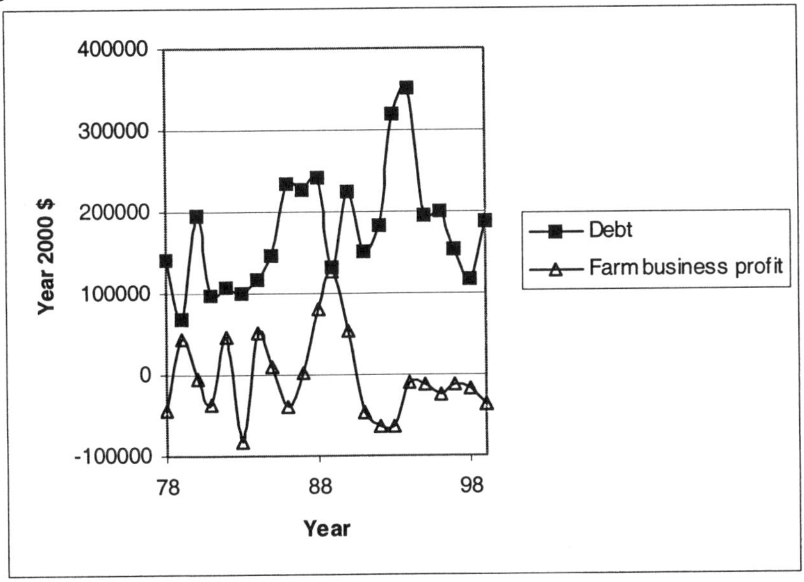

Figure 13.2: Pastoral financial trends. Source: ABARE (2000)

Conservationists have noted the severe impacts of pastoralism on some native species (Table 13.1). The main production strategy of pastoralists has been to develop artificial water points (bores and dams), to the extent that around 80% of the region is now within 2km of surface water. Before pastoral settlement only the areas adjacent to perennial rivers and springs had year round surface water. Thus most biota in the region evolved with only episodic grazing. Landsberg *et al.* (1997) found in a sample survey of national rangelands using distance from artificial water as a proxy for grazing pressure that the frequency of occurrence of 15–38% of species from a range of plant and animal taxa declined under grazing. The frequency of between 10% and 33% of species appeared to increase under grazing. Frequencies of the rest were apparently unaffected. In terms of resilience theory (discussed later), permanent surface water is the slowly changing variable that determines the intensity and duration of grazing pressure on the rangelands, thus species composition. Conservationists understand that to reverse trends in biodiversity this slow variable must be managed by the selective closure of water points.

Table 13.1: Changes to the status of central and western division vertebrates (the study region is the Western Division)

Numbers of species:	Mammals	Birds	Reptiles	Amphibians
Recorded since 1788	88	329	117	28
Regionally extinct by early 1990s	27	6	0	0
Nationally extinct by early 1990s	7	0	0	0
Under threat in early 1990s	27	60	48	8

Sources: Dickman (1994), Sadlier (1994) and Smith and Smith (1994).

The first National Park was not gazetted in the region until 1967. The area now occupied by Parks, and Forest and Nature Reserves is 3.5%. This small area cannot include the range of species, ecological communities and processes now present in the region, nor does it provide secure habitat for endangered species.

Conservationists and some researchers argue that land degradation caused by pastoralism threatens the sustainability of the region for future pastoralism. Pastoralists mainly deny this; they are supported by other researchers. The inconclusive evidence is reviewed in Baker *et al.* (1999), and summarized in Table 13.2.

Mining has been a powerful economic contributor to the region since the discovery of gold in adjacent Victoria in 1851. The direct impact of gold upon the region was through the demand for meat to feed a human population that was exploding in response to the wealth from gold. Pastoralists hastened to occupy Aboriginal land in the region from which to produce the meat. Mining of copper, silver, lead, zinc and opals began in the late 19th Century and continues. Major towns and infrastructure (slowly changing variables) developed around the metal deposits, and urban populations increased to dilute the political influence of the pastoralists.

Minerals have become supremely important to the economy of the region (NIER, 1999). Mines occupy tiny areas of land from which they generate great wealth and employment. Their off-site impacts are upon groundwater and infrastructure. In this region the minerals industry enjoys fairly peaceful relationships with Aboriginal peoples, agro-pastoralists and conservationists, probably because its mines are generally long established. Expected mining of mineral sands deposits may sharpen these conflicts. For Aboriginal peoples the issue is what they see as the violation of the earth by mining. Conservationists are concerned about water use and loss of species locally. Agro-pastoralists are in competition with mining companies for ground water. Because minerals are owned by the State, not the landholder, the latter may lose land and amenity to new mines.

Tourism has played only a small part so far in the history of the region. With the decline of pastoral opportunities in the region, and of 'wilderness' around the globe, tourism may have an important future.

These histories influenced our objectives, which are outlined next.

Table 13.2 Summary of the impact of pastoralism on the rangeland types (adapted from Baker et al., 1999)

Rangeland Type	% of region	Decrease of desirable perennial grasses	Increase of undesirable plant species	Active soil erosion	Condition **
Belah and Bluebush	16	Little	Some	Little	Resilient
Bimble box–pine	7	Extreme	Some	Some	Poor
Downs Country	12	Extreme	Extreme	Large	Very poor
Gidgee and Brigalow	2	Some	Potentially extreme	Some	Resilient
Mallee	9	Extreme	Extreme	Extreme	Very poor
Mitchell Grass Plains	1	Little	Little	Some	Productive and resilient
Mulga	32	Large	Extreme (woody weeds)	Large	Reasonable but potentially unstable
Northern Floodplains	13	Extreme	Some (exotics)	Some	Productive but potentially unstable
Saltbush Plains	4	Large	Extreme (woody weeds)	Some	Poor
Southern Grasslands	1	Some	Some (exotics)	Little	Modified but resilient and productive
Southern Riverine Woodlands	3	Extreme	Some (exotics)	Little	Modified but resilient

*Desirable plant species may include perennial woody plants that provide soil stability, protection for annual grasses and forbs, drought fodder and diversity to cope with drought (i.e., function).

** Condition is used here to assess the status of the rangeland after adverse impacts caused by pastoralism based on a landscape function viewpoint.

13.3 PROJECT OBJECTIVES

Objectives were to:

a) develop visions of sustainable land use for the region, and a strategy for achieving them by:
   - *explaining past and current determinants of land use;*
   - *generating alternative scenarios of land use, representing the values of groups of stakeholders, and assessing their likely ecological, social and economic outcomes;*
   - *designing the institutions needed to support change towards sustainable land uses;*
   - *developing a phased strategy for implementation.*

b) establish some nationally-applicable principles, social processes and methods for translating broad policies on sustainable resource use into operational programs, and communicate them widely.

Our methods were built to achieve these objectives. Before discussing methods we need to consider the political economic background to the region, as our interpretation of this affected the design of methods.

13.4 POLITICAL ECONOMY, LAND USE, SOCIAL CHANGE AND RESILIENCE

In this section we describe our use of political economy and resilience theory. The section is based on Abel (1999), Pritchard and Sanderson (2002), Magee *et al.* (1989), Godden (1997) and Gunderson and Holling (2002). These theories were used to explain why and how the pattern of land use in the region changed and therefore how future change might be guided towards increased sustainability.

Institutions are formed and changed through political economic processes. The pattern of land use in the region is the result of land attributes interacting with past and present institutional arrangements, costs, prices and other values. We envisage a political economic system in which the aim of political parties is to be elected to government. They compete with other parties to supply policies (including changes to laws) or public investments (e.g. in infrastructure) that will earn sufficient votes to win elections. Parties need operational funds in order to compete with other parties for votes. They get these funds from lobby groups, mainly industries and trade unions, in exchange for policies that satisfy the needs of the lobbyists. In this simple model the outcome in terms of institutional change or public investment is the result of trade-offs by political parties among the competing aims of voters and lobbyists. The simple model is complicated by the ability of politicians and the media to manipulate perceptions and to control what is included on political agendas. Complexity is also added by bureaucracies that have their own aims and resources, but are not elected, by local governments and by legally and informally constituted community resource management groups. Moreover, freedom to offer policies and changes to laws is limited by pre-existing institutions established by similar processes in the past, such as the State and Federal constitutions, and land tenure. No one group guides the evolutionary path of this system, and the competitive behavior of the groups can change the system towards configurations that are inefficient in the sense of their failure to satisfy the value sets of any stakeholders. This view of the social-ecological system reinforces the assumption that institutional change is likely to be an effective way of influencing the evolutionary patch of the system, and its resilience.

Resilience is the capacity of a system to undergo disturbances and persist (Gunderson and Holling, 2002). The higher the resilience, the greater the sustainability. Resilience must be defined in relation to a purpose (Walker and Abel, 2002). A system that is resilient for one purpose may not be resilient for another – in this region, resilience for the production of wool does not coincide at all precisely with resilience for the conservation of native species.

Holling and Gunderson (2002) hypothesize that resilience varies over time as a system evolves. A project such as ours that is attempting to enhance resilience for a number of purposes simultaneously may well be more effective if it operates at a time when the resilience of the system for some purposes is falling, rather than when that system is gaining in resilience. This can be explained in terms of the 'adaptive cycle'.

Holling and Gunderson (2002) have noted the tendency of social-ecological systems (SESs) to pass through four stages which they call the 'adaptive cycle' (Figure 13.3). The first is one of collapse, which in this region corresponds with the demise of the pre-European Aboriginal system. The second is re-organization, which is mirrored by the early activities of pastoralists as they established a new system with different objectives from the Aboriginal system it replaced. In this stage the course of the SES is indeterminate, opportunities are open, and it could move in a number of different directions. In the third, or growth stage, pastoralism consolidated its hold on land and political influence and reduced the range of land use options (e.g. National Parks and nature reserves), before entering what we call a monopoly stage. We see the social-ecological system as being now near the sunset of this stage. It has undergone a number of small collapses in the past when wool price and rainfall decreases have coincided, but the general course has been maintained. The monopoly stage is characterized by legal and organizational complexity, unresolved land use conflicts, and inability to take up options and break social gridlock.

Holling and Gunderson (2002) hypothesize that the resilience of a system will increase following collapse and reorganization, peak during the early monopoly stage, and fall again in the late monopoly stage as the system becomes 'over-connected', and as options are suppressed by monopolies. A project such as ours may be well placed to influence the onset of change if it operates at a time when the resilience of the dominant resource use system is waning, because at this time resilience theory predicts the system is vulnerable to disturbance, including the disturbance of new ideas.

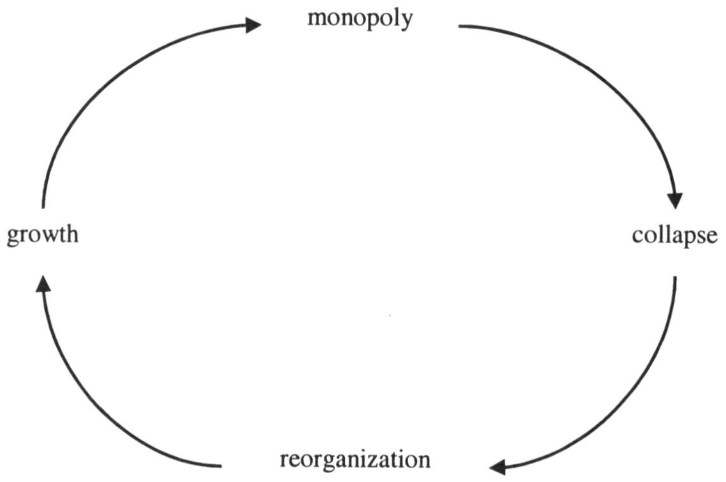

Figure 13.3: The adaptive cycle

According to resilience theory, evidence that the current system is moving into a late monopoly stage is provided by its institutional arrangements, which are:

> ....a complex mixture of ancient and modern, which has seen layer built upon layer [of legislation], with no attempt to conduct any fundamental re-evaluation. The approach of government in recent years has been to deal with specific problems relating to the older legislation as and when they have arisen by ad hoc amendments, rather than carrying out a fundamental reform. For the most part, the new breed of holistic 'environmental law' has been simply superimposed on outdated sectoral and segmented natural resources law. (Abel et al., 1999, p. 71).

The current institutional arrangements are slowly changing variables that affect relationships among bureaucracies, stakeholders and ecosystems, thus the resilience of the system. Hence our focus upon them. Relationships among stakeholders are already changing (Table 13.3). In addition, the region is facing major uncertainties with impending climatic and global economic changes. Stakeholders have much adaptive capacity. This is demonstrated, for example, by the ability of pastoralists to cope with highly variable rainfall and prices, and of Aboriginal peoples to adapt to the invasion and alienation of their lands. Institutional impediments, however, prevent the expression of these capacities in new and constructive ways. Our project was designed to become part of a process of social change, and it was intended to influence the direction of change by contributing to the resolution of conflicts, realization of win-win opportunities, the release of adaptive capacity and the building of resilience.

Table 13.3: Political influences of stakeholders

	Relative Strength Now	Postulated Reasons for Relative Strength	Estimated 20 Year Trend	Reason for Trend
Conservationists	Moderate	- Conservationist paradigm clashes with dominant developmental one - Perceived as anti-growth, so anti-jobs (limits their influence)	Growing	- Concerns about sustainability of resource use - Species losses - Support from urban middle class - Support from scientists
Pastoral Industry	Strong	- Australian icon status - the self sufficient pioneers - Economic importance in the past - Uniformity of rural voting - Voting systems and electoral boundaries	Weakening	- Perceived as damaging the land by urban people - Perceived as subsidized - Low and diminishing economic importance - Decline in ratio of rural/urban voters - High average age
Aboriginal People	Weak	- Loss of land and culture - 'Protection' policies in the past - Low numbers	Growing	- Political activism - Growth in numbers - International support - Some urban middle class support

The key considerations for our choice of methods emerging from the history of the region and its political economy, and from resilience theory, are that:

- the main stakeholders value the same land in very different ways, and are in conflict over how it should be used and by whom;
- to be sustainable, land use patterns should be suited to land attributes, be politically feasible, address conflicts, provide employment and be economically viable;
- change in policies, laws and organizational structures is probably the most effective way of influencing land use change; however, existing organizations have vested interests in the present arrangements, and change will involve conflict;
- an effective approach to addressing and resolving conflicts among stakeholders and among government agencies requires a sharing of perceptions, searches for win-win opportunities, and negotiations of trade-offs;
- in terms of the adaptive cycle, the region is in the monopoly stage, so when change comes it may happen rapidly; the project is an attempt to both initiate change, and influence the direction of change through major modifications of institutions.

These considerations influenced our choice of approach. We used participative research because 'no one is in charge'. A participative approach is also justified by 'mental model' theory. This shows that humans will not readily accept change on the basis of spoken or written information alone, especially if it threatens their own interests. They need experiences that are convincing, and opportunities to analyse new propositions deeply (Abel *et al.*, 1998). We designed stakeholder and policy workshops to provide these experiences and opportunities. We also justified a participative approach on the assumption that it is more likely to lead to implementation of proposals because it gives ownership of issues to particular groups (Abel *et al.*, 2002).

13.5 METHODS

Figure 13.4 shows the structure of the project. It comprises a stakeholder process, a policy process and sustainability analysis. The three processes ran in parallel. Workshops and subsequent mail and telephone were the media through which stakeholders and policy makers interacted with researchers.

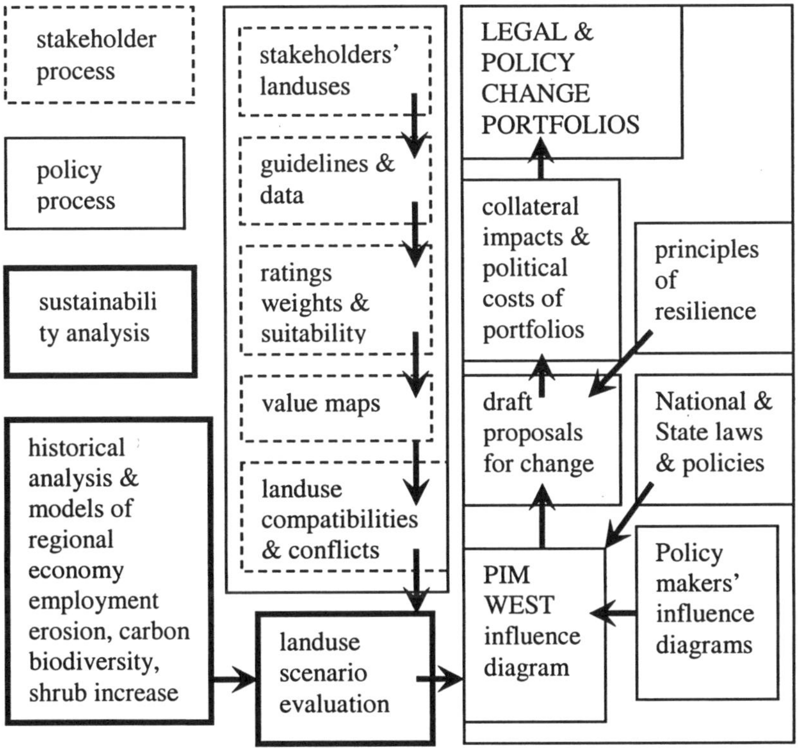

Figure 13.4: The structure of the project

13.5.1 Stakeholder Process

13.5.1.1 Recruitment of stakeholders

Stakeholder and policy processes both involved researchers and participants as co-researchers. To recruit pastoral representatives to the process we began with the two Catchment Management Committees with jurisdictions that span the region. Both had sponsored our research grant application. We followed a network of contacts until quotas were reached for the number of representatives in each stakeholder group, based on resources and transaction costs. The groups were broader in outlook on most issues than the population they represented. We see this as an advantage. We added people practicing novel land uses by networking from a business enterprise centre. Their enterprises included kangaroo and goat harvesting and processing, sandalwood, olives and crayfish culture. Mining stakeholders were found for us through the industry's representative body, and the regulatory agency. In seeking Aboriginal participants we looked for people likely to be influential and innovative, rather than representative. As a consequence these stakeholders had long experience of working for Aboriginal rights. Representatives were sought from each of the main language groups, with advice from Aboriginal people and others who had worked with them. To recruit conservationists we approached non-governmental conservation organizations, and the National Parks and Wildlife Service. We aimed for a mix of rural and urban interests, since urban conservationists regard themselves as rangeland stakeholders. Tourism stakeholders were found through the Tourism Commission. All were associated with local government. Thirty-six stakeholders worked with the project over four years.

13.5.1.2 Eliciting stakeholders' priorities and values

The stakeholder process was based on the SIRO-MED approach (Ive and Cocks, 1999). Facilitating understanding of each other's mental models was a major objective in exploring conflicts. Early workshops were a vehicle for providing social interactions, building networks, communicating ideas, and teaching researchers about land use and society in the region. Understanding among the various stakeholder groups was developed through discussion of the various conflicts separating them. Plenary and feedback discussions, shared meals and accommodation helped communication. Stakeholders identified 55 land uses. A few were dominant existing uses. Most were minor uses or are not yet practiced.

For each land use, stakeholders developed guidelines to match land uses to land attributes. Examples for pastoralism are in Table 13.4. Stakeholders contributed deep and extensive knowledge. The value of land for a land use was given different interpretations by the groups: suitability or profitability for agro-pastoral and tourism enterprises; probability of sites of importance to Aboriginal people; suitability for different types of conservation management; and probability of minerals exploration or mining activity in the future.

Table 13.4: Pastoral stakeholders' WINLUPIS guidelines for allocating mapping units to sheep and cattle production

Guidelines - as far as possible allocate mapping units:	Data requirements
where pastures are dominated by desirable short perennial to sheep.	Area of perennial grasses Area of mixed grasses
where pastures are dominated by chenopods to sheep	Proportion of mapping unit carrying chenopod shrubland
Where woody cover is low to sheep	Shrub present
away from predator-prone areas to sheep	Distance to non-dingo protected land Distance to river
with access to labor, schools, off-farm employment and general services to sheep	Distance to urban limits
with access to labor, schools, off-farm employment and general services to cattle	Distance to urban limits
that have low-lying country to cattle	Low lying ground
that have belah-lignum country to cattle	Lignum and belah present
where pasture dominated by tall perennial grass to cattle	Area of perennial grasses
that are flat or with gentle slopes to sheep	Slope
that are flat or with gentle slopes to cattle	Slope
have fodder trees and shrubs to sheep	Fodder trees present
with low erosion risk to sheep	Proportion of low erosion risk land within land system
with low erosion risk to cattle	Proportion of low erosion risk land within land system
with high mean annual rainfall to sheep	Index for mean annual rainfall
with high mean annual rainfall to cattle	Index for mean annual rainfall
with low rainfall variability to sheep	Rainfall variability index
with low rainfall variability to cattle	Rainfall variability index
with high percentages of winter rainfall to sheep	Index for percent winter rainfall
with high percentages of winter rainfall to cattle	Index for percent winter rainfall
with high percentages of summer rainfall to sheep	Index for percent summer rainfall
with high percentages of summer rainfall to cattle	Index for percent summer rainfall
that are subject to floodwater passage to sheep	Percentage of mapping unit subject to floodwater passage
that are subject to floodwater passage to cattle	Percentage of mapping unit subject to floodwater passage

13.5.1.3 Mapping stakeholders' values

Between workshops, a land classification adaptable to mapping of all the potential land uses was developed by researchers using an existing land systems classification (Walker, 1991). Using the SIRO-MED approach and supporting WINLUPIS software (Figure 13.4, and Ive and Cocks, 1999), researchers wrote rules for deriving suitability ratings. A rule relates a land attribute to the rating of each mapping unit in relation to that attribute. Ratings could be categorical or quantitative. Quantitative rules could provide suitabilities as levels, or as continuous functions. The guidelines and ratings, and the scale of the mapping units told researchers what data were needed (Figure 13.4). One hundred and eighty layers of spatial data were required representing vegetation, soil and landform, infrastructure, heritage values, climatic information and land use. They are described on the project website. These layers provided a crucial bio-physical template for the expression of stakeholders' values through the guidelines.

Mapping units, ratings and data were provided to subsequent workshops where they were modified as requested by stakeholders. Ratings were weighted by stakeholders and combined in WINLUPIS to produce suitability scores, defined as:

$$\text{Suitability score} = \text{sum } r_{ijk} \cdot w_k$$

Where:

r_{ijk} = the rating representing the ability of the mapping unit i (i = 1, I) to satisfy guideline k (k = 1, K) when assigned land use j (j = 1, J);

w_k = the guideline importance weighting required by stakeholders to gain acceptable achievement for guideline k;

- ratings are defined to span an interval of 0 to 1, the higher the value the more highly valued the mapping unit is from the perspective of the guideline;
- suitability scores were standardized across land uses to allow for the different numbers of guidelines applied to the various land uses;
- the suitability score represented each land unit's relative value for a specified land use.

This linear combination approach is regarded by Hogarth (1980) as robust.

The resulting land value map represented a stakeholder group's vision of how a land use and its value might be dispersed across the region in the future. These maps were used to inform policy makers of the spatial variability of land use values. Two out of 55 possible examples are shown in Figures 13.5 and 13.6.

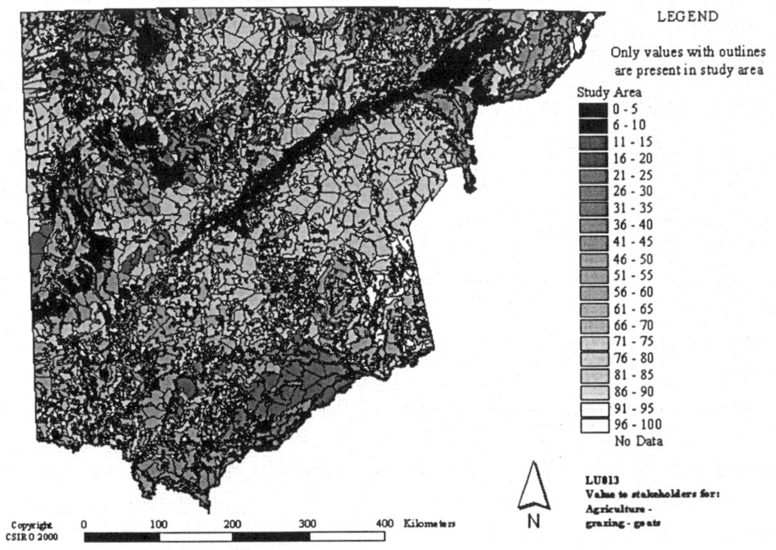

Figure 13.5: Value of the land for goat production, according to pastoral stakeholders

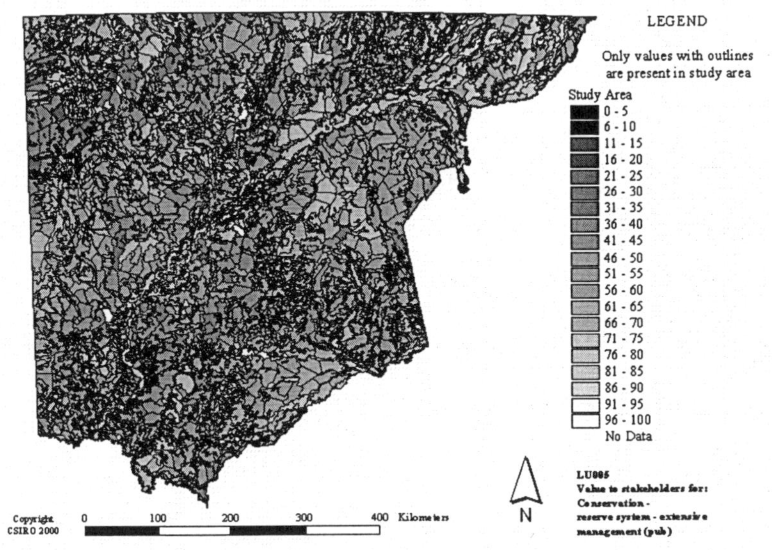

Figure 13.6: Value of the land for conservation under extensive management, according to conservation stakeholders

Some stakeholders showed strategic behavior in weighting guidelines. One group, for example, was uncomfortable with the strong gradients of suitability that emerged from the application of their own guidelines. This was due to the influence of rainfall. In the opinion of the researchers this was an accurate reflection of the suitability of the land for the stated purpose, but this stakeholder group reduced the weight on rainfall to achieve a more homogeneous surface. We believe they did so in order to 'claim' most of the region. This strategy would probably have put them at a disadvantage had the negotiating capabilities of WINLUPIS been used in the stakeholder process.

Suitability for conservation is especially difficult to estimate because there is no objective way of measuring biodiversity value (Margules and Redhead, 1995). It is not species richness, because that may exclude rare and endangered species. Besides, a site may be rich in species because it is not vulnerable to species loss, and may therefore have a low conservation priority. Moreover, it is not in practice possible to enumerate the distribution and abundance of species over a heterogeneous region spanning 325,000 km^2 (many of the invertebrate species are yet to be named, and the soil biota are largely unknown). Our conservation stakeholders therefore used the concepts of representatives and vulnerability to derive guidelines (Margules and Redhead, 1995). The representativeness of a mapping unit is the extent to which the genes, species or ecological communities it contains are already present in other mapping units. We used range sub-types (aggregations of land system types) as a proxy for biodiversity, and one of our stakeholders' guidelines was to allocate a proportion of each of the 20 categories to conservation. The target for representativeness was set at around 15% of each category.

Supporting guidelines allocated mapping units to conservation reserves according to their vulnerability to threatening processes (soil erosion, rabbits, kangaroos, goats, pigs), and their vulnerability because of their small size or wide separation from other, similar mapping units. The guidelines allocated mapping units to three categories of conservation reserve according to the intensity of management indicated by their vulnerability (in accordance with IUCN recommendations). Mapping units containing rare or endangered species, rivers, wetlands and dry lakes were allocated preferentially to conservation because stakeholders saw these as especially important for conservation.

13.5.1.4 Comparing stakeholders' values

WINLUPIS provides a platform for developing a socially acceptable and efficient land use pattern. It does this by:

- comparing the suitability scores of all land uses of all stakeholder groups on all mapping units;
- allocating to each mapping unit the land use with the highest suitability score;
- enabling stakeholder groups to change weightings and negotiate an acceptable compromise.

In this project we did not pursue the negotiation because the stakeholder representatives did not have sufficient authority from their peers to do so. The information for negotiations will be available for future use.

13.5.1.5 Land use conflicts and compatabilities

WINLUPIS has some capability to address multiple land uses, but this was not used in our project. It was more important for policy makers to understand the extent of compatibility and conflict among land uses than it was for them to see a compromise multiple land use pattern. Stakeholders used a matrix for recording mutual compatibilities of land uses. They were asked 'Which pairs of land uses could co-exist in the same space, at the same time or sequentially, without impacting on their values and performance?'. They listed levels of compatibility:

- fully compatible land uses from the same stakeholder group;
- fully compatible land uses from different stakeholder groups;
- potentially compatible with negotiated conditions;
- incompatible;
- agreement not reached on compatibility.

Eighty percent were compatible to some extent, a strong indication that multiple land use was both possible, and desired by stakeholders. This information was fed into the policy workshops. Land use value maps can be used to identify areas where conflict is likely to be greatest due to coexistence of high values for incompatible land uses.

13.5.2 Scenarios of Sustainable Futures

With the National Institute for Economic and Industry Research, and working outside the workshops, we developed social and economic scenarios for the region (NIEIR, 1999). The scenarios are: continuation of the present arrangements – the baseline scenario; increase in agriculture; increase in government spending in the Region (including nature conservation); increase in mining; increase in tourism. Some of the outputs are summarized in Figures 13.7a,b. We also modeled the biophysical consequences of each scenario in terms of changes in soil erosion hazard, ability to store carbon, and capacity to implement a representative conservation reserve system. In addition we explored the consequences of climatic change for land use in the region (Howden *et al.*, 2001).

Outputs from these activities were synthesized and fed into the policy process (Figure 13.4) to encourage policy makers to think about the economic, employment and environmental consequences of land use changes. Outputs from the stakeholder process were likewise introduced to the policy makers to improve their understanding of stakeholders' values, the conflicts and compatibilities among them, and the potential for win-win solutions.

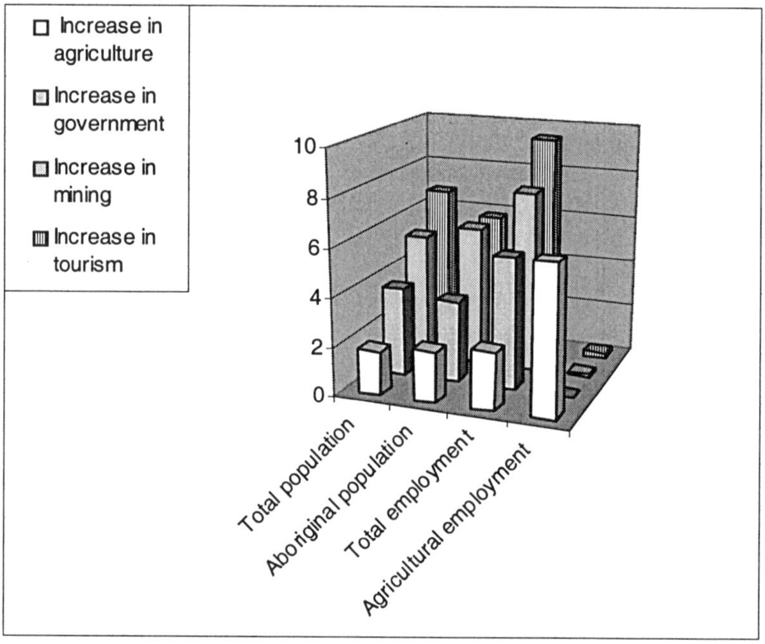

Figure 13.7a: Percentage change in population and employment in the four scenarios relative to baseline scenario (see text).

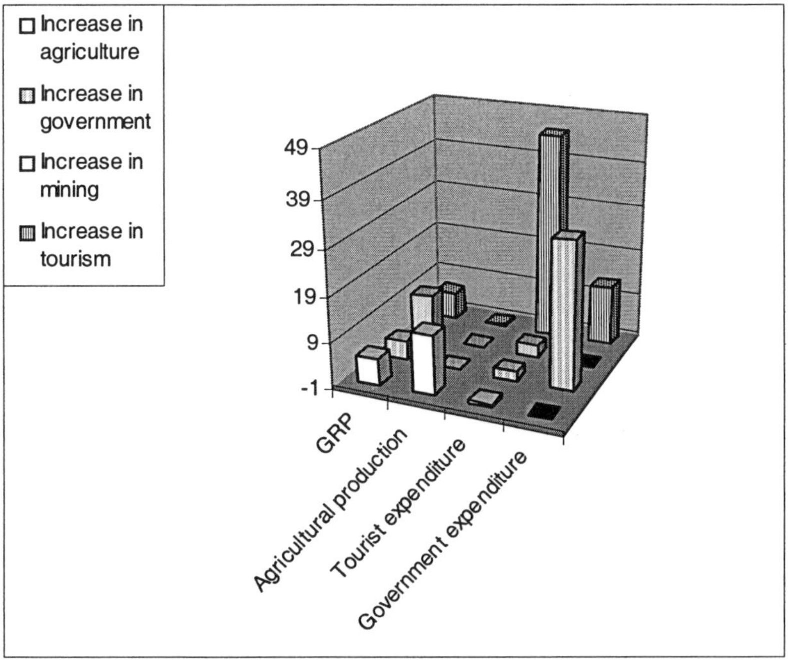

Figure 13.7b: Percentage change in gross regional product (GRP), agricultural production and expenditure in the four scenarios relative to baseline scenario (see text).

13.5.3 Policy process

To choose whom to invite to the policy workshops we made matrices of organizations against stakeholder groups at three scales – local, regional, State and Federal – that reflect the hierarchy of decentralized decision making. In the cells we entered the names of organizations that mediated the interests of a particular stakeholder group at a particular scale. From this list we selected organizations with most influence on land use. Within each organization we identified the person most appropriate to discuss and influence changes in policies and laws affecting land allocation. We assumed that if a sufficient number of policy makers at sufficiently high levels attended our workshops, their mental models would be influenced, and some would become champions for change. We involved 150 policy makers from 45 organizations during the course of the project. Up to 60 participants attended a single policy workshop.

At the workshops policy makers were exposed to resilience theory, and these principles of regional resilience (Gunderson and Holling, 2002):

- Humans are part of nature, adaptive agents in a social-ecological system, not external controllers of a bio-physical system.
- Institutions should be designed to allow just enough disturbance to maintain resilience; they should allow just enough external support to prevent a system crash, but without discouraging internal adaptation.
- Organizational memory is necessary to guide regions through recovery following disturbance, just as land managers' memories guide recovery at local scale.
- Capacity for learning from past disturbances must be built and maintained in land managers and in organizations, thus enhancing the ability to anticipate and adapt to future disturbances.
- Innovation and diversity should be fostered within societies and land uses to provide a wide range of options when conditions change.
- We should accept a level of redundancy in infrastructure, technology and institutions – if part fails there are backups.

Policy makers were encouraged to incorporate these principles into their proposals for change (Figure 13.4). We do not claim that full use was made of the principles.

Policy makers divided into groups, each representing the interests of a stakeholder group. There they built influence diagrams of the factors affecting the allocation of land to their stakeholder group. Workshops were held at a regional city and in the State capital to capture these different perspectives. Researchers subsequently searched the literature and laws, and incorporated their findings with the policy makers' influence diagrams into the policy model PIMWEST using Vensim (Ventana Systems, 1999). PIMWEST is a large influence diagram showing the positive and negative influences of laws, policy, economy, society and environment on land uses across the region. Relationships among factors were extracted as causal trees showing the influences that enhance or hinder the land use interests of each stakeholder group (Figure 13.8). During the workshops participants began to propose institutional changes to enhance the interests of each stakeholder group, minimize conflict and promote resilience. Outputs from the stakeholder process were used to seed ideas for change, for example towards multiple land use.

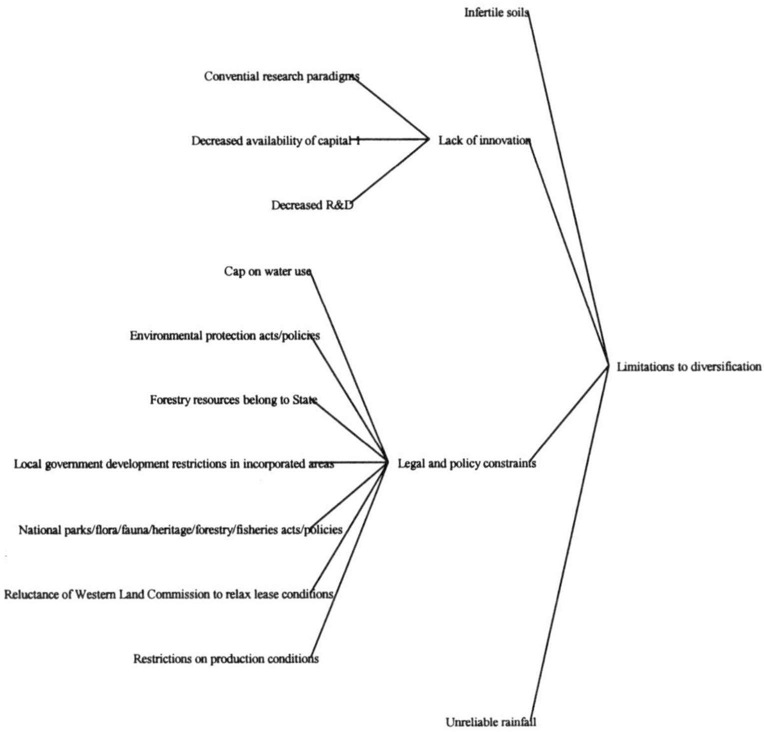

Figure 13.8: Factors limiting the ability to innovate: a causal tree from PIMWEST. Note: this tree shows 3 levels of causation. The causal chains can be extended to the limits of the data entered into the model. For example, the actual legislation that restricts innovation is listed in the fourth level, which is not shown here.

Inevitably we, as researchers, were influenced by our own values and analyses, and by the stakeholder process, and when we saw opportunities to insert suggestions, particularly those involving win-win opportunities, we felt free to do so, on condition our proposals were labeled as originating from the researchers. No objections were made. Two main criticisms were made about the process. First was its political nature, because it analysed the distribution of power and resources. Second, we were criticized because we invited as stakeholders people who had few or no property rights in the region, including Aboriginal peoples, conservationists and the tourism industry.

Policy makers developed five portfolios for institutional change, one for each stakeholder group. Interactions among proposals were captured by asking each group of policy makers to evaluate the effects on their own stakeholders of proposals emanating from other policy groups. To evaluate the political feasibility of proposals, policy makers were asked to estimate their influence on rural and urban votes. Champions for each proposal were identified, with a view to future implementation. One hundred and sixty proposals were developed. They are summarized in the Appendix.

13.6 IMPLEMENTATION, ADOPTION AND LESSONS LEARNED

We assumed that the experience provided by participation in the workshops would influence the mental models of policy makers. To test this influence we administered the same questionnaire to participants at the beginning of the project and again at the end. We also asked participants to identify a 'twin', a person who in their view held similar views and performed similar functions. Twins also completed the questionnaire at the beginning and end of the project. The questionnaire was about the relationships among policies and laws, land use and sustainability. There was no detectable change over time in the responses of our participants, nor was there any detectable difference between twins and participants. The variation between individuals was high. Job turnover among policy makers during the project left us with only 22 pairs, so small sample size may have masked any effects. More important, the fact that turnover occurred may have reduced our impact. Turnover is a reality of working over a four-year period.

The project has influenced policy making substantially within the region. The two Catchment Management Boards that span the region were our community sponsors for the funding application. We have built strong links through our workshops. They are now preparing catchment plans to which we are contributing ideas, information and outputs from this project, including the design of a conservation reserve system. We are also working with pastoralists, non-governmental organizations and State and Federal agencies to develop a pilot stewardship scheme that would complement the conservation reserve system. Representatives of these organizations were among our stakeholders and policy makers, and establishing such a scheme was a key recommendation.

During the project, the New South Wales Government called for a review of policies and laws for the region (Kerin, 2000). We worked closely with the Review, to our mutual benefit. Despite this, we do not claim great success at the State scale. The section on political economy, land use and social change alludes to an explanation. Our understanding is this. The main agencies operate in the region under separate legislation and pursue their own policies.

These do not add up to a coherent strategy for the region. The agencies are in competition for power and resources, and have insufficient incentive to cooperate. Each operates within its own regional sub-divisions, and the boundaries are mutually incompatible. In addition, three separate community committees have been established to develop separate strategies for managing catchments, rivers and vegetation. Furthermore, local government operates under its own legislation, and within its own boundaries. These arrangements were not devised by a madman, they evolved over time through the accumulation of multiple incremental changes, each one more-or-less rational. The unwieldy structure that now over-bears the rangelands is very poorly adapted to promote or even to allow adaptive changes by landholders. The Premier's Department has been developing a coordinating role that may break the gridlock, but meanwhile radical proposals generated in the policy workshops have not been championed from within the agencies, at least partly because of multiple disincentives. An incomplete list includes: the likelihood of retribution if departmental policies are challenged; reluctance to put at risk departmental influence, resources and jobs; the absence of rewards for innovation and problem solving; and the fact that all the major problems identified fall across rather than within departmental portfolios.

Another factor contributing to stalemate is Native Title. Pastoralists and bodies that represent them do not wish to use a process that is available under the legislation for negotiating Native Title rights with Aboriginal groups. They prefer to take each case to court as it arises. The matter is made worse because under the legislation three parties need to participate in negotiations – the landholder, the claimants, and the State government. The agency representing government has not promoted negotiations because there may be a risk of compensation claims in the future.

In retrospect the project objectives were too ambitious. The SES is displaying the inertia predicted by the Holling and Gunderson (2002) model for a system in the monopoly stage. The factors that are likely to end the stalemate are summarized in Table 13.3. In terms of resilience theory, we hypothesize that the slowly changing variables listed in the column 'Postulated Reasons for Relative Strength' need to cross threshold values before the social-ecological system can accommodate radical change. Meanwhile our communication strategy is spreading information about the potential for win-win changes through multiple land use, Native Title land use agreements and stewardship agreements, for example (Appendix), and we have made the portfolios of change accessible to potential champions.

The main method of communication and adoption has been through our policy maker and stakeholder networks. They have been receiving our Newsletters. We have a booklet and the website. The latter will be emulated in a CD-ROM for stakeholders with restricted Internet access. Stakeholders and policy makers will be equipped to promote change as opportunities arise, and arise they will.

ACKNOWLEDGEMENTS

We are deeply grateful to our stakeholders and policy makers for providing the foundations for the project. We are also indebted to these people for their work, support and ideas: Phil Price, Russel Harland, Steve Marsden, Peter Clark, Brian Walker, Buzz Holling, Richard Dunworth, David Farrier, John Kerin, Mandy Yialeloglou, Nick Nicholls, Ross O'Shea, Steve Orr, Tim Ferraro, Brendon Baker, Margaret Cawsey, Guy Barnett, Bob Pressey, Mike Fleming, Carla Mooney, Peter Walker, Cathy Mobbs, Terry Mazzer, David Tongway, Ken Hodgkinson, Richard Price, Steve Morton, Dick Condon, Guy Fitzhardinge, Dee Murdoch, Jen Shearing, Vanessa Chewings, Dennis Barber, John Carter and Monica van Wesween. Community sponsors were the Western and the Lower Murray Darling Catchment Management Committees, (now Boards), respective chairs Jenny McLellan and Ron Rees. Three anonymous referees provided excellent comments that enabled us to improve the original draft. The work was funded by CSIRO Sustainable Ecosystems, Land and Water Australia, and the NSW Department of Land and Water Conservation.

APPENDIX: SUMMARY OF PROPOSALS FROM THE POLICY PROCESS

ABORIGINAL PORTFOLIO:

- *Increase Aboriginal influence on regional planning and development* through: recognition of customary law; and increased Aboriginal representation on planning and decision-making bodies.
- *Resolve conflict between Aboriginal peoples and landholders* through: changes in Native Title law; Indigenous Land Use Agreements (ILUAs); co-management of pastoral leases; and better representation of Aboriginal people with local connections to land.
- *Resolve conflicts among Aboriginal peoples* through: ILUAs; changes in Land Rights and Native Title law; and recognition of tribal areas.
- *Increase Aboriginal ownership or control of land and water* through: lease purchases; co-management of more National Parks and reserves; and allocation of water rights.

- *Enhance Aboriginal access to land and water* through: ILUAs; changes to access and liability law; access agreements; and recognition of traditional rights to hunt and fish.
- *Improve protection of Aboriginal cultural sites* by: changes to National Parks and Wildlife Act and Environmental Planning and Assessment Act; and through Aboriginal heritage management agreements.

AGRO-PASTORAL PORTFOLIO

- *Promote diversification of land uses* through: rebuilding extension services; adding aquaculture extension; easing restrictions where land is capable of more intensive uses; 'one-stop-shop' development approval; production cooperatives; giving timber rights to leaseholders; and stewardship payments for conservation.
- *Increase investment at property level* by enhancing security of tenure and freeholding of leases.
- *Eliminate debt* by property amalgamation and other measures.
- *Improve roads* by establishing a legal road network, and resolving trespass and liability issues.
- *Promote sustainable management of land* by: amending land law so it specifies lessee's rights and responsibilities, management outcomes, voluntary nature conservation agreements, incentives, and periodic changes of lease conditions.

NATURE CONSERVATION PORTFOLIO

- *Promote sustainable use of natural resources across the region* through: 1) legislation that sets strategic social, environmental and economic outcomes; establishes natural resource policies, plans and programs; integrates natural resource and catchment management policies; sets out rights and obligations of resource users, environmental controls and incentives; and establishes participatory planning processes; 2) taxation, grant, user-pays, and incentive measures; 3) linking restructuring policy to land suitability and conservation value; 4) better representation of conservation interests on advisory committees; 5) incentives and levies on uncapped bores; selective removal of water licenses; 6) formal land condition and trend reports required when a lease is sold.
- *Establish a comprehensive, adequate and representative reserve system (CAR)* by: formalizing a policy under the law; acquiring public reserves and encouraging conservation on private land through taxes, rents and stewardship payments; linking the CAR to the Conservation Trust; biodiversity accreditation linking consumers, retailers and producers; a biodiversity credits scheme; debt-for-conservation swaps; conservation on leases purchased by mining companies to get water rights.

- *Clarify the priorities of National Parks and Public Reserves*: conservation of native biodiversity should be first, conservation of heritage second, and public enjoyment the third priority.

MINERALS INDUSTRY PORTFOLIO

- *Improve participatory planning* by better representation of stakeholder interests on permanent advisory committees.
- *Promote regional development* through: favoring approvals for value-adding development projects; a simplified and improved development approval process; tax incentives for new mines and for employment of local people.
- *Improve rehabilitation* by setting the level of the post-mining land rehabilitation bond according to past record of the company.
- *Promote multiple and sequential land use* through: State Conservation Reserves that contribute to a CAR reserve system, but where exploration and mining are allowed; and a development approval process that favors multiple and sequential use.
- *Promote efficient water use* through: user charges; water trading; incentives for water exploration; and review of current water licenses.

TOURIST INDUSTRY PORTFOLIO

- *Expand the tourism industry in the region* while maintaining quality by developing a carrying capacity strategy and the means of implementation; supporting the establishment of a CAR reserve system; and accreditation of operators.
- In addition to these institutional changes the tourism representatives proposed improvements to roads, communications, transport and accommodation; Aboriginal cultural tourism; and development of eco-tourism.

RESEARCHERS' PROPOSAL FOR RESRUCTURING REGIONAL ORGANISATIONS

- State agency, local government and community committee arrangements are complex, over-redundant and often unconnected. Simplification is needed. To encourage policy makers to think beyond current arrangements, researchers proposed incorporation of community-based resource management committees and local governments with the existing catchment management boards to form two Regional Councils. The Councils would co-ordinate strategic socio-economic and natural resource planning for the region. They would undertake development approvals, collect rates and provide traditional local government services. Local representatives would be included in many of the functions currently provided by resource management committees. Relevant State agencies would staff the technical and regulatory sections of the Councils, with appropriate transfers of those agencies' budgets to the Councils.

NOTES

1. Website: Sustainable Use of Rangeland in the 21st Century: www.cse.csiro.au/nsw_rangelands

References

ABARE (2000), *ASPIRE: Agricultural Data Base*, Australian Bureau of Agricultural and Resource Economics, Canberra.

Abbot, J., R. Chambers, C. Dunn, T. Harris, P.G. de Merode, E.J. Townsend, and D. Weiner (1998), 'Participatory GIS: opportunity or oxymoron?', *Participatory Learning and Action Notes*, **33**: 27–34.

Abel, N. (1999), 'Resilient rangeland regions', in D. Eldridge and D. Freudenberger (eds), *People and Rangelands: Building the Future*, VIth International Rangeland Congress, Townsville, Australia, pp. 21–30.

Abel, N., M. Gachugu, A. Langston, D. Freudenberger, M. Howden, and S. Marsden (2002), 'Policies, planning and institutions for sustainable resource use: a participatory approach', in A.C. Grice and K.C. Hodgkinson (eds). *Global Rangelands: Progress and Prospects*, Wallingford, CAB International, in press.

Abel, N, H. Ross, and P. Walker (1998), 'Mental models in rangeland research, communication and management', *The Rangeland Journal*, **20**: 77–91.

Abel, N., D. Farrier, B. Tatnell, and C. Mooney (1999), 'A rangeland enmeshed: the legal and administrative framework of the Western Division of New South Wales', Report to the Western Lands Review, Department of Land and Water Conservation, Sydney.

Ajzen, I. (1985), 'From intentions to actions: A theory of planned behaviour', in J. Kuhl and J. Beckmann (eds) *Action-Control: From Cognition to Behaviour*, Heidelberg, Germany: Springer, pp. 11–39.

Ajzen, I. (1988), *Attitudes, Personality and Behaviour*, Milton Keynes, England, Open University Press

Ajzen, I. (1991), 'The theory of planned behavior', *Organizational Behavior and Human Decision Processes*, **50**: 179–211.

Alberti, M. and P. Waddell (2000), 'An integrated urban development and ecological simulation model', *Integrated Assessment*, **1**: 215–27.

Alonso, W. (1968), 'Predicting best with imperfect data', *Journal of the American Institute of Planners*, **34**: 248–55.

Anderies, J. (1998), 'Culture and human agro-ecosystem dynamics: the Tsembaga of New Guinea', *Journal of Theoretical Biology*, **192**: 515–530.

Anderies, J.M., M.A. Janssen, and B.H. Walker (2002), 'Grazing management, resilience and the dynamics of a fire driven rangeland system', *Ecosystems*, **5**: 23–44.

Andreoni, J. and J.H. Miller (1995), 'Auctions with artificial adaptive agents', *Games and Economic Behavior*, **10**: 39–64.

Antona, M., F. Bousquet, C. Le Page, J. Weber, A. Karsenty, and P. Guizol (1998), 'Economic theory of renewable resource management: a multi-agent system approach', in J. Sichman, R. Conte and N. Gilbert (eds), *Multi-Agent Systems and Agent-Based Simulation*, Lecture Notes in Artificial Intelligence, Paris (France): Springer-Verlag, pp. 61–78.

Aquino (d'), P., O. Barreteau, M. Etienne, S. Boissau, S. Aubert, B. F. and C. Le Page (2002), 'The role playing games in a ABM participatory modeling process: outcomes from six different experiments carried out in the last five years', International Society for Ecological Economics, thematic session on Models, Role Games and Negotiations, Sousse, Tunisia.

Archer, S., C. Scifres, C.R. Bassham and R. Maggio (1988), 'Autogenic succession in subtropical savanna: conversion of grassland to thorn woodland', *Ecological Monographs*, **58**: 111–27.

Arrow, K.J. and A.C. Fisher (1974), 'Environmental preservation, uncertainty, and irreversibility', *Quarterly Journal of Economics*, **88**: 312–19.

Arthur, W.B. (1994), 'Complexity in economic theory', *American Economic Review*, **84**: 406–11.

Ash, A. and D.M. Stafford Smith (1996), 'Evaluating stocking rate impacts in rangelands: animals don't practice what we preach', *Rangelands Journal*, **18**: 216–43.

Attonaty, J.M., M.H. Chatelin, and J.C. Poussin (1990), 'L'évolution des méthodes et langages de simulation', in J.L. Lemoigne (ed.), *Modélisation systémique et système agraire*, Paris : INRA Editions, pp.119–34.

Axelrod, R. (1984), *The Evolution of Cooperation,* New York: Basic Books.

Axelrod, R. (1997), 'Advancing the art of simulation in the social sciences', in R. Conte, R. Hegselmann, and P. Terna (eds), *Simulating Social Phenomena*, Lecture Notes in Economics and Mathematical Systems, **456**: 21–40.

Axtell, R. and J. Epstein (1994), 'Agent-based modeling: understanding our creations', *Bulletin of the Santa Fe Institute* (Winter): 28–32.

Baas, N.A. and C. Emmeche (1997), 'On Emergence and Explanation', Santa Fe Institute Working Paper 97-02-008, Santa Fe, NM.

Babin, D. and A. Bertrand (1998), 'Managing pluralism: subsidiarity and patrimonial mediation', *Unasylva*, **49**: 19–25.

Baker, B., G. Barnett, M. Doherty, A. Gordon, A. Kearns, T. Mazzer, A. Sarre, and B. Tatnell (1999), 'Study D: Biophysical – status and trends of the land and water systems of the Western Division', Report to Western Lands Review, CSIRO Wildlife and Ecology, Canberra, Australia.

Balmann, A. (1997), 'Farm-based modelling of regional structural change', *European Review of Agricultural Economics*, **25**(1): 85–108.

Balmann, A. (1999), 'Path dependence and the structural development of family farm dominated regions', *IX European Congress of Agricultural Economists*, Organized Session Papers, Warsaw, Poland, 24-28 August,, 263–84.

Balmann, A. (2001), 'Modeling land use with multi-agent systems. Perspectives for the analysis of agricultural policies', *IIFET Conference 'Microbehavior and Macroresults'*, proceedings, Corvallis.

Balmann, A., B. Czasch, and M. Odening (2002), 'Employment and Efficiency of Farms in Transition: An Empirical Analysis for Brandenburg', *Berlin 2000, XXIVth International Conference of Agricultural Economists, Proceedings*, forthcoming.

Balmann, A. and C. Hilbig (1998), 'Zur Identifikation von Pfadabhängigheiten in hochdimensionalen dynamischen Systemen: Eine Anwendung multivariater Analyseverfahren auf simulierte Agrarstrukturentwicklungen', Working paper 47, Wirtschafts- und Sozialwissenschaften an der Landwirtschaftlich-Gärtnerischen Fakultät der Humboldt-Universität zu Berlin.

Balmann, A., H. Lotze, and S. Noleppa (1998a), 'Agrarsektormodellierung auf der Basis "typischer Betriebe". Teil 1: Eine Modellkonzeption für die neuen Bundesländer', *Agrarwirtschaft*, **47**(5): 222–30.

Balmann, A., H. Lotze and S. Noleppa (1998b), 'Modelling Agricultural Sectors based on Virtual Farm Structures: Effects of the "Agenda 2000" in the New Federal States of Germany', Working Paper, Humboldt University Berlin. <http://www.agrar.hu-berlin.de/wisola/fg/abl/ALFONS/Mavis07.PDF>.

Bandura, A. (1977), *Social Learning Theory*, Englewood Cliffs, NJ: Prentice Hall.

Bandura, A. (1986), *Social Foundations of Thought and Action: A Social Cognitive Theory*, Englewood Cliffs, NJ: Prentice Hall.

Barreteau, O. and F. Bousquet (2000), 'SHADOC: a multi-agent model to tackle viability of irrigated systems', *Annals of Operations Research*, **94**: 139–62.

Barreteau, O., F. Bousquet, and J.-M. Attonaty (2001), 'Role-playing games for opening the black box of multi-agent systems: method and lessons of its application to Senegal River Valley irrigated systems', *Journal of*

Artificial Societies and Social Simulation **4**:. <http://www.soc.surrey.ac.uk/JASSS/4/2/5.html>

Barreteau, O. (2002), 'Joint use of role games and models: a review. Introduction to the thematic session', International Society for Ecological Economics, thematic session on Models, Role Games and Negotiations, Sousse, Tunisia.

Bascompte, J. and R.V. Solé (1995), 'Rethinking complexity: modelling spatiotemporal dynamics in ecology', *Trends in Ecology and Evolution*, **10**: 361–66.

Batty, M. (2000), 'Less is more, more is different: complexity, morphology, cities, and emergence', *Environment and Planning B*, **27**: 167.

Baudry, J. (1993), 'Landscape dynamics and farming systems: Problems of relating patterns and predicting ecological changes', in R.G.H. Bunce, L. Ryszkowski and M.G. Paoletti (eds), *Landscape ecology and agroecosystems*, Ann Arbor, Lewis Publishers: pp. 21–39.

Bellman, R. (1957), *Dynamic Programming*, Princeton, NJ: Princeton University Press.

Ben Zur, H. and S.J. Breznitz (1981), 'The effect of time pressure on risky choice behavior', *Acta Psychologica*, **47**: 89–104.

Berdegue, J.A., M. Installe, Ch. Duque, R. Garcia and X. Quezada (1989), 'Application of a simulation software to the analysis of a peasant farming system', *Agricultural Systems*, **30**: 317–34.

Berger, J.O. (1986), *Statistical Decision Theory and Bayesian Analysis*, New York: Springer-Verlag.

Berger, T. (2001), 'Agent-based spatial models applied to agriculture. A simulation tool for technology diffusion, resource use changes, and policy analysis', *Agricultural Economics*, **25**: 245–60.

Berk, R.A. (1994), 'Uncertainty in the construction and interpretation of mesoscale models of physical and biological processes', in P. M. Groffman and G. E. Likens (eds), *Integrated Regional Models*, New York: Chapman & Hall, pp. 50–64.

Berkes, F. and C. Folke (eds) (1998), *Linking Ecological and Social Systems*, Cambridge: Cambridge University Press.

Berkes, F. (1997). 'New and not-so new directions in the use of the commons: co-management,' *The Common Property Resource Digest. Quaterly Publication of the International Association for the Study of Common Property*, **42**: 5–7

Berlyne, D.E. (1966), 'Curiosity and exploration', *Science*, **152**: 25–33.

Bian, L. (1997), 'Multiscale nature of spatial data in scaling up environmental models', in D. A. Quattrochi and M. F. Goodchild (eds), *Scale in Remote Sensing and GIS*, New York: Lewis Publishers, pp. 13–26.

Binmore, K. (1998), 'The complexity of cooperation by Robert Axelrod', *Journal of Artificial Societies and Social Simulation*, **1**(1): <http://jasss.soc.surrey.ac.uk/1/1/review1.html>
Blume, L. (1993), 'The statistial mechanics of strategic interaction', *Games and Economic Behavior*, **4**: 387–424.
Blume, L. (1995), 'The statistical mechanics of best-response strategy revision', *Games and Economic Behavior*, **11**: 111–45.
Boissau, S. (in prep.), unpublished doctoral dissertation, Wageningen University, expected 2004.
Bond, A.H. and L. Gasser (eds) (1988), *Readings in Distributed Artificial Intelligence*, San Mateo, CAl.: Morgan Kaufmann.
Bonnefoy, J.L., C. Le Page, J. Rouchier and F. Bousquet (2000), 'Modelling spatial practices and social representations of space using multi-agent systems', in G. Ballot and G. Weisbuch, *Applications of Simulation to Social Sciences*, Hermes: pp. 155–68.
Borrini-Feyerabend, G., M. Farvar, J.C. Nguiringuiri and V.A. Ndangang (2000), 'Co-Management of natural resources: organising, negotiating and learning by doing', Heidelberg (Germany).
Box, G.E.P., W.G. Hunter and J.S. Hunter (1978), *Statistics for Experimenters. An Introduction to Design, Data Analysis, and Model Building*, New York: John Wiley & Sons.
Bousquet, F., I. Bakam, H. Proton and C. Le Page (1998), 'Cormas: Common-pool resources and multi-agent systems', *Lecture Notes in Artificial Intelligence*, **1416**: 826–38.
Bousquet, F., O. Barreteau, C. Le Page, C. Mullon, and J. Weber (1999), 'An environmental modelling approach: the use of multi-agent simulations', in F. Blasco (ed.), *Advances in environmental and ecological modeling*, Paris, Elsevier: pp. 113–22.
Bousquet, F. and D. Gautier (1998), 'Comparaison de deux approches de modélisation des dynamiques spatiales par simulation multi-agents: les approches spatiales et acteurs', CyberGéo **89:** <http://www.cybergeo.presse.fr/modelis/bousquet/bousquet.htm>
Bousquet, F., C. LePage, I. Bakam and A. Takforyan (2001), 'Multi-agent simulations of hunting wild meat in a village in eastern Cameroon', *Ecological Modelling*, **138**: 331–46.
Bradbury, R.H. (1998a), 'Science, social science and the Dig Tree'. in I. Falk (ed.), *International Symposium on Learning Communities*, Regional Sustainability and the Learning Society, pp. 40–44.
Bradbury, R.H. (1998b), 'Sustainable development as a subversive issue', *Nature and Resources*, **34**: 7–11.
Bradbury, R.H. (1999), 'Just what is science anyway?', *Nature and Resources*, **35**: 9–11.

Bradbury, R.H., L.S. Hammond, R.E. Reichelt, and P.C. Young (1984), 'Prediction and explanation as a dialectical pair in ecology', *Search* **15**: 220–22.

Brandes, W., G. Recke, and T. Berger (1997), *Produktions– und Umweltökonomik – Band 1*, Stuttgart: Ulmer.

Breiman, L., J.H. Freidman, R.A. Olshen and J.H. Stone (1984), *Classification and Regression Trees*, New York: Chapman & Hall.

Breitschuh, G. (2000), *Kostenvergleich bei der pflanzlichen Produktion zwischen Ost und West*, Thüringische Landesanstalt für Landwirtschaft Jena.

Burton, M. (1994), 'The irrigation management game: a role playing exercise for training in irrigation management', *Irrigation and Drainage Systems*, **7**: 305–48.

Camazine, S., J-L. Deneubourg, N.R. Franks, J. Sneyd, G. Theraulaz and E. Bonabeau (2001), *Self-Organization in Biological Systems*, Princeton, NJ: Princeton University Press.

Cangelosi, D. and A. Parisi (eds) (2001), *Simulating the Evolution of Language*, London: Springer-Verlag.

Cardenas, J.C. and E. Ostrom (2001), 'What do people bring into the game? How norms help overcome the tragedy of the commons', 4eme journées d'Economie de l'environnement, Toulouse, INRA.

Caro, T.M. (1986), 'The functions of stotting: a review of the hypotheses', *Animal Behavior*, **34**: 649–62.

Carpenter, S.R., W. Brock, and P. Hanson (1999), 'Ecological and social dynamics in simple models of ecosystem management', *Conservation Ecology*, 3(2): 4 <http://www.consecol.org/vol3/iss2/art4>

Casari, M. and C.R. Plott (2000), 'Keeping an eye on your neighbors: agents monitoring and sanctioning one another in a common-pool resource environment', Social Science Working Paper no. 1072, California Institute of Technology, Pasadena, August 2000.

Castella, J.C., S. Boissau, H. Lan Anh and O. Husson (2001a), 'Enhancing communities' adaptability to a rapidly changing environment in Vietnam uplands: the SAMBA role-play', Sustaining Upland Development in Southeast Asia: Issues, Tools and Institutions for Local Natural Resource Management, Makati, Metro Manila, Philippines, SANREM CRSP/ Southeast Asia.

Castella, J.C., S. Boissau, T.N. Trung and D.D. Quang (2001b), 'How allocation of lowlands affects upland agricultural dynamics in Northern Vietnam mountains: From systems analysis to multi-agent model', Hanoi, Vietnam Agricultural Science Institute.

CBFMO (1998), 'Community mapping, manual for resource management', Quezon City, Philippines, Environmental Science for Social Change, 45 pages.

Chapman, G.P. (1987), 'Gaming simulations and systems analysis: two factions of the truth', *Journal of Applied Systems Analysis*, **14**: 3–15.

Chattoe, E. and N. Gilbert (1998), 'A basic simulation of information diffusion', <http://wwwlisc.clermont.cemagref.fr/ImagesProject/>

Checkland, P. (1981), *Systems Thinking, Systems Practice*, Chichester: John Wiley.

Chichilnisky, G. (2000), 'An axiomatic approach to choice under uncertainty with catastrophic risks', *Resource and Energy Economics*, **22**: 221–31.

Chou, Y.H. (1993), 'Map resolution and spatial autocorrelation', *Geographical Analysis*, **23**: 228–46.

Christensen, N.L., A.M. Bartuska, J.H. Brown, S.A. Carpenter, S. D'Antonio, R. Francis, J.F. Franklin, J.A. MacMahon, R.F. Noss, D.J. Parsons, C.H. Peterson, M.G. Turner, and R.G. Woodmansee (1996), 'The report of the ecological society of America Committee on the scientific basis for ecosystem management', *Ecological Applications*, **6**: 665–91.

Clark, C.W. (1973), 'The economics of overexploitation', *Science*, **189**: 630–4.

Clark, C.W. (1990), *Mathematical Bioeconomics: The Optimal Management of Renewable Resources*, New York: J. Wiley.

Clark, C.W., F.H. Clarke and G.R. Munro (1979), 'The optimal exploitation of renewable resource stocks: Problems of irreversible investment', *Econometrica*, **47**: 25–47.

Cockes, K.D. and J.R. Ive (1996), 'Mediation support for forest land allocation: the SIRO-MED system', *Environmental management*, **20**: 41–52.

Conte, R. (1999), 'Social intelligence among autonomous agents', *Computational & Mathematical Organization Theory*, **5**(3): 203–28.

Conte, R. and C. Castelfranchi (1995), *Cognitive and Social Action*, London: UCL Press.

Conte, R., R. Hegselmann, and P. Terna (eds) (1997), 'Simulating Social Phenomena', *Lecture Notes in Economics and Mathematical Systems*, **456**, Berlin: Springer.

Cooper, D.J., J.H. Kagel, W. Lo, and Q.L. Gu (1999), 'Gaming against managers in incentive systems: experimental results with Chinese students and Chinese managers', *American Economic Review*, **89**: 781–804.

Costanza, R. and M. Ruth (1998). 'Using dynamic modeling to scope environmental problems and build consensus', *Environmental management*, **22**: 183–95

Coughenour, M.B. (1991), 'Spatial components of plant-herbivore interactions in pastoral, ranching, and native ungulate ecosystems', *Journal of Range Management*, **44**: 530–42.

Coughenour M.B. (1992), 'Spatial modelling and landscape characterisation of an African pastoral ecosystem: a prototype model and its potential use for modelling drought', in D.H. McKenzie D.E. Hyatt and V.J. McDonald (eds) *Ecological Indicators*, Vol 1, London and New York: Elsevier Applied Science, pp. 787–810.

Crawford, S.E.S. and E. Ostrom (1995), 'A grammar of institutions', *American Political Science Review*, **89**: 582–600.

Cridland, S. and D.M. Stafford Smith (1993), 'Development and dissemination of design methods for rangeland paddocks which maximise animal production and minimise land degradation', Western Australian Department of Agriculture Miscellaneous Publication, 42/93, 79.

Cumming, D.H.M. (1991), 'Multispecies systems and rural development in southern Africa: opportunities, constraints and challenges', Project Report 1. World Wide Fund for Nature, Multispecies Animal Production Systems Project, Harare, Zimbabwe.

Cumming, D.H.M. and T.J.P. Lynam (1997), 'Landuse changes, wildlife conservation and utilisation and the sustainability of agro-ecosystems in the Zambezi Valley', World Wide Fund for Nature, Harare, Zimbabwe.

Cutshall, C.R. (1989), 'A socio-economic baseline survey of community households. Masoka/Kanyurira Ward', Centre for Applied Social Sciences, University of Zimbabwe, Harare, Zimbabwe.

Dare, W. (2002), 'Role playing games in negotiation on irrigated systems: between play and reality. Proposal of a methodology', International Society for Ecological Economics, thematic session on Models, Role Games and Negociations, Sousse, Tunisia.

Dasgupta, D. (ed.) (1999), *Artificial Immune Systems and Their Applications*, Springer-Verlag.

Dasgupta, P. (1982), *The Control of Resources*, Cambridge, MA: Harvard University Press.

Dasgupta, P. and G. Heal (1974), 'The optimal depletion of exhaustible resources', *Review of Economic Studies*, **41**: 3–28.

Dasgupta, P.S. and. M. Heal (1979), *Economic Theory and Exhaustible Resources*, Cambridge: Cambridge University Press.

Davis, M.B. (1993), *Old Growth in the East*, Richmond, VT: The Cenozoic Society.

Dawkins, R. (1976), *The Selfish Gene*, Oxford and New York: Oxford University Press.

Day, R.H. (1963), *Recursive Programming and Production Response*, Amsterdam: North-Holland.

Deadman, P.J. (1997), 'Modeling individual behavior in common pool resource management experiments with autonomous agents', PhD thesis, School of renewable natural resources, University of Arizona.

Deadman, P.J. (1999), 'Modelling individual behaviour and group performance in an intelligent agent-based simulation of the tragedy of the commons', *Journal of Environmental Management*, **56**: 159–72.

Deadman, P.J., E. Schlager and R. Gimblett (2000), 'Simulating common pool resource management experiments with adaptative agents employing alternate communication routines', *Journal of Artificial Societies and Social Simulation*, **3**(2) <http://www.soc.surrey.ac.uk/JASSS/3/2.2.html>

DeAngelis, D.L. and L.J. Gross (eds) (1992), *Individual-Based Models and Approaches In Ecology*, New York: Chapman & Hall.

De Boer, B. (1997), 'Generating vowels in a population of agents', in P. Husbands and I. Harvey (eds), *Proceedings of the Fourth European Conference on Artificial Life*, MIT Press, pp. 503–10.

Deffuant, G. (2001), 'Final report of the IMAGES project (CT 2092)', 192, http://wwwlisc.clermont.cemagref.fr/ImagesProject/

Deffuant, G., D. Neau, F. Amblard, and G. Weisbuch. (2000), 'Mixing beliefs among interacting agents', *Advances in Complex Systems*, **3**: 87–98.

DeLong, J.B., A. Shleifer, and R.J. Waldmann (1990), 'Noise traders and risk in financial markets', *Journal of Political Economy*, **98**: 703–38.

Dennett, D.C. (1995), *Darwin's Dangerous Idea*, New York: Simon & Schuster.

Dickman, C.R. (1994), 'Native mammals of Western New South Wales: past neglect, future rehabilitation?', in D. Lunney, S. Hand, P. Reed, and D. Butcher (eds), *Future of the Fauna in Western New South Wales*, Royal Zoological Society of NSW, Mosman, pp. 81–92.

Dieckmann, U., R. Law, and J.A.J. Metz (eds) (2000), *The Geometry of Ecological Interactions: Simplifying Spatial Complexity*, Cambridge: Cambridge University Press.

Dudley, D. (1993), 'Essays on individual behavior in social dilemma environments: An experimental analysis', PhD Thesis, Department of Economics, Indiana University, Indiana.

Dudzinski, M.L., H.J. Schuh, D.G. Wilcox, H.G. Gardiner and J.G. Morrissey (1978), 'Statistical and probabilistic estimators of forage conditions from grazing behaviour of merino sheep in a semi-arid environment', *Applied Animal Ethology*, **52**: 93–109.

Duffy, J. (2001), 'Learning to speculate: experiments with artificial and real agents', *Journal of Economic Dynamics and Control*, **25**: 295–319.

Durrett, R. and S. Levin (1998), 'Spatial aspects of interspecific competition', *Theoretical Population Biology*, **53**: 30–43.

Eastman, R. (1999), *Guide to GIS and Image Processing*, Worcester, MA: Clark University.

Ellison, G. (1993), 'Learning, social interaction and co-ordination', *Econometrica*, **61**: 1047–71.

Elwell, H.A. (1980), 'Design of safe rotational systems', Department of Conservation and Extension (AGRITEX), Harare, Zimbabwe.

Epstein, J.M. (1999), 'Agent-based models and generative social science', *Complexity*, **4**(5): 41–60.

Epstein, J.M. and R. Axtell (1996), *Growing Artificial Societies: Social Science from the Bottom Up*, Washington: Brookings Institution Press/ MIT Press Cambridge, MA.

EU Commission (1999), 'Agenda 2000: Strengthening the Union and preparing enlargement', Retrieved from Internet on 20.03.02 <http://europe.eu.int/comm/agenda2000/index_en.htm>.

Evans, T.P., G.M. Green, and L.A. Carlson (2001a), 'Multi-scale analysis of landcover composition and landscape management of public and private lands in Indiana', in A. Millington, S. Walsh, and P. Osborne (eds), *Remote Sensing and GIS Applications in Biogeography and Ecology*, Kluwer Publications, pp. 271–87.

Evans, T. P., A. Manire, F. De Castro, E. Brondizio and S. McCracken (2001b), 'A dynamic model of household decision making and parcel level landcover change in the Eastern Amazon', *Ecological Modeling*, **143**: 95–113.

Feeny, D., F. Berkes, B. J. McCay, and J. M. Acheson (1990), 'The tragedy of the commons: twenty-two years later', *Human Ecology*, **16**: 1–19.

Fehr, E. and S. Gächter (1998), 'Reciprocity and economics: The economic implications of Homo Reciprocans', *European Economic Review*, **42**: 845–59.

Ferber, J. (1995), *Les Systèmes Multi-Agents: Vers une intelligence collective*, Paris : InterEditions.

Ferber, J. (1999), *Multi-Agent Systems: an Introduction to Distributed Artificial Intelligence*, Reading, MA: Addison-Wesley.

Festinger, L. (1954), 'A theory of social comparison processes', *Human Relations*, **7**: 117–40.

Firbank, L.G. (1993), 'The implications of scale on the ecology and management of weeds', in R.G.H. Bunce, L. Ryszkowski and M.G. Paoletti (eds), *Landscape ecology and agroecosystems*, Ann Arbor: Lewis Publishers, 91–103.

Fiske, S.T. and S.E. Taylor (1991), *Social Cognition, 2nd. Edn,*. New York: McGraw-Hill.

Fitzgibbon, C.D. and J.H. Fanshawe (1988), 'Stotting in Thomson gazelles - an honest signal of condition', *Behavioral Ecology Sociobiology*, **23**(2): 69–74.

Fogel, D., K. Chellapilla and P. Angeline (1999), 'Inductive reasoning and bounded rationality reconsidered', *IEEE Transactions on Evolutionary Computation*, **3**: 142–6.

Forman, R.T. and M. Godron (1986), *Landscape Ecology*, New York: John Wiley & Sons.

Fox, J. (1998), 'Mapping the commons: the social context of spatial information technologies', *The Common Property Resource Digest: Quaterly Publication of the International Association for the Study of Common Property*, **45**: 1–4

Fresco, L.O. (1986), 'Cassava in shifting cultivation. A systems approach to agricultural technology development in Africa', Royal Tropical Institute, Wageningen, The Netherlands.

Friedel, M.H., B.D. Foran and D.M. Stafford Smith (1990), 'Where the creeks run dry or ten feet high: pastoral management in arid Australia', *Proceedings of Ecological Society of Australia*, **16**: 185–94.

Frohn, R. (1998), *Remote Sensing for Landscape Ecology: New Metric Indicators for Monitoring Modeling, and Assessment of Ecosystems*, Boca Raton: Lewis Publishers.

Frost, P., E. Medina, J.-C. Menaut, O. Solbrig, M. Swift, and B. Walker (eds) (1986), 'Responses of savannas to stress and disturbance: a proposal for a collaborative programme of research', *Biology International*, Special Issue **10**.

Funtowicz, S.O. and J.R. Ravetz (1994), 'The worth of a songbird: ecological economics as a post normal science', *Ecological Economics*, **10**: 197–207.

Gallant, S.I. (1993), *Neural Network Learning and Expert Systems*, Cambridge, MA: MIT Press.

Gardner, M. (1970), 'The fantastic combinations of John Conway's new solitaire game life', *Scientific American*, **223**: 120–3.

Gibson, E.J. (1988), 'Exploratory behaviour in the development of perceiving, acting, and the acquiring of knowledge', *Annual Review of Psychology*, **39**: 1–41.

Giddens, A. (1987), *Social Theory and Modern Sociology*, Cambridge: Polity Press.

Gilbert, N. and R. Conte (eds) (1995), *Artificial Societies*, UCL Press.

Gilbert, N. and K.G. Troitzsch (1999), *Simulation for the Social Scientist*, London: Open University Press.

Giles, R.H., Jr. and M.K. Trani (1999), 'Key elements of landscape pattern measures', *Environmental Management*, **23**: 477–81.

Gilruth, P.T., S.E. Marsh and R. Itami (1995), 'A dynamic spatial model of shifting cultivation in the highlands of Guinea, West Africa', *Ecological Modelling*, **79**: 179–97.

Gintis, H. (2000a), *Game Theory Evolving*, Princeton, NJ: Princeton University Press.

Gintis, H. (2000b), 'Beyond Homo economicus: evidence from experimental economics', *Ecological Economics*, **35**: 311–22.

Godden, D.P. (1997), *Agricultural and Resource Policy: Principles and Practice*, Melbourne: Oxford University Press.

Goldberg, L.R. (1990), 'An alternative "Description of Personality": The Big-Five Factor Structure', *Journal of Personality and Social Psychology*, **59:** 1216–29.

Goldenberg, J., B. Libai, and E. Muller (1999), *Modelling new product growth through cellular automata: is consumers' heterogeneity a barrier to aggregate analysis*, Israel Institute for Business Research, Tel Aviv University, Tel Aviv, Israel.

Gonzalez, R. (2000), 'Platforms and terraces: bridging participation and GIS in joint learning for watershed management with the Ifugaos of the Philippines', PhD., ITC-Wageningen University.

Goodall, D W (1967), 'Computer simulation of changes in vegetation subject to grazing', *Journal of the Indian Botanical Society*, **46**: 356–62.

Gordon, D. (1999), *Ants at Work: How an Insect Society is Organized*, New York/London: W.W. Norton & Company.

Gordon, H.S. (1954), 'The economic theory of a common property resource: the fishery', *Journal of Political Economy*, **62**: 124–42.

Granovetter, M. (1973), 'The strength of weak ties', *American Journal of Sociology*, **83**: 1444–65.

Granovetter, M. (1978), 'Threshold models of collective behavior', *American Journal of Sociology*, **88**: 1360–80.

Granovetter, M. (1985), 'Economic action and social structure: the problem of embeddedness', *American Journal of Sociology*, **91**: 481–510.

Greenblat, C.S. and R.D. Duke (1981), *Principles and Practices of Gaming-Simulation,* Sage.

Grenney, W. J., D.A. Bella, and H.C. Curl (1973), 'A theoretical approach to interspecific competition in phytoplankton communities', *American Naturalist*, **107**: 405–25.

Grumbine, R.E. (1994), 'What is ecosystem management?', *Conservation Biology*, **8**: 27–38.

Grumbine, R.E. (1997), 'Reflections on "What is ecosystem management?"', *Conservation Biology*, **11**: 41–7.

Gunderson, L. and C.S. Holling (eds) (2002), *Panarchy: Understanding Transformations in Human and Natural Systems*, Washington, DC: Island Press.

Hanson, J .D, J .W. Skiles and W. J. Parton (1988), 'A multi-species model for rangeland plant communities', *Ecological Modelling*, **44**: 89–123.

Happe, K., K. Kellermann and A. Balmann (2002), *Model Documentation.* <http://www.alfons-balmann.de/agripolis>.

Hardin, G. (1968), 'The tragedy of the commons', *Science*, **162**: 1243–8.

Hare, M., J. Heeb and C. Pahl-Wostl (2002), 'The symbiotic relationship between role playing games and model development: A case study in participatory model building and social learning for sustainable urban water management', International Society for Ecological Economics, thematic session on Models, Role Games and Negotiations, Sousse, Tunisia

Hayek, F. (1945), 'The use of knowledge in society', *American Economic Review,* **35**: 519–30.

Hegselmann, R. (1998), 'Modeling social dynamics by cellular automata', in W.B.G. Liebrand, A. Nowak, and R. Hegselmann (eds), *Computer Modeling of Social Processes*, London: Sage Publications, pp. 37–64.

Helmcke, B., (1996), *Zur Bedeutung der betrieblichen Flächenausstattung für die Produktionskosten im Marktfruchtbau*, Shaker, Aachen.

Henry, C. (1974), 'Investment decisions under uncertainty: the irreversibility effects', *American Economic Review*, **64**: 1006–12.

Heuvelink, G. and P. Burrough (1993), 'Error propagation in logical cartographic modelling using Boolean methods and continuous classification', *International Journal of Geographical Information Systems,* **3**: 231–46.

Hightower, R.R, S. Forrest, and A.S. Perelson (1995), 'The evolution of emergent organization in immune system gene libraries', in L.J. Eshelman (ed.), *Proceedings of the Sixth International Conference on Genetic Algorithms*, San Francisco, CA: Morgan Kaufman, pp. 344–50.

Hine, D.W. and R. Gifford (1996), 'Individual restraint and group efficiency in commons dilemmas: The effects of two types of environmental uncertainty', *Journal of Applied Social Psychology,* **26**: 993–1009.

Hofmeyr, S.A. and S. Forrest (2000), 'Architecture for an artificial immune system', *Evolutionary Computational Journal*, **7**: 45–68.

Hofmeyr, S.A. (2001), 'An interpretative introduction to the immune system', in I. Cohen and L. Segel (eds), *Design Principles for the Immune System and other Distributed Autonomous Systems*, Oxford" Oxford University Press, pp. 3–26.

Hofstadter, D.R. (1979), *Gödel, Escher, Bach: An Eternal Golden Braid,* Harmondsworth: Penguin Books.

Hogarth, R.M. (1980), *Judgement and Choice*, Chichester: Wiley.

Hogeweg, P. and B. Hesper (1983), 'The ontogeny of the interaction structure in bumble bee colonies: A MIRROR model', *Behavioural Ecology and Sociobiology*, **12**: 271–83.

Hogeweg, P. (1988), 'Cellular Automata as a paradigm for ecological modeling', *Applied Mathematics and Computation*, **27**: 81–100.

Holland, J.H. (1992), 'Complex adaptive systems', *Daedalus*, **121**: 17–30.

Holland, J.H. (1995), *Hidden Order*, New York: Addison Wesley.

Holland, J.H. (1998), *Emergence: From Chaos to Order*, Oxford: Oxford University Press.

Holland, J.H. (2001), Exploring the evolution of complexity in signaling networks, *Complexity*, **7**(2): 34–45.

Holling, C.S. (1973), 'Resilience and stability of ecological systems', *Annual Review of Ecology and Systematics,* **4**: 1–23.

Holling, C.S. (1978), *Adaptive Environmental Assessment and Management*: London, John Wiley.

Holling, C.S. (1986), 'The resilience of terrestrial ecosystems: local surprise and global change', in W.C. Clark and R.E. Munn (eds), *Sustainable Development of the Biosphere*, Cambridge: Cambridge University Press, pp. 292–317.

Holling, C.S. and L.H. Gunderson (2002), 'Resilience and adaptive cycles', in L. Gunderson and C.S Holling (eds), *Panarchy: Understanding Transformations in Human and Natural Systems*, Washington, DC: Island Press, pp. 25–62.

Howden, S.M., A.C. Langston, N. Abel, S. Marsden, P.J. Reyenga, and W. Tatnell (2001), 'Global change impacts on the Western Division of New South Wales', Working Paper Series 01/1, Australian Greenhouse Office, Canberra, Australia.

Hudson, W.D. and C.W. Ramm (1987), 'Correct formulation of the Kappa coefficient of agreement', *Photogrammetric Engineering and Remote Sensing*, **53**: 421–2.

Huet, S. and G. Deffuant (2001), 'Modèle Allier Calcul de l'impact économique de la conversion à l'agriculture biologique', <http://wwwlisc.clermont.cemagref.fr/ImagesProject/defaultDef.htm>

Hutt, C. (1970), 'Specific and diversive exploration', in H.W. Reese and L.P. Lipsitt (eds), *Advances in Child Development and Behaviour, Vol 5*. New York: Academic Press, pp. 120–72.

Hyden, G. (1986), 'Invisible economy of smallholder agriculture', in J.L. Moock (ed.), *Understanding Africa's Rural Households and Farming Systems*, London: Westview Press, pp. 11–35.

IIASA (1999), 'Cleaner air for a cleaner future: controlling transboundary air pollution', *Options*, International Institute for Applied System Analysis.

Illius, A.W. and T.G. O'Connor (1999), 'On the relevance of nonequilibrium concepts to arid and semiarid grazing systems', *Ecological Applications*, **9**: 798–813.

Ive, J. and D. Cocks (1999), 'SIRO-MED: bringing information-based mediation support to natural resource allocation', *Australasian Dispute Resolution Journal*, **10**: 165–72.

Jacob, F. (1977), 'Evolution and tinkering', *Science*, **196**: 1161–6.

Jager, W. (2000), 'Modelling consumer behaviour', PhD thesis, University of Groningen,. <http://docserver.ub.rug.nl/eldoc/dis/ppsw/w.jager/>

Jager, W., M.A. Janssen and C.A.J. Vlek (1999), 'Consumats in a commons dilemma: Testing the behavioural rules of simulated consumers', COV report no. 99-01. Groningen: Centre for Environment and Traffic Psychology, University of Groningen, <http:// www.bdk.rug.nl/ medewerkers/ w.jager/ simpap.pdf>

Jager, W., M.A. Janssen, H.J.M. De Vries, J. De Greef, and C.A.J. Vlek (2000), 'Behaviour in commons dilemmas: Homo economicus and Homo psychologicus in an ecological-economic model', *Ecological Economics*, **35**: 357–79.

Jager, W., M.A. Janssen and C.A.J. Vlek (2002), 'How uncertainty stimulates over-harvesting in a resource dilemma: three process explanations', *Journal of Environmental Psychology*, 22: 247-63.

James, C.D., J. Landsberg and S.R. Morton (1999), 'Provision of watering points in the Australian arid zone: a review of effects on biota', *Journal of Arid Environments*, **41**: 87–121.

Jansen, V.A.A and A.M. de Roos (2000), 'The role of space in reducing predator-prey cycles', in U. Dieckmann, R. Law and J.A.J. Metz (eds), *The Geometry of Ecological Interactions: Simplifying Spatial Complexity*, Cambridge: Cambridge University Press: pp. 183–201.

Janssen, M.A. (2001), 'An immune-system perspective on ecosystem management', *Conservation Ecology*, **5**(1): <http://www.consecol.org/vol5/iss1/art13>

Janssen, M.A. and S.R. Carpenter (1999), 'Managing the resilience of lakes: a multi-agent modelling approach', *Conservation Ecology*, **3**(2): <http://www.consecol.org/vol3/iss2/art15>.

Janssen, M.A. and W. Jager (2001) 'Co-evolution of cognitive strategies and the environment', Paper presented at SIMSOC5, Kazimierz Dolny, Poland, 21-24 September.

Janssen, M.A. and E. Ostrom (2002), 'Adoption of a new regulation for the governance of common-pool resources by a heterogeneous population', in J.M. Baland, P. Bardhan and S. Bowles (eds), *Inequality, Cooperation and Environmental Sustainability*.

Janssen, M.A. and D.W. Stow (2002), *Ecosystem Management and the Evolution of Rules: An immune system perspective*, in review.

Janssen, M.A., B.H. Walker, J. Langridge, and N. Abel (2000), 'An adaptive agent model for analysing co-evolution of management and policies in a complex rangelands system', *Ecological Modelling,* **131**: 249–68.

Jerne, N.K. (1973), 'The immune system', *Scientific American*, **229**(1): 52–60.

Jervis, R. (1998), *System Effects*, Princeton, NJ: Princeton University Press.

Jiang B. (1999), 'SimPed: simulating pedestrian flows in a virtual urban environment', *Journal of Geographic Information and Decision Analysis*, **3**: 21–30.

Kanyurira Wildlife Committee (1989), 'Wildlife land use plan Kanyurira Ward, Dande Communal Area', WWF Multispecies Project, Harare, Zimbabwe.

Keeling, M., B. Howard and S. Pacala (2000), 'Reinterpreting space, time lags, and functional responses in ecological models', *Science*, **290**: 1758–62.

Kelley, H.H. and J.W. Thibaut (1978), *Interpersonal Relations: A Theory of Interdependence,* New York: Wiley.

Kerin, J. (2000), 'Western Lands Review, Final Report', Department of Land and Water Conservation, Sydney.

Kimble, G.H.T. (1951), 'The inadequacy of the regional concept', in L.D. Stamp and S.W. Woodridge (eds), *London Essays in Geography*, London: Longman.

Kirby, S., and J.R. Hurford (2001), 'The emergence of linguistic structure: an overview of the iterated learning model', in A. Cangelosi and D. Parisi (eds) *Simulating the Evolution of Language*, London: Springer-Verlag, pp. 121–47.

Kirschke D., M. Odening, R. Doluschitz, Th. Fock, K. Hagedorn, D. Rost, and H. von Witzke (1998), *Weiterentwicklung der EU-Agrarpolitik - Aussichten für die neuen Bundesländer*, Kiel.

Kiser, L.L. and E. Ostrom (1982), 'The three worlds of action: a meta-theoretical synthesis of institutional approaches', in E. Ostrom (ed.) *Strategies of Political Inquiry*, Beverly Hills, CA: Sage, pp. 179–222.

Kleingarn, A. (2002), 'Anpassungskosten von Agrarpolitikänderungen. Erstellung und Kalibrierung der Datengrundlage für ein räumlich-dynamischen Multiagentenmodell der Region Hohenlohe', Diploma-Thesis, Humboldt University, Berlin.

Klepper, O. (1997), 'Multivariate aspects of model uncertainty analysis: tools for sensitivity analysis and calibration', *Ecological Modelling*, **101**: 1–13.

Kohler, T.A. and G.J. Gumerman (eds) (2000), *Dynamics in Human and Primate Societies: Agent-based Modeling of Social and Spatial Processes*, Oxford: Oxford University Press.

Koontz, T. (2001), 'Money talks – but to whom? Financial v. non-financial motivations in land-use decisions', *Society and Natural Resources*, **14**: 51–65.

Krummel, J.R., R.H. Gardner, G. Sugihara, and R.V. O'Neill (1987), 'Landscape patterns in a disturbed environment', *Oikos*, **48**: 321–4.

KTBL (1997), *KTBL - Taschenbuch Landwirtschaft*, 18, Edition, Munich.

Kuhlman, D.M. and A. Marshello (1975), 'Individual differences in game motivation as moderators of preprogrammed strategic effects in prisoner's dilemma', *Journal of Personality and Social Psychology*, **32**: 922–31.

Kummer, D. and C.H. Sham (1994), 'The causes of tropical deforestation: a quantitative analysis and case study from the Philippines', in K. Brown and D.W. Pearce (eds), *The Causes of Tropical Deforestation*, London: University College of London Press, pp. 146–58.

Kwasnicki, W. (1999), 'Evolutionary economics and simulation', in T. Brenner (ed.), *Computational Techniques for Modelling Learning in Economics*, Boston, MA: Kluwer Academic Publishers, pp. 3–44.

Lam, N. and D.A. Quattrochi (1992), 'On the issues of scale, resolution, and fractal analysis in the mapping sciences', *Professional Geographer*, **44**: 88–98.

Landesanstalt für Landwirtschaft des Landes Brandenburg (2001), *Datensammlung für die Betriebsplanung und die betriebswirtschaftliche Bewertung landwirtschaftlicher Produktionsverfahren im Land Brandenburg*, Vol. 2, Teltow/Ruhlsdorf.

Lane, D. (1993), 'Artificial worlds and economics', *Journal of Evolutionary Economics*, (Parts One and Two), **3**: 89–107, 177–97.

Lange, R.T., A.D. Nicolson and D.A. Nicolson (1984), 'Vegetation management of Chenopod rangelands in South Australia', *Australian Rangelands Journal*, **6:** 46–54.

Landsberg, J.J. (1987), 'Book review', Forest *Ecology and Management*, **18**: 161–3.

Landsberg, J, C. James, S. Morton, T. Hobbs, J. Stol, A. Drew, and H. Tongway (1997), 'Effects of artificial sources of water on rangeland biodiversity', Final report to the Biodiversity Convention and Strategy Section of the Biodiversity Group, Environment Australia, CSIRO Wildlife and Ecology, Lyneham.

Lansing, J.S. and J.N. Kremer (1993), 'Emergent properties of Balinese water temple networks: Coadaptation on a rugged fitness landscape', *American Anthropologist*, **95**: 97–114.

Lansing, J.S. (2000), 'Anti-chaos, common property and the emergence of cooperation', in T.A. Kohler and G.J. Gumerman (eds), *Dynamics in Human and Primate Societies*, Oxford: Oxford University Press, pp. 207–24.

Lansing, J.S. (2002), '"Artificial Societies" and the Social Sciences', Santa Fe Institute Working paper 02-03-11, Santa Fe, NM.

Law, A.M. and W.D. Kelton (1991), *Simulation Modeling and Analysis*, New York: McGraw-Hill, Inc.

Levin, S.A. (1998), 'Ecosystems and the biosphere as complex adaptive systems', *Ecosystems*, **1**: 131–6.

Levin, S.A. (1999), *Fragile Dominion*, Cambridge, MA: Perseus Publishing.

Levins, R. (1970), 'Complex systems', in C.H. Waddington (ed.), *Towards a theoretical biology*, 3, Drafts, Edinburgh: Edinburgh University Press, pp. 73–88.

Li, B.-L. (2000), 'Fractal geometry applications in description and analysis of patch patterns and patch dynamics', *Ecological Modelling*, **132**: 33–50.

Li, H. and J.F. Reynolds (1997), 'Modeling effects of spatial pattern, drought, and grazing on rates of rangeland degradation: A combined Markov and cellular automaton approach', in D.A. Quattrochi and M.F. Goodchild (eds), *Scale in Remote Sensing and GIS*, New York: Lewis Publishers, pp. 211–30.

Liebrand, W.B.G. (1984), 'The effect of social motives, communication and group size on behaviour in a N-person multi-stage mixed-motive game', *European Journal of Social Psychology*, **14**: 239–64.

Lindsey A. (1997), 'Walking in wilderness', in M.T. Jackson (ed.), *The Natural Heritage of Indiana*, Bloomington, IN: Indiana University Press, pp. 113–23.

Lindsey, A., A.B.W. Crankshaw and A.S. Qadir (1965), 'Soil relations and distribution map of the vegetation of presettlement Indiana', *Botanical Gazette*, **126**: 155–63.

Linster, B. (1992), 'Evolutionary stability in the repeated prisoners' dilemma played by two-state Moore machines', *Southern Economic Journal*, **58**: 880–903.

Low, A. (1986), *Agricultural Development in Southern Africa. Farm-Household Economics and the Food Crisis*, London: James Currey.

Low, B., E. Ostrom, C. Simon, and J. Wilson (2002), 'Redundancy: How does it influence the management of natural resources?', in F. Berkes, C. Folke and J. Colding (eds), *Navigating Nature's Dynamics: Responding to Change in Social-Ecological Systems*, Cambridge: Cambridge University Press, forthcoming.

Ludwig, J.A., D.J. Tongway, D.O. Freudenberger, J.C. Noble, and K.C. Hodgkinson (eds) (1997), *Landscape Ecology: Function and Management: Principles from Australia's Rangelands*, Melbourne: CSIRO Publishing.

Ludwig, J.A., J.A. Wiens and D.J. Tongway (2000), 'A scaling rule for landscape patches and how it applies to conserving soil resources in savannas', *Ecosystems,* **3**: 84–97.

Lynam, T.J.P. (1993), 'Agroecosystem analysis in a semi-arid, smallscale farming system of Zimbabwe: A systems approach', Unpublished PhD Dissertation, Michigan State University, East Lansing, Michigan, USA.

Lynam, T.J.P., F. Bousquet, P. d'Aquino, O. Barreteau, F. Chinembiri, and B. Mombeshora (2002), 'Adapting science to adaptive managers: Spidergrams, belief models and multi-agent systems modelling', *Conservation Ecology*, **5**(2): <http://www.consecol.org/vol5/iss2/art24>

Macy, M. and J. Skvoretz (1998), 'The evolution of trust and cooperation between strangers: a computational model', *American Sociological Review*, **63**: 638–60

MacWhinney, B. (1998), 'Models of the emergence of language', *Annual Review of Psychology*, **49**: 199–277.

Maes, P. (1995), 'Modelling adaptive autonomous agents', in C.G. Langton (ed.), *Artificial Life: an overview*, Cambridge/London: MIT Press, pp. 135–62.

Magee, S.P., W.A. Brock and L. Young (1989), *Black Hole Tariffs, and Endogenous Policy Theory*, Cambridge: Cambridge University Press.

Mäler, K.G. (1974), *Environmental Economics*, Baltimore: John Hopkins Press.

Manson, S.M. (2000), 'Agent-based dynamic spatial simulation of land-use/cover change in the Yucatán peninsula, Mexico', Fourth International Conference on Integrating GIS and Environmental Modeling (GIS/EM4), Banff, Canada.

Manson, S.M. (2001), 'Simplifying complexity: a review of complexity theory', *Geoforum*, **32**: 405–14.

Marschall, L.A. (1994), 'A world that slipped away', *The Sciences*, **34**: 45–6.

Margules, C. and T.D. Redhead (1995), 'Guidelines for using the BioRap Methodology and Tools', CSIRO, Canberra, Australia.

Martin, R.B. (1986), 'Communal areas management program for indigenous resources (CAMPFIRE)', Department of National Parks and Wildlife Management, Branch of Terrestrial Ecology, Government of Zimbabwe, Harare, Zimbabwe.

Mason, C.F. and O. Philipps (1997), 'Mitigating the tragedy of the commons through cooperation: an experimental evaluation', *Journal of Environmental Economics and Management*, **34**: 148–72.

Maturana, H. (1996), 'Le langage et la nature de la communication humaine', *JFIADSMA 96*, Paris: Hermes.

McCabe, K., D. Houser, L. Ryan, V. Smith, and T. Trouard (2001), 'A functional imaging study of cooperation in two-person reciprocal exchange', *Proceedings of the National Academy of Sciences,* **98** (20): 1832–5.

McClelland, J.L. and D.E. Rumelhart (eds) (1986), *Explorations in Parallel Distributed Processing*, Cambridge, MA: MIT Press.

McClintock, C.G. (1978), 'Social values: their definition, measurement and development', *Journal of Research and Development in Education*, **12**: 121–37.

McClintock, C.G. and W.B.G. Liebrand (1988), 'The role of interdependence structure, individual value orientation and other's strategy in social decision making: A transformational analysis', *Journal of Personality and Social Psychology,* **55**: 396–409.

McKeon, G.M., K.A. Day, S.M. Howden, J.J. Mott, D.M. Orr, W.J. Scattini, and E.J. Weston (1990), 'Northern Australian savannas: management for pastoral production', *Journal of Biogeography*, **17**: 355–72.

Meadows, D. and D. Meadows (1993), *Fish Banks News*, Durham, NH, USA, Fish Banks Ltd and Laboratory for Interactive Learning University of New Hampshire.

Meentemeyer, V. and O.E. Box (1987), 'Scale effects in landscape studies', in M.G. Turner (ed.), *Landscape Heterogeneity and Disturbance*, New York: Springer-Verlag, pp. 15–34.

Mehrota, K, C.K. Mohan and S. Ranka (1997), *Elements of Artificial Neural Networks*, Cambridge, MA: MIT Press.

Mermet, L. (1993), *La nature comme jeu de société*, L'Harmattan.

Mertens, B. and E.F. Lambin (2000), 'Land-cover-change trajectories in Southern Cameroon', *Annals of the Association of American Geographers*, **90**: 467–94.

Messick, D.M. and C.G. McClintock (1968), 'Motivational basis of choice in experimental games', *Journal of Experimental Social Psychology*, **4**: 1–25.

Miller, J.H. (1998), 'Active nonlinear tests (ANTs) of complex simulation models', *Management Science*, **44**: 820–30.

Milne, J.A., A.R. Sibbald, K.D. Farnsworth, and C.P.D. Birch (1999), 'A model for predicting the impact of grazing animals on animal production and vegetation changes in temperate grasslands and rangelands', in D. Eldridge and D. Freudenberger (eds), *People and Rangelands. Proceedings of VIth International Rangeland Congress,* Townsville, Australia, pp. 872–3.

Ministerium Ländlicher Raum des Landes Baden-Württemberg (MLR) (2000), *Landwirtschaftliche Betriebsverhältnisse und Buchführungsergebnisse*, Wirtschaftsjahr 1998/99, Vol. 48.

Mintzberg, H., D. Raisinghani, and A. Theoret (1979), The structure of unstructured decision processes', *Administrative Science Quarterly*, **21**: 246–75.

Mitchell, M. (1996), *An Introduction to Genetic Algorithms*, Cambridge, MA: MIT Press.

Montgolfier, J. and J.M. Natali (1987), *Le patrimoine du futur, approches pour une gestion patrimoniale des ressources naturelles*, Paris: Economica.

Muller, J.P. (1996), 'The design of intelligent agents: a layered approach', *Lectures notes in Artificial Intelligence*, **1177**, Springer-Verlag.

Nelson, J. (1998), 'Indiana's forests: past, present, and future', *Woodland Steward*, *7*(3) <http://www.inwoodlands.org>

Nettle, D. (1999), *Linguistic Diversity*, Oxford: Oxford University Press.

NIEIR (1999), 'The Western Division of New South Wales: its human geography and economic prospects', Report by the National Institute for Economic and Industry Research to the Sustainable Use of Rangelands in the 21st Century Project, and the Western Lands Review, Department of Land and Water Conservation, Sydney.

Noble, I.R. (1975), 'Computer simulations of sheep grazing in the arid zone', Unpublished PhD. Thesis, University of Adelaide.

Noble, J.C. (1997), 'The delicate and noxious scrub: CSIRO studies on native tree and shrub proliferation in the semi-arid woodlands of Eastern Australia', CSIRO Australia.

Nowak, M.A., N.L. Komarova and P. Niyogi (2001), 'Evolution of Universal Grammar', *Science*, **291**: 114–18.

Noy-Meir, I. (1975), 'Stability of grazing systems: an application of predator-prey graphs', *Journal of Ecology*, **63**: 459–81.

Ollagnon, H. (1989), 'Une approche patrimoniale de la qualité du milieu naturel,' *Du rural à l'environnement, la question de la nature aujourd'hui*, ARF, l'Harmattan.

Ollagnon, H. (1991), 'Vers une gestion patrimoniale de la protection et de la qualité biologique des forêts', *Forest, trees and people Newsletter*, **3**: 2–35.

Openshaw, S. (1977), 'A geographical study of scale and aggregation problems in region-building, partitioning, and spatial modelling', *Transactions of the Institute of British Geographers*, **2**: 459–72.

Openshaw, S., C. Wymer and A. Craft (1987), 'A mark I geographic analysis machine for the automated analysis of point data sets', *International Journal of Geographical Information Systems*, **1**: 335–58.

Ostrom, E. (1990), *Governing the Commons*, Cambridge: Cambridge University Press.

Ostrom, E. (1995), 'Designing complexity to govern complexity', in S. Hanna and M. Munasinghe (eds), *Property Rights and the Environment*, Beijer International Institute of Ecological Economics and The World Bank, pp. 33–46.

Ostrom, E. (1999), 'Coping with tragedies of the commons', *Annual Review of Political Science*, **2**: 493–535

Ostrom, E., R. Gardner and J. Walker (1994), *Rules, Games and Common-Pool Resources*, The University of Michigan Press.

Ostrom, E. (2000), 'Collective action and the evolution of social norms', *Journal of Economic Perspectives*, **14**(3): 137–58.

Pacala, S.W. and S.A. Levin (1997), 'Biologically generated spatial pattern and the coexistence of competing species', in D. Tilman and P. Kareiva (eds), *Spatial Ecology: The Role of Space in Population Dynamics and Interspecific Interactions,* Princeton, NJ: Princeton University Press, pp. 204–32.

Palmer, R.G., W.B. Arthur, J.H. Holland, B. LeBaron, and P. Tayler (1994), 'Artificial economic life: a simple model of a stockmarket', *Physica D*, **75**: 264–74.

Parker, D., S. Manson, M.A. Janssen, M. Hoffmann, and P. Deadman (2001), 'Multi-agent systems for the simulation of land-use and land-cover change: a review', CIPEC Working Paper CW-01-05. Bloomington: Center for the Study of Institutions, Population, and Environmental Change, Indiana University, Bloomington.

Parks, P. (1995), 'Explaining 'Irrational' Land-use: Risk Aversion and Marginal Agricultural Land', *Journal of Environmental Economics and Management*, **28**: 34–47.

Perelson, A.S. (1989), 'Immune Network Theory', *Immunological Review*, **10**: 5–36.

Peter, G. (1993), 'Eine Ermittlung der langfristigen Durchschnittskostenkurve von Marktfruchtbetrieben anhand des "economic engineering" Ansatzes', Dissertation, University of Göttingen.

Peterson, J.H. Jr. (1991), 'CAMPFIRE: a Zimbabwean approach to sustainable development and community empowerment through wildlife utilization', Centre for Applied Social Sciences, Harare, Zimbabwe.

Pickup, G. (1985), 'The erosion cell – a geomorphic approach to landscape classification in range assessment', *Australian Rangeland Journal*, **7**: 114–21.

Pickup, G. and V.H. Chewings (1988), 'Estimating the distribution of grazing and patterns of cattle movement in a large arid zone paddock – an approach using animal distribution models and Landsat imagery', *International Journal of Remote Sensing*, **9**: 1469–90.

Pierson, F.B., K.E. Spaeth and D.H. Carlson (1999), 'Spur 2000: decision support system for assessing and managing rangelands' in D. Eldridge and D. Freudenberger (eds), *People and Rangelands. Proceedings of VIth International Rangeland Congress*, Townsville, Australia, pp.840–1.

Pingle, M. and L. Tesfatsion (2001), 'Non-employment benefits and the evolution of cooperation: experiments with real and computational agents', ISU Economic Report 55.

Piveteau, V. (1995), *Prospective et territoire: apports d'une réflexion sur le jeu*, Cemagref éditions.

Pontius, R.G.J. (2000), 'Quantification error versus location error in comparison of categorical maps', *Photogrammetric Engineering and Remote Sensing*, **66**: 1011–16.

Pritchard, L. Jr. and S.E. Sanderson (2002), 'The dynamics of political discourse in seeking sustainability', in L. Gunderson and C.S. Holling (eds), *Panarchy: Understanding Transformations in Human and Natural Systems*, Washington DC: Island Press, pp. 147–69.

Rappaport, R.A. (1968), *Pigs for the Ancestors: Ritual in the Ecology of a New Guinea People*, Yale University Press.

Regierungsbezirk Mittelfranken (2000), 'Deckungsbeiträge, Variable Kosten, AKh-Bedarf der wichtigsten landwirtschaftlichen Produktionsverfahren, incl. Sonderkulturen', Retrieved from Internet on 28.8.01,<http://www.regierung.mittelfranken.bayern.de/wir_f_s/wissensw/landwirt/db2000.pdf>

Reynolds, J.F. and J. Wu (1999), 'Do landscape structural and functional units exist?', in J.D. Tenhunen and P. Kabat (eds), *Integrating Hydrology, Ecosystem Dynamics, and Biogeochemistry in Complex Landscapes*, John Wiley, pp. 274–96.

Reynolds, P. (1991), *Dance civet cat: Child labour in the Zambezi Valley*, Harare, Zimbabwe: Boabab Books.

Robinson, A. (1995), *The Story of Writing: Alphabets, Hieroglyphs & Pictograms*, London: Thames & Hudson.

Robinson, W.S. (1950), 'Ecological correlations and the behavior of individuals', *American Sociology Review*, **15**: 351–57.

Robinson, J. (1994), 'Land-use and land-cover projections', in W.B. Meyer and B.L. Turner II (eds), *Changes in Land-Use and Land-Cover: A Global Perspective*, Cambridge: Cambridge University Press, pp. 73–92.

Rogers, E.M. (1983), *Diffusion of Innovations*, New York: Free Press.

Röling, N. (1996), 'Towards an interactive agricultural science', *European Journal of Agricultural Education and Extension*, **2**(4): 35–48.

Rothman, D., J. Robinson and D. Biggs (2001), 'Signs of life: linking indicators and modelling in the context of QUEST', SCOPE project on Integrated Adaptive Ecological and Economic Modeling and Assessment.

Rouchier, J. and F. Bousquet (1998), 'Non-merchant economy and multi-agents systems, an analysis of structuring exchanges', *Lecture Notes in Artificial Intelligence*, **1534**: 11–124

Rouchier, J., F. Bousquet, M. Requier-Desjardins, and M. Antona (2001a), 'A multi-agent model for transhumance in North Cameroon', *Journal of Economic Dynamics and Control*, **25**: 527–59.

Rouchier, J., M. O'Connor and F. Bousquet (2001b), 'The creation of a reputation in an artificial society organised by a gift system', *Journal of Artificial Societies and Social Simulation* **4**(2): <http:// www.soc.surrey.ac.uk/JASSS/4/2/8.html>

Ruxton, G.D. and L.A. Saravia (1998), 'The need for biological realism in the updating of cellular automata models', *Ecological Modelling*, **107**: 105–12.

Sadlier, R.A. (1994), 'Conservation status of the reptiles and amphibians in the Western Division of New South Wales – an overview', in D. Lunney, S. Hand, P. Reed, and D. Butcher (eds), *Future of the Fauna in Western New South Wales*, Mosman: Royal Zoological Society of NSW, pp. 161–8.

Sampson, G. (1985), *Writing Systems*, Stanford, CA: Stanford University Press.

Sargent, T.J. (1993), *Bounded Rationality in Macroeconomics*, Oxford: Clarendon Press.

Sauget, N. and G. Balent (1993), 'The diversity of agricultural practices and landscape dynamics: the case of a hill region in the southwest of France', in R.G.H. Bunce, L. Ryszkowski and M.G. Paoletti (eds), *Landscape Ecology and Agroecosystems*, Ann Arbor: Lewis Publishers, pp. 113–29.

Saura, S. and M. Millan (2001), 'Sensitivity of landscape pattern metrics to map spatial extent', *Photogrammetric Engineering and Remote Sensing*, **67**: 1027–36.

Schelling, T.C. (1971), 'Dynamic models of segregation', *Journal of Mathematical Sociology*, **1**: 143–86.

Schwartz, P. (1991), *The Art of the Long View*, New York: Doubleday.

Sell, J. and Y. Son (1997), 'Comparing public goods with common pool resources: three experiments', *Social Psychology Quarterly*, **60**(2): 118–37.

Senft, R.L., L.R. Rittenhouse and R.G. Woodmansee (1983), 'The use of regression models to predict spatial patterns of cattle behavior', *Journal of Range Management*, **36**: 553–7.

Shanteau, J. (1990), 'Psychological characteristics of agricultural experts: applications to expert systems', A. Weiss (ed.), *Climate and Agriculture: Systems Approaches to Decision Making*, Lincoln, University of Nebraska Press, pp. 163–79.

Sherman, H. and R. Schultz (1998), *Open Boundaries: Creating Business Innovation through Complexity*, New York: Perseus Books.

Shiller, R.J. (1981), 'Do stock prices move too much to be justified by subsequent changes in dividends?', *American Economic Review*, **71**: 421–36.

Simon, H.A. (1947), *Administrative Behavior*, New York: Macmillan.

Simon, H.A. (1955), 'A behavioral model of rational choice', *Quarterly Journal of Economics*, **69**: 99–118.

Simon, H.A. (1959), 'Theories of decision making in economics and behavioral science', *American Economic Review*, **49**: 253–83.

Simon, H.A. (1976), *Administrative Behavior: A Study of Decision Making Processes in Administrative Organizations*, New York: Harper.

Simon, H.A. (1986), 'The failure of armchair economics', *Challenge*, **29**(5): 18–25.

Singh, I., L. Squire, and J. Strauss (eds) (1986), *Agricultural Household Models. Extensions, Applications and Policy*, Baltimore: World Bank/The Johns Hopkins University Press.

Smith, J., T.R. Mitchell and L.R. Beach (1982), 'A cost-benefit mechanism for selecting problem-solving strategies: some extensions and empirical tests', *Organizational Behavior and Human Performance*, **29**: 370–96.

Smith, P. and J. Smith (1994), 'Historical change in the bird fauna of Western New South Wales – ecological patterns and conservation implications', in D. Lunney, S. Hand, P. Reed, and D. Butcher (eds), *Future of the Fauna in Western New South Wales*, Royal Zoological Society of NSW, Mosman, pp. 123–48.

Sober, E. and D. Sloan Wilson (1999), *Unto Others: The Evolution and Psychology of Unselfish Behavior*, Harvard University Press.

Sokal, A.D. and J. Bricmont (1998), *Fashionable Nonsense: Postmodern Intellectuals' Abuse of Science*, New York: St Martin's Press

Stafford Smith, D.M. (1984), 'Behavioural Ecology of Sheep in the Australian Arid Zone'. Unpublished Ph.D. Thesis, Australian National University, Canberra: Australia.

Stafford Smith, D.M. (1988), 'Modeling: three approaches to predicting how herbivore impact is distributed in rangelands', New Mexican Agriculture Station Regional Resources Report 628.

Stafford Smith, D.M. and B.D. Foran (1990), 'RANGEPACK: the philosophy underlying the development of a microcomputer-based decision support system for pastoral land management', *Journal of Biogeography*, **17**: 541–6.

Steels, L. (1995), 'The artifical life roots of artifical intelligence', in C.G. Langton (ed.), *Artificial Life: an overview*, Cambridge/London: MIT Press, pp. 75–110.

Steels, L. (1997), 'The synthetic modeling of language origins', *Evolution of Communication Journal*, **1**(1): 1–34.

Steels, L. (2000), 'The emergence of grammar in communicating autonomous robotic agents', in W. Horn (ed.), *Proceeding of ECAI 2000*, Amsterdam: IOS Publishing.

Steins, N.A. and V.M. Edwards (1999), 'Platforms for collective action in multiple-use common-pool resources', *Agriculture and Human Values*, **16**: 241–55.

Stevenson, G.G. (1991), *Common Property Economics: A General Theory and Land Use Applications*, Cambridge: Cambridge University Press.

Stroustrup, B. (1998), *The C++ Programming Language*, Redwood City, CA: Addison-Wesley

Tada, T. (1997), 'The immune system as a supersystem', *Annual Review in Immunology*, **15**: 1–13.

Terna, P. (1998), 'Simulation tools for social scientists: building agent based models with Swarm', *Journal of Artificial Societies and Social Simulation*, **1**(2) <http://jasss.soc.surrey.ac.uk/1/2/4.html>

Tesfatsion, L. (2001), 'Introduction to the special issue on agent-based computational economics', *Journal of Economic Dynamics and Control*, **25**: 281–93.

Thaler, R.H. (1992), *The Winners' Curse: Paradoxes and Anomalies of Economic Life*, Princeton, NJ: Princeton University Press.

Thébaud, O. and B. Locatelli (2001), 'Modelling the emergence of resource-sharing conventions: an agent-based approach', *Journal of Artificial Societies and Social Simulation* **4**(2): <http://www.soc.surrey.ac.uk/JASSS/4/2/3.html>.

Tobias, R. (2002), 'Empirical test of a psychological based simulation of collective action', Paper presented at the First Lake Arrowhead Conference on Computational Social Science: Agent-based Modelling in the Social Sciences, 9 to 12 May.

Tobler, W.R. (1979), 'Cellular Geography', in S. Gales and G. Olsson (eds), *Philosophy in Geography*, Dordrecht, Netherlands: D. Reidel Publishing Company, pp. 379–86.

Tongway, D.J. and J.A. Ludwig (1990), 'Vegetation and soil patterning in semi-arid mulga lands of eastern Australia', *Australian Journal of Ecology*, **15**: 23–34.

Turner, M.G., R. Costanza and F. Sklar (1989), 'Methods to evaluate the performance of spatial simulation models', *Ecological Modelling*, **48**: 1–18.

Turner, M.G. and R.H. Gardner (1991), *Quantitative Methods in Landscape Ecology*, New York: Springer.

Turner, M.G., D.N. Wear and R.O. Flamm (1996), 'Land ownership and land-cover change in the Southern Appalachian Highlands and the Olympic Peninsula', *Ecological Applications*, **6**: 1150–72.

Tversky, A. (1969), 'Intransitivity of preferences', *Psychological Review,* **76**: 31–48.

Tversky, A., (1972), 'Elimination by aspects: a theory of choice', *Psychological Review,* **79**: 281–99.

Tversky, A. and D. Kahneman (1974), 'Judgement and uncertainty: Heuristics and biases', *Science*, **185**: 1124–31.

Tversky, A. and D. Kahneman (1981), 'The framing of decisions and the psychology of choice', *Science*, **211**: 453–8.

Valente, T.W. (1995), *Network Models of the Diffusion of Innovations*, Cresskill, NJ: Hampton Press.

Van Asselt, M., J. Mellors, N. Rijkens-Klomp, S. Greeuw, K. Molendijk, P. Beers and P. van Notten (2001), 'Participatory methods in integrated assessment', International Center for Integrative Studies, Maastricht, The Netherlands.

Van de Koppel, J., M. Rietkerk and F.J. Weissing (1997), 'Catastrophic vegetation shifts and soil degradation in terrestrial grazing systems', *Trends in Ecology and Evolution*, **12**: 352–6.

Van Lange, P.A.M. (1999), 'The pursuit of joint outcomes and equality in outcomes: An integrative model of social value orientation', *Journal of Personality and Social Psychology,* **77**: 337–49.

Van Lange, P.A.M., W.B.G. Liebrand, D.M. Messick, and H.A.M. Wilke (1992), 'Social dilemmas: the state of the art. Introduction and literature review', in W.B.G. Liebrand, D.M Messick and H.A.M. Wilke (eds), *Social Dilemmas. Theoretical Issues and Research Findings,* Oxford/New York: Pergamon Press: pp. 3–28.

Varela, F., E. Thompson and E. Rosch (1993), *L'inscription Corporelle de l'Esprit, Sciences Cognitives et Expérience Humaine*, Seuil.

Ventana Systems (1999), *Vensim 4 Modelling Guide*, Ventana Systems Inc., Harvard.

Veregin, H. (1995), 'Developing and testing of an error propagation model for GIS overlay operations', *International Journal of Geographical Information Systems*, **9**: 595–619.

Vitousek, P.M., H.A Mooney, J. Lubchenco, and J.M. Melillo (1997), 'Human domination of earth's ecosystems', *Science*, **277**: 494–9.

Waldrop, M. (1992), *Complexity: The Emerging Science at the Edge of Order and Chaos*, New York: Simon & Schuster.

Walker, P. (1991), 'Land systems of Western New South Wales', Technical Report Number 25, Soil Conservation Service of New South Wales, Sydney.

Walker, B.H. D. Ludwig, C.S. Holling, and R.M. Peterman (1981), 'Stability of semi-arid savanna grazing systems', *Journal of Ecology*, **69**: 473–98.

Walker, B.H. and N. Abel (2002), 'Resilient rangelands: adaptation in complex systems', in L. Gunderson and C.S. Holling (eds), *Panarchy: Understanding Transformations in Human and Natural Systems*, Washington: Island Press, pp. 293–314.

Wallsten, T.S. and C. Barton (1982), 'Processing probabilistic multidimensional information for decisions', *Journal of Experimental Psychology: Learning, Memory, and Cognition*, **8**: 361–84.

Walters, C. (1986), *Adaptive Management of Renewable Resources*, MacMillan Publishing.

Walters, C.J. and R. Hilborn (1978), 'Ecological optimization and adaptive management', *Annual Review of Ecology and Systematics*, **9**: 157–88.

Watts, D.J. (1999), *Small Worlds, The Dynamics of Networks between Order and Randomness*, Princeton, NJ: Princeton University Press.

Weber, J. (1998), 'Perspectives de gestion patrimoniale des ressources renouvelables', in P. Lavigne Delville (ed.), *Quelle Politique Foncière en Afrique Rurale?*, Paris, Karthala-Coopération Française, pp. 534–52.

Weber, J. and D. Bailly (1993), 'Prévoir c'est gouverner', *Natures, sciences, sociétés*, **1**: 59–64.

Weisbuch, G. (1991), *Complex Systems Dynamics*, Santa Fe Institute Publications: Westview Press.

Weisbuch, G. and G. Boudjema (1999), 'Dynamical aspects in the Adoption of Agri-environmental Measures', *Advances in Complex Systems*, **2**: 11–36.

Weiss, G. (ed.) (1999), *Multiagent Systems: A Modern Approach to Distributed Artificial Intelligence*, MIT Press.

Welch, P.D. (1983), 'The statistical analysis of simulation results', in S.S. Lavenberg (ed.), *Computer Performance Modeling Handbook*. New York: Academic Press, pp. 267–329.

Westoby, M., B.H. Walker and I. Noy-Meir (1989), 'Opportunistic management for rangelands not at equilibrium', *Journal of Range Management*, **42**: 266–74.

Wiegart, R.G. (1988), 'Holism and reductionism in ecology: Hypotheses, scale and systems models', *Oikos*, **53**: 267–9.

Wiens, J.A. (1989), 'Spatial scaling in ecology', *Functional Ecology*, **3**: 385–97.

Wiens, J.A., J.F. Addicott, T.J. Case, and J. Diamond (1986), 'Overview: The importance of spatial and temporal scale in ecological investigations', in J. Diamond and T.J. Case (eds), *Community Ecology*, New York: Harper & Row, pp. 145–53.

Wiggins, J.S. (1980), 'Circumplex models of interpersonal behavior', in L. Wheeler (ed.), *Review of Personality and Social Psychology* (Vol. 1) Greenwich, CT: JAI Press, pp. 265–94.

Wilson, J., B. Low, R. Costanza and E. Ostrom (1999), 'Scale misperceptions and the spatial dynamics of a social-ecological system', *Ecological Economics*, **31**: 243–57.

Wooldridge, M. (1999), 'Intelligent Agents', in G. Weiss (ed.), *Multiagent Systems: A Modern Approach to Distributed Artificial Intelligence,* MIT Press, pp. 27–77.

Wright, P. (1974), 'The harassed decision maker: time pressures, distractions, and the use of evidence', *Journal of Applied Psychology,* **39**: 555–61.

Yates, F. (1937), 'The design and analysis of factorial experiments', Imperial Bureau of Soil Science, Harpenden, Herts, England.

Young, P. (1998), *Individual Strategy and Social Structure: An Evolutionary Theory of Institutions*, Princeton, NJ: Princeton University Press.

Young, P. (1999), 'Diffusion in Social Networks', CSED Working Paper No. 2, Washington DC: Brookings Institute.

ZMP (2001), *Neuausrichtung der Gemeinschaftsaufgabe,Verbesserung der Agrarstruktur und des Küstenschutzes' (GAK): Zielsetzung und Inhalte der Änderungen*, Retrieved from Internet on 28.8.01. <www.zmp.de/oekomarkt/vbereich/pages/planak.pdf>

Index